Advances in Mathematical Fluid Mechanics

Series Editors

Advances in Mathematical Fluid Mechanics is a forum for the publication of high quality monographs, or collections of works, on the mathematical theory of fluid mechanics, with special regards to the Navier-Stokes equations. Its mathematical aims and scope are similar to those of the *Journal of Mathematical Fluid Mechanics*. In particular, mathematical aspects of computational methods and of applications to science and engineering are welcome as an important part of the theory. So also are works in related areas of mathematics that have a direct bearing on fluid mechanics.

The monographs and collections of works published here may be written in a more expository style than is usual for research journals, with the intention of reaching a wide audience. Collections of review articles will also be sought from time to time.

Mathematical Fluid Mechanics

Recent Results and Open Questions

Jiří Neustupa
Patrick Penel
Editors

Springer Basel AG

Editors' addresses:

Jiří Neustupa
Department of Technical Mathematics
Faculty of Mechanical Engineering
Czech Technical University
Karlovo nám. 13
121 35 Praha 2
Czech Republic

e-mail: neustupa@marian.fsik.cvut.cz

Patrick Penel
Département de Mathématiques
Université de Toulon et du Var
B.P. 132
83957 La Garde Cedex
France

e-mail: penel@univ-tln.fr

2000 Mathematical Subject Classification 76D05, 35Q30; 76M10, 76N10, 35B32

A CIP catalogue record for this book is available from the
Library of Congress, Washington D.C., USA

Deutsche Bibliothek Cataloging-in-Publication Data
Mathematical fluid mechanics : recent results and open questions / Jiří Neustupa ;
Patrick Penel, ed. - Basel ; Boston ; Berlin : Birkhäuser 2001
(Advances in mathematical fluid mechanics)
ISBN 978-3-0348-9489-0 ISBN 978-3-0348-8243-9 (eBook)
DOI 10.1007/978-3-0348-8243-9

Originally published by Birkhäuser Verlag in 2001
Softcover reprint of the hardcover 1st edition 2001
Printed on acid-free paper produced from chlorine-free pulp. TCF ∞

ISBN 978-3-0348-9489-0

9 8 7 6 5 4 3 2 1

www.birkhäuser-science.com

Contents

Foreword vii

Claude Bardos:
What Use for the Mathematical Theory of the Navier–Stokes Equations 1

Markus Bause, John G. Heywood, Antonín Novotný and Mariarosaria Padula:
An Iterative Scheme for Steady Compressible Viscous Flow, Modified to Treat Large Potential Forces 27

Catherine Cherfils–Clérouin, Olivier Lafitte and Pierre–Arnaud Raviart:
Asymptotic Results for the Linear Stage of the Rayleigh–Taylor Instability ... 47

Eduard Feireisl:
Recent Progress in the Mathematical Theory of Viscous Compressible Fluids 73

Miloslav Feistauer:
Numerical Methods for Compressible Flow 105

Susan Friedlander and Alexander Shnirelman:
Instability of Steady Flows of an Ideal Incompressible Fluid 143

Jiří Fürst, Michal Janda and Karel Kozel:
Finite Volume Solution of 2D and 3D Euler and Navier–Stokes Equations 173

John G. Heywood:
On a Conjecture Concerning the Stokes Problem in Nonsmooth Domains ... 195

Hideo Kozono:
On Well–Posedness of the Navier–Stokes Equations 207

Jiří Neustupa and Patrick Penel:
Anisotropic and Geometric Criteria for Interior Regularity of Weak Solutions to the 3D Navier–Stokes Equations 237

Appendix 267

List of Authors 269

Foreword

This book represents an attempt to describe some recent developments in mathematical fluid mechanics, in the hope of stimulating further consideration and future research. In the process of collecting the series of theoretical and numerical papers presented here, we wished to choose several important fields with their own essential questions and problems, and to present a range of different approaches and views on them.

Mathematical modeling and numerical simulation in the mechanics of fluids are topics of great importance in theory as well as in technical applications. We are still far from understanding all properties of the used models, or from being in a position to justify their correctness. Thus they represent a long-term challenge for scientists and engineers. Related questions and problems have been studied for many years, and they remain as attractive and as important as ever.

The model which is usually called the Navier–Stokes system of equations was independently proposed by H. Navier in 1822 and by G. H. Stokes in 1845. Since that time it has become one of the most popular nonlinear models in the theory of partial differential equations. This model exhibits many features which are also typical for other nonlinear partial differential equations, and so its qualitative analysis often has a backward influence on the general theory of partial differential equations itself. The Navier–Stokes equations formed a starting point for the construction of many other mathematical models of various special types of flows, and many discrete models used in numerical computations are also based on them. However, the coincidence of their solutions with the behaviour of fluids is still a matter of discussion, and this concerns especially the properties of solutions in "small scales" (regularity or "blow ups") and the correspondence between the solutions and turbulent flows.

Deep mathematical analysis of the Navier–Stokes model was initiated by Jean Leray and his remarkable papers in 1933. The model has a high degree of generality and universality, and its analysis has over the past seventy years led to a powerful, structured theory which involves many beautiful approaches and ideas, and also enables further motivating questions to be formulated. Many new and interesting results have appeared, especially in recent decades, due to the great progress in the theory of function spaces and in functional analysis. The still–unsolved problems include, e.g., the regularity and the uniqueness of weak solutions to the Navier–Stokes equations in three space dimensions, and questions of turbulence.

The following chapters of this book concern the general theory of incompressible Navier-Stokes equations:

– C. Bardos inserts the Navier–Stokes equations into a hierarchy of a larger system of models. The models start with kinetic theory and Hamiltonian systems with a high number of particles, and they end with models of turbulence used in engineering.

– J. Heywood describes the Xie conjecture concerning one important inequality in the theory of the Stokes and Navier–Stokes equations. He also shows that the desired inequality would follow from another conjecture that seems to be more approachable. J. Heywood also presents some partial results and observations from the efforts to prove this new conjecture.

– H. Kozono discusses uniqueness, regularity and stability of weak and strong solutions to the Navier-Stokes equations. He also deals with various estimates in the BMO–Hardy spaces and with Sobolev embeddings in the Besov space and he proves a criterion on regularity and break–down of the solutions.

– J. Neustupa and P. Penel give a survey of qualitative results from the theory of the Navier–Stokes equations, stressing the question of the regularity of weak solutions. They prove two regularity criteria: the first uses different assumptions about various components of velocity, while the second is closely related to the question what types of deformations of infinitely small volumes of the fluid "support" regularity and what types "work" for the "blow up".

The next two chapters concern the general theory of the Navier–Stokes equations for compressible fluids:

– E. Feireisl gives a detailed description of the mathematical model for the viscous gas flow, based on the compressible Navier–Stokes equations. He explains the notions of distributional, renormalized and weak solutions and he proves some recent results on a-priori estimates, compactness, global existence and the long-time behaviour of solutions to the Navier–Stokes equations for compressible fluid flow.

– M. Bause, J. Heywood, A. Novotný and M. Padula begin with theoretical aspects of the compressible Navier–Stokes equations, and they study the properties of certain iterative schemes. They finally stimulate the construction of a new possible numerical method.

The transition from laminar flow to turbulence is explained in mathematical models by the arising of instabilities in the solutions of governing equations. The next two chapters deal with this important theme:

– S. Friedlander and A. Shnirelman direct their attention to recent developments and open questions concerning instabilities in ideal fluid flows. They argue that in an appropriate sense almost all steady flows of an inviscid incompressible fluid are unstable. However they distinguish between different kinds of instability and they also show the importance of the norm in which the growth of disturbances is measured.

– C. Cherfils–Clérouin, O. Lafitte and P. A. Raviart present a mathematically rigorous analysis of the linear Rayleigh–Taylor instability for some simple model flows which occur in connection with inertial confinement fusion. They explain the effect of a smooth density profile and convection velocity on the rate of growth of instabilities.

The next two chapters deal with numerical methods in mathematical fluid mechanics. The increasing importance of these methods is so obvious that it needs no further comment.

– M. Feistauer presents the definition of main initial–boundary value problems for inviscid and viscous compressible flows, and he gives a brief review of their known properties. He goes on to describe numerical methods for solving them. He deals especially with the finite-volume method for inviscid flows and the combined finite–volume and finite–element method for viscous flows.

– J. Fürst, M. Janda and K. Kozel deal with modern finite volume schemes for numerical solution of the Euler and Navier–Stokes equations for transonic flow problems. They discuss the TVD theory for first order and higher order schemes and, among others, they present two new 2D finite volume schemes. They show the extension of the TVD MacCormack method and a finite volume composite scheme to a method which solves 3D flows through channels and turbine cascades.

These last two chapters are both accompanied by numerous applications and examples of results of concrete calculations.

The editors wish to thank all the authors for their contributions. Moreover, they wish to express their special gratitude to two of their colleagues and friends: to Giovanni Paolo Galdi who suggested that the editors should take on the task of putting this book together, and provided encouragement from the beginning, and to Adelia Sequeira, who organized the Thematic Term on Theoretical and Computational Fluid Dynamics at the International Mathematical Centre in Coimbra in 1999 and created such a pleasant and stimulating working environment that the first idea about a book emerged there.

We wish to finish by pointing out that the variety of mathematical problems appearing in connection with the mechanics of fluids is so broad that no single book can encompass all the advances made in this field. We intentionally restricted ourselves in the choice of themes mainly to the topics discussed during the 1st International French–Czech Conference on Mathematical Fluid Mechanics held in Marseille (Centre International de Rencontres Mathématiques de la Société Mathématique de France) from March 20–24, 2000.

The chapters are ordered alphabetically according to the name of the first author. The book is addressed firstly to specialists. However, in order to make it also more readable for doctoral students and young researchers, we have appended a brief list of books which provide an explanation of the backgrounds to most of the topics discussed here.

April, 2001

Jiří Neustupa, Patrick Penel

What Use for the Mathematical Theory of the Navier-Stokes Equations

Claude Bardos

Abstract. This contribution is a review of mathematical results concerning the Navier–Stokes equations. The reader should be aware that it is far from being complete and that it has been organized according to the taste and field of interest of the writer.

Compared with the needs of physicists, engineers and even people who design large numerical methods, the contribution of mathematics seems to be very limited. Therefore, I try to show that even in this limitation, mathematics contributes seriously to the understanding of the problem.

Eventually both for physical reason and to reach a better understanding, it seems adequate to insert the Navier–Stokes equations in a large hierarchy of equations. This hierarchy starts from the Hamiltonian systems with large number of particles and ends with models of turbulence used in engineering. Emphasis is put on this hierarchy and the relations existing between its various steps.

1. Introduction: Scalings, hierarchies and formal derivations

There are several good reasons why the study of the Navier Stokes and others closely related equations has been in the center of the activities of mathematicians for more than two centuries. It started probably with Euler and involved the contributions of such diverse personalities as, Leray, Kolmogorov, Arnold and others... The Navier Stokes equations are perfectly well defined mathematical objects and are paradigms of nonlinear equations. The solutions exhibit in their behaviour most of the characteristics of genuinely non linear phenomena.

Compared with the needs for practical applications in engineering sciences success have been very limited. However the results that have been obtained contribute to our understanding of the program and this is the main idea that I would like to describe in these notes.

Fluid mechanics was in the range of our capacity of observations since the beginning of modern science. It is usual, as it is done in the book of Uriel Frisch, [23] to quote some notes written by Leonardo da Vinci about turbulence in fluids. Furthermore the mechanic of fluids has been used as a model for description of phenomena that in the 18 and 19 centuries were much more mysterious, like electromagnetism (cf. Helmholtz who also made important contributions in vortex theory). Eventually the study of fluid mechanics contributed in an essential way, with the work of Boltzmann and Maxwell to the understanding of the notion of atoms.

There is no question about the validity of the equations. Nothing has to be discovered concerning the intimate nature of the physic. The present issues are *how to describe the phenomena with the convenient equations, how to compute these phenomena, how to use and visualize the results in spite of their complexity.*

The equations involve some physical parameters and turn out to be relevant when these parameters have certain value. Therefore as an introduction it is natural to consider a "chain" of equations, hoping and, this is often the case, that the next equation will become relevant when the structure of the phenomena becomes to complicated to be computed by the previous one. The Navier Stokes equations appear to be one of the main pieces this chain:

I Hamiltonian system of particles,

$\Downarrow$

II Boltzmann equation,

$\Downarrow$

III Navier Stokes equations,

$\Downarrow$

IV Models of turbulence.

Each step is deduced from the previous one with the introduction of hierarchy of equations and a process of closure which in some case leads to the appearance of irreversibility.

According to the classical Newton law, the evolution of N particles is an hamiltonian system defined in phase space $I\!R^{3N} \times I\!R^{3N}$:

$$H_N(x_1, x_2, \dots, x_N, v_1, v_2, \dots, v_N) = \sum_{1 \le i \le N} \frac{|v_i|^2}{2} + \sum_{1 \le i < j \le N} V(|x_i - x_j|), \quad (1.1)$$

N is the Avogadro number, of the order of 10^{24}. Furthermore one introduces σ, the range of action of the interacting potential V (or the diameter of the molecules when instead of (1.1) one uses the dynamic of elastic collisions for the evolution of the system).

To realize a connection between {I} and {III}, Boltzmann and Maxwell had the idea of introducing a function $f(x, v, t)$ which describes the density of particles in the vessel which at the point x and at the time t do have the velocity v and which is a solution of the so called Boltzmann equation:

$$\partial_t f + v \nabla_x f = C(f)\,. \quad (1.2)$$

In (1.2) the left hand side represents the evolution of the function f under the sole action of the proper velocity v of the particles. The right hand side is a collision operator which model the interaction between the particles.

Formally (i.e. without proof of convergence) one proceeds as follows.

First for the connection between {I} and {II} one introduces the density function

$$f_N(x_1, x_2, \dots, x_N, v_1, v_2, \dots, v_N)$$

which describes the probability of having at time t the first molecule at the point x_1 with velocity v_1, the second at the point x_2 with velocity v_2 and so on. This function is a solution of the Liouville equation:

$$\partial_t f_N + \{H_N, f_N\} = 0 . \tag{1.3}$$

Assume that the particles are indistinguishable which means that at time $t = 0$ and therefore at any time t and for any permutation σ of the set $\{1, 2 \dots N\}$ one has:

$$\begin{aligned} & f_N(x_1, x_2, \dots, x_N, v_1, v_2, \dots, v_N, t) \\ = \;& f_N(x_{\sigma(1)}, x_{\sigma(2)}, \dots, x_{\sigma(N)}, v_{\sigma(1)}, v_{\sigma(2)}, \dots, v_{\sigma(N)}, t) . \end{aligned} \tag{1.4}$$

Consider for

$$N \to \infty, \sigma \to 0 \quad N\sigma^2 \to \lambda \tag{1.5}$$

the limit of the first marginal:

$$\begin{aligned} f(x, v, t) &= \lim f_N^1(x, v, t) \\ &= \int f_N(x, x_2, \dots, x_N, v, v_2, \dots, v_N, t)\, dx_2\, dx_3 \dots dx_N\, dv_1\, dv_2 \dots dv_N \end{aligned} \tag{1.6}$$

and prove that this function is a solution of the Boltzmann equation. However to analyze this limit one integrates (1.3) with respect to the variables $x_2, x_3, \dots, x_N$, $v_2, v_3, \dots, v_N$ and obtain, using (1.4), an equation of the form:

$$\partial_t f_N^1 + L_N^1 f_N^1 = M_N^1 f_N^2 \tag{1.7}$$

with

$$\begin{aligned} f_N^2 &= f_N^2(x, x_2, v, v_2, t) \\ &= \int f_N(x, x_2, \dots, x_N, v, v_2, \dots, v_N, t)\, dx_3\, dx_4 \dots dx_N\, dv_3\, dv_4 \dots dv_N \end{aligned} \tag{1.8}$$

and L_N^1 and M_N^2 denoting some convenient operators. To analyze the second marginal f_N^2 one integrates (1.7) and obtain an equation for the third marginal f_N^3 defined in a similar manner. Eventually one has a hierarchy (called the BBGKY hierarchy) of N equations

$$\partial_t f_N^l + L_N^l f_N^l = M_N^l f_N^{l+1} \quad 1 \le l \le N-1, \quad \partial_t f_N^N + \{H_N, f_N^N\} = 0 \tag{1.9}$$

for the marginals:

$$\begin{aligned} f_N^l &= \int f_N(x, x_2, \dots, x_N, v, v_2, \dots, v_N, t)\, dx_{l+1}\, dx_{l+2} \dots dx_N \\ &\qquad\qquad dv_{l+1}\, dv_{l+2} \dots dv_N, \quad 1 \le l \le (N-1) \\ f_N^N &= f_N \,. \end{aligned}$$

Letting N go to infinity one obtains an infinite hierarchy which is called the Boltzmann hierarchy and formally written as:

$$\partial_t f^l + L^l f^l = M^l f^{l+1} \,. \tag{1.10}$$

Then one observes that if $f(x, v, t)$ is a solution of the corresponding Boltzmann equation

$$f^l(x, x_2, \dots, x_l, v, v_2, \dots, v_l, t) = \prod_1^l f(x_l, v_l, t) \tag{1.11}$$

produces a solution of the hierarchy (1.10). An uniqueness argument (of the Cauchy Kowalewskaya type) plus the fact that the initial data are assumed to be factorized leads to the conclusion that

$$f(x, v, t) = \lim f_N^1(x, v, t) \,. \tag{1.12}$$

is a solution of the Boltzmann equation.

Going from {II} to {III} is simpler and probably the part of the theory which is by now the best established both at a formal and at a rigorous level. One observes that the collision operator of the Bolztmann equation satisfies the following invariance (inherited from the underlying Liouville equation)

$$\int_{\mathbb{R}^3_v} C(f)\; \Xi(v)\, dv \;=\; 0\,, \text{ for } \Xi(v) = 1\,,\ v_1\,,\ v_2\,,\ v_3 \text{ and } |v|^2 \tag{1.13}$$

and the entropy condition

$$\int_{\mathbb{R}^3_v} C(f)\; \log f(v)\, dv \le 0 \tag{1.14}$$

with equality if and only if f as a function of v is a Maxwellian ie is given by a formula of the following type:

$$f(v) = \frac{\rho}{(2\pi\theta)^{d/2}}\, e^{-\frac{|v-u|^2}{2\theta}} \,. \tag{1.15}$$

In (1.15) d is the dimension of the space, ρ is the macroscopic density, u the macroscopic velocity and θ the macroscopic temperature.

The derivation of {II} from {I} corresponds to a regime where:

$$\lambda = \lim N\sigma^2 \quad 0 < \lambda < \infty \,. \tag{1.16}$$

This number has the dimension of the inverse of a length which is called the mean free path or the Knudsen number. On the other hand the total volume occupied by the gas is of the order of $N\sigma^3$ therefore in the above derivation this volume is very small. The term *rarefied gas* is used in this context and to go from {II} to {III} one should let $\lambda = \epsilon^{-1}$ go to infinity. Therefore the Boltzmann equation is rescaled according to the formula:

$$\partial_t f_\epsilon + v\nabla_x f_\epsilon = \frac{1}{\epsilon} C(f_\epsilon) \,. \tag{1.17}$$

The quantities

$$\begin{aligned} \rho_f &= \int f_\epsilon(x,v,t)\, dv\,, \\ \rho_f u_f &= \int v\, f_\epsilon(x,v,t)\, dv\,, \\ \rho_f\Big(\frac{|u_f|}{2} + \frac{d}{2}\theta_f\Big) &= \int \frac{|v|^2}{2} f_\epsilon(x,v,t)\, dv \end{aligned}$$

define the macroscopic density momentum internal energy and the temperature.

At the level of the equation (1.17) their computation would require the knowledge of higher moment according to the formulas

$$\partial_t \int f_\epsilon(x,v,t)\, dv + \nabla_x \int v f_\epsilon(x,v,t)\, dv = 0\,, \tag{1.18}$$

$$\partial_t \int v f_\epsilon(x,v,t)\, dv + \nabla_x \int v\otimes v f_\epsilon(x,v,t)\, dv = 0\,, \tag{1.19}$$

$$\partial_t \int |v|^2 f_\epsilon(x,v,t)\, dv + \nabla_x \int v\,|v|^2 f_\epsilon(x,v,t)\, dv = 0\,, \tag{1.20}$$

and eventually as in the first derivation an infinite hierarchy of moments. However due to the relaxation property contained in the equation (1.14) the fact that ϵ goes to zero forces f_ϵ to become a maxwellian and this leads to an explicit computation of the moments in term of ρ, u and θ.

In such a way not only the compressible Euler equation but all the equations of macroscopic fluid dynamic (for perfect gazes) can be deduced (with some other convenient scaling) from the Boltzmann equation ([7],...,[9], [1]).

Much more difficult and completely unsolved questions arise for the relation between {III} and {IV}. It corresponds to situations where the macroscopic fluid becomes turbulent and when some type of averaging is necessary for quantitative or qualitative results. In spite of being the very end of the chain, this step shares in common some points with the previous one.

It is an averaging process and the "turbulent model" starts to be efficient when the original Navier Stokes are outside the reach of direct numerical simulations.

In this averaging appears a hierarchy of moments which has been studied "per se" (cf section (3) and [24]). However this is not enough for the following reasons:

There is up to now no well defined notion of equilibrium and relaxation to this equilibrium with something that would play the role of the entropy as it appears in the derivation of {III} from {II} and not even an indirect proof as it is done in the derivation of {II} from {I} by a uniqueness argument (which does not fully explain how things happen).

The parameters that would lead to turbulent phenomena are not so clearly identified as in the previous steps of the hierarchy. In some sense they are less universal and more local.

In conclusion there is up to now no case where a proof (even formal) of the validity of a derivation of {IV} from {III} is available. The arguments when they exist relies on phenomenological considerations and engineering experiments.*In spite of this lack of justification such equations are the one which are used to design the airplanes in which you fly!*

It is an experimental fact (not a theorem) that no new mathematical results cannot be obtained at the level n of the chain of equations without the knowledge of its counterpart at the level $n+1$. A tentative explanation would be that the equations at the level n contains in their asymptotic behaviour the properties of the equation at the level $n+1$. However as said above the derivation of the model of turbulence is not for the time being accessible by first principles from the macroscopic equations and this may be a reason why theorems at the level of the macroscopic equation remain incomplete. Even the macroscopic equations are the cornerstone of the theory and this is the object of the next section. where comments will be made on the following issue:

1) The existence of a smooth solution of the compressible Euler equation for "short time" before the appearance of singularities due to the generations of shocks.
2) The existence of a weak solution of the incompressible Navier Stokes equation.

In agreement with the results one observes that the derivation from {I} to {II} has been fully proved by O.Lanford [28] for the hard spheres model. But this prove is valid only for short time. The derivation of the compressible Euler equation from the Boltzmann equation has also been proven for short time ([37], [52] and...). Eventually with the introduction of a convenient scaling one derive with proofs which are almost complete the Leray weak solution from the renormalized solution introduced by diPerna and Lions for the Boltzmann equation. In some cases such derivation is valid with no restriction on time or smallness assumption on the initial data and under "physical hypothesis". [7], [8], [9], [10], [34] and [45].

2. Stabilities and instabilities of solutions of macroscopic equations

The stability properties of the macroscopic fluid dynamic equations depend on the size intrinsic parameters which describe the average state of the fluid. In particular are involved the Reynolds number

$$Re = \frac{\text{Characteristic speed of the Fluid} \times \text{ Characteristic length of the flow}}{\text{viscosity}} \tag{2.21}$$

and the Mach number:

$$Ma = \frac{\text{Characteristic speed of the Fluid}}{\text{Sound speed}} \tag{2.22}$$

In an adimensional form this leads to the considerations of the following three closely related equations:

The compressible Euler equation which involves the density, the velocity (or moment), and the temperature (or internal energy)

$$\rho\,, u\,, m = \rho u\,,\, \theta \quad \text{and} \quad e = \rho\{\tfrac{1}{2}|u|^2 + \tfrac{3}{2}\theta\}\ : \tag{2.23}$$

$$\partial_t \rho + \nabla_x \cdot (\rho u) = 0\,, \tag{2.24}$$

$$\rho(\partial_t + u \cdot \nabla_x)u + \nabla_x(\rho\theta) = 0\,, \tag{2.25}$$

$$\tfrac{3}{2}\rho(\partial_t + u \cdot \nabla_x)\theta + \rho\theta\nabla_x \cdot u = 0\,. \tag{2.26}$$

The incompressible Euler equation:

$$\partial_t u + \nabla(u \otimes u) + \nabla p = 0, \nabla u = 0\,. \tag{2.27}$$

and the incompressible Navier Stokes equation

$$\partial_t u + \nabla(u \otimes u) - \nu\Delta u + \nabla p = 0,\ \ \nabla u = 0\,. \tag{2.28}$$

The incompressible Euler equation is formally deduced from the incompressible Navier Stokes equation as the viscosity goes to zero or rather with a convenient change of scale as the Reynolds Number goes to infinity. The incompressible Euler equation is deduced from the compressible one as the Mach number goes to zero.

As a consequence stability results for these three equations should be closely related. The compressible Euler equation is the "prototype" of non linear hyperbolic problem. Classical theorems (The Cauchy Lipschitz Theorem) shows that the problem is well posed for a finite time with smooth initial data in $H^s(\mathbb{R}^n)$ with $s > \frac{n}{2} + 1$. However after a finite time singularities appears This correspond to the generation of shocks by compression. A mathematical formulation of this process can be found in Sideris ([49]).

However written in conservation form the compressible Euler equation *should* admits weak solutions which satisfy an entropy condition and do correspond to the propagation of shock or rarefaction waves. Such a result is up to now out of reach. It has been proven by a very clever method in *one space dimension* by Glimm in (1965). Since this time it has not been seriously improved. *Among other things it uses stability in the space of functions with bounded variation. Such a space seems to be well adapted to the description of solutions with discontinuities. However it was proven by Littman than in more than one space variable the acoustic equation (which is a linearized version of the compressible Euler equation is not well posed*

J. Rauch extended the result to the Euler equation itself [43]. *Therefore there is no hope of extending the method of Glimm to more than one space variable dimension.*

The situation is even more complex for the incompressible Euler equation in three space variables. The proof of the local in time existence of smooth solution say for initial data in $H^s(I\!R^n)$ with $s > \frac{n}{2}+1$ can be done along the same line as in the case of the compressible with the introduction of pseudodifferential operators; in a less formal way that was done by Lichtenstein in (1925)[31]. On the other hand due to the incompressibility, the appearance of singularities is a much more subtle problems which up to now remains basically open. According to the physical intuition this would be the increase of vorticity that would be responsible for the appearance of singularities. Along this line the most basic results is probably the one due to Beale Kato and Majda [13] which goes as follow:

Theorem 2.1. *(Beale, Kato, Majda): Let $u \in C^0([0,T[; H^3(I\!R^3\backslash \mathbf{Z}^3))$ be a solution of the three dimensional incompressible Euler equation. Suppose that there exists a time T_* such that the solution cannot be continued up to $T = T_*$ and assume that T_* is the first such time. Then one has for $\omega(x,t) = \nabla \times u(x,t)$,*

$$\int_0^{T_*} \|\omega(t)\|_{L^\infty}\, dt = \infty\,, \tag{2.29}$$

and in particular

$$\limsup_{t\uparrow T_*} \|\omega(t)\|_{L^\infty} = \infty. \tag{2.30}$$

The above theorem and the absence of proof for the appearance of singularities should be completed by the following observation due to Lions and DiPerna (cf. [33] page 150-153.)

Theorem 2.2. *In three space variables* 3 *and for any p (the entropy corresponds to $p = 2$) there exist no function*

$$\phi(Z,t), \quad \lim_{Z\to 0} \phi(Z,t) = 0$$

such that the vorticity of the incompressible Euler equation satisfies the estimate:

$$\|\omega(t)\|_{L^p} \le \phi(\|\omega(0)\|_{L^p}, t)\,. \tag{2.31}$$

The proof is done by construction of examples and the simplest one which emphasize the role of the frequencies is obtained in the torus $\mathbf{T} = (\frac{I\!R}{2\pi})^3$. Observe that the vector field

$$\big(U(y), 0, W(x,y,z,t)\big)\,, \quad W(x,y,z,t) = W(x - tU(y), y, z, 0)\big)$$

is an explicit solution of the incompressible Euler equation with zero pressure. In particular one can consider

$$U(y) = \sin ky, \quad V(y) = 0, \quad W(x,y,z,t) = \sin l(x - t\sin(ky)) \tag{2.32}$$

which gives for the vorticity a time increment which is frequency dependent.

$$\begin{aligned}\omega(x,y,t) &= \big(-tkl\cos l(x-t\sin(ky))\cos(ky),\\ &\qquad -l\cos(x-t\sin(ky)),\ -k\cos(ky)\big)\,.\end{aligned} \tag{2.33}$$

Theorem 2.2 has for our purpose several important consequences.

It shows the existence of solutions that may be do not blow up in finite time but with vorticity which increases arbitrarily fast. Such phenomena may be present in practical turbulence. *The name "quasisingularities " should be appropriate.* The existence of the quasisingularities reduces seriously the hope of proving, as it was tried by several authors, the appearance of blow up by numerical simulations. *Numerical simulations cannot discriminate between quantities that blow in a finite time and quantities that increase fast enough to produce an overflow in finite time.* In the mean time this also shows that apriori estimates (other than the one already known) for the incompressible Navier Stokes equation and that would be independent of the Reynolds number (or the viscosity) have very few chance if any to exist, if they would exist they would lead to similar estimates for $\nu = 0$ which may contradict the statement of the above theorem. Personally I do not believe that the solutions of the incompressible Euler or Navier Stokes equation blows up but it may well be that there are no other general estimates than the one presently found.

Multiplying by u the incompressible Navier Stokes equation:

$$\partial_t u + u\nabla u - \nu\Delta u + \nabla p = 0, \quad \nabla\cdot u = 0 \quad \text{in } \Omega\times I\!R_t^+, \tag{2.34}$$

$$u(x,t) = 0 \quad \text{on } \partial\Omega\times I\!R_t^+, \tag{2.35}$$

$$u(x,0) = u_0(x), \quad u_0(x)\in(L^2(\Omega))^3,\ \nabla\cdot u_0 = 0,\ u_0(x)\cdot n(x) \text{ on } \partial\Omega, \tag{2.36}$$

one obtains (with the hypothesis that the solution is smooth) the local and global "formal" energy relations:

$$\partial_t\frac{|u|^2}{2} + \nabla\Big(u\Big\{\frac{|u|^2}{2}+p\Big\}\Big) + \nu|\nabla u|^2 - \Delta\frac{|u|^2}{2} = 0 \tag{2.37}$$

and

$$\partial_t\int_\Omega \frac{|u|^2}{2}\,dx + \nu\int_\Omega |\nabla u|^2\,dx = 0 \tag{2.38}$$

which were at the origin of Leray's proof of the existence of a weak solution which satisfies the estimate:

$$\int_\Omega \frac{|u(x,t)|^2}{2}\,dx + \nu\int_0^t\int_\Omega |\nabla u(x,s)|^2\,dx\,ds \le \int_\Omega \frac{|u_0(x)|^2}{2}\,dx\,. \tag{2.39}$$

The followers of Leray tried very hard to improve this result and to obtain with some assumption regularity assumption the existence of a smooth solution which then would, easily be shown to be unique and stable with respect to the initial data. *Even if they are very incomplete the present results are of practical importance because in some sense they show where and how "quasi singular solu-*

tions" may not be well controlled or what is sufficient to imply that the solution is indeed well controlled.

With functional analysis one shows (cf. [47], [50] and others) that a weak solution which is in $L^r(0,T;(L^s(\Omega))^3)$ with smooth initial data is in fact a smooth solution provided (r,s) satisfy the relation:

$$r \in [2,+\infty[, \quad s \in]3,\infty[, \quad \frac{2}{r}+\frac{3}{s} \leq 1 \tag{2.40}$$

Refining the notion of weak solutions, V.Scheffer [46], L.Caffarelli, R.Kohn and L.Nirenberg [15] and H.Sohr and W.von Wahl [50] proved that it is possible to construct global in time weak solutions which satisfy for any $T>0$ the following relations:

$$u \in L^\infty(0,T;(H_0^1(\Omega))^3), \quad p \in L^{\frac{5}{3}}(0,T) \tag{2.41}$$

and

$$\partial_t \frac{|u|^2}{2} + \nabla\Big(u\Big\{\frac{|u|^2}{2}+p\Big\}\Big) + \nu|\nabla u|^2 - \Delta\frac{|u|^2}{2} \leq 0\,. \tag{2.42}$$

Such solutions carry the name of *suitable weak solutions*. Notice that for weak solutions or suitable weak solutions the equality in (2.37) and (2.39) is not proven. Solutions are not smooth enough to validate multiplications or integrations by part.

If it exist the set of singularities is of Hausdorff measure 1 in space time. The proof started from the work of J.Leray himself continuing with the contribution of V.Scheffer and culminating with the result of L.Caffarelli, R.Kohn and L.Nirenberg [15]. A more direct and simpler proof has been provided by Fanghua Lin [32], using the pressure estimate of H.Sohr and W.von Wahl [50]. *From a practical point of view these theorems evaluate the measure (small) of the set where a quasi singular solution may have large vorticity*

In two space variables (cf below) the situation concerning the finite time stability of the solution is much better. This implies that to become large (out of control and so on) the $3d$ solutions have to be genuinely three dimensional (in the language of engineers the flow is not *laminar*) this means that there should be no direction in which the velocity should be small and and that the vorticity should in arbitrarily small neighborhood of a singular point "visit" all directions.

The complement of this observation leads to a series of theorems quoted below.

Theorem 2.3. *J.Neustupa, A.Novotný and P.Penel* [40]*: Consider a suitable weak solution u to the Navier stokes equation in a domain of $\Omega\times]0,T[$ assume that there exist a sub domain $D\times]T_1,T_2[\subset \Omega\times]0,T[$ and a space direction $e\in S^2(\mathbb{R}^3)$ such that*

$$u\cdot e_{|D\times]T_1,T_2} \in L^r(]T_1,T_2[,L^s(D)) \text{ with } r\in[4,\infty[,\ s\in]3,\infty[,\ 2/r+3/s\leq \tfrac{1}{2} \tag{2.43}$$

(compare with (2.40) then u is smooth in $D\times]T_1,T_2]$.

Theorem 2.4. *P. Constantin and Ch. Fefferman* [20]*: Write the Navier Stokes equation in term of the vorticity:*

$$\partial_t \omega + u \cdot \nabla_x \omega - \nu \Delta_x \omega = \omega \cdot \nabla_x u \tag{2.44}$$

and introduce the quantity:

$$\xi(x,t) = \frac{\omega(x,t)}{|\omega(x,t)|} \quad \text{if} \quad \omega(x,t) \neq 0\,, \tag{2.45}$$

$$\xi(x,t) = 0 \quad \text{if} \quad \omega(x,t) = 0\,. \tag{2.46}$$

i) Then for any Leray solution of the 3d Navier Stokes equation one has:

$$\int_0^T dt \int_{\{(x,t);|\omega(x,t)|>\Omega\}} |\nabla_x \xi(x,t)|^2\, dx\, dt \leq \frac{C}{\nu\Omega} \int_{\mathbb{R}^3} \{|u_0(x)|^2 + |\nabla \times u_0(x)|\}\, dx. \tag{2.47}$$

ii) Assume that the direction of the vorticity of u is uniformly lipschitz with respect to x when the modulus of this vorticity is large; this means that there exist two positive finite constants C and ρ such that one has

$$\forall (x,y,t) \in (\mathbb{R}^3)^2 \times \mathbb{R}_t^+ \ \{|\omega(x,t)| > \Omega \text{ and } |\omega(y,t)| > \Omega$$
$$\Longrightarrow |\sin\phi(x,y,t)| \leq \frac{|x-y|}{\rho}. \tag{2.48}$$

with $\phi(x,y,t)$ denoting the angle of the two vectors $\omega(x,t)$ and $\omega(y,t)$. Then the vorticity, if bounded for $t = 0$, in $L^2(\mathbb{R}^3)$ remains bounded in the same space for $t \geq 0$ and therefore the solution is "regular".

With other more sophisticated ingredients the proof of the theorem 2.3 uses the fact that some regularity on ω_3 can be controlled by the regularity on u_3 according to the equation (2.44) which gives:

$$\partial_t \omega_3 + u \cdot \nabla_x \omega_3 - \omega \cdot \nabla_x u_3 - \nu \Delta_x \omega_3 = 0 \tag{2.49}$$

the fact that the regularity of the pair (u_3, ω_3) can be transferred to the other velocity component according to the formula:

$$\partial_{x_1} u_1 + \partial_{x_2} u_2 = -\partial_{x_3} u_3 \tag{2.50}$$

$$\partial_{x_1} u_2 - \partial_{x_2} u_1 = \omega_3\,. \tag{2.51}$$

The proof of the theorem 2.4 uses the explicit representation of the right hand side of the equation (2.44) according to the formulas:

$$|(\omega \cdot \nabla_x u, \omega)(x,t)|$$
$$= |\omega(x,t)|^2 \left\{ \frac{3}{4\pi} \text{P.V.} \int \det(\hat{y}, \xi(x), \xi(x+y,t))\, |\omega(x+y,t)| \frac{dy}{|y|^3} \right\} \tag{2.52}$$

Remark. *The significance of the relation (2.47) in the theorem 2.4 is: in regions of high vorticity the direction of the vorticity is regular in an averaged sense, but uniformly with respect to the initial data, and with a $\frac{1}{\nu}$ dependence with respect to the viscosity.*

The significance of the assertion ii) is that for singularities (or loss of control of the regularity) to appear need both large (in modulus) vorticity and large oscillations of the direction of this vorticity.

The theorem (2.4) has its counterpart at the level of the Euler equation (a similar result can be proven but with the extra assumption that the velocity is bounded in L^∞.

Furthermore the significance of both theorems is that the loss of control of regularity may appear only if the velocity and the vorticity of the fluid oscillates in all directions and this emphasizes the role of the dimension 3, in agreement with the fact (cf. section 4) that in two dimensions the fluid remains smooth. It seems important to observe that it is the volume element:

$$\det(\hat{y}, \xi(x), \xi(x+y,t))$$

which drive the system to singularities. Whenever such volume element remains smooth say of the order of $|y|^\alpha$ the fluid remains smooth. This may also be a diagnostic of "turbulent" phenomena.

Eventually one should observe similarities (I have no idea how to use such similarities) between the term involved in the proof of the theorem 2.4

$$\text{P.V.} \int \det(\hat{y}, \xi(x), \xi(x+y,t)) |\,\omega(x+y,t)|\, \frac{dy}{|y|^3}$$

with the helicity

$$\mathcal{H} = \int \omega(x,t) \wedge u(x,t)\, dx = \frac{1}{4\pi} \iint (\hat{y}, \xi(x), \xi(x+y,t))\, \frac{dy}{|y|^2}\, dx\, dy \tag{2.53}$$

which is an invariant of the flow describing in particular the asymptotic crossing number (cf [2] page 141)

Up to now the fact that "big" vortices should stabilize the phenomena has not been fully used. Some preliminary results can be found in the contributions of Babin Mahalov and Nicolaenko [4] and [5]. The idea goes as follow. Write the Euler or Navier Stokes equation in the form:

$$\partial_t u + u \wedge \omega - \nu \Delta u + \nabla_x \Big(\frac{|u|^2}{2} + p \Big) = 0 \tag{2.54}$$

Then according to ([13]) singularities would appear when the vorticity becomes infinite therefore as a first step in the analysis one considers in the equation (2.54) situations where at time $t = 0$

$$\omega(x,0) = \frac{\Omega}{\epsilon} + \tilde{\omega}_0 \tag{2.55}$$

with Ω constant and ϵ small. Introduce the Poincaré propagator defined by the equations

$$\partial_t \psi + \frac{\Omega}{\epsilon} \wedge \psi - \nu \Delta \psi + \nabla p = 0, \quad \nabla \psi \equiv 0, \tag{2.56}$$

$$\psi(x,0) = \psi_0(x), \quad \psi(x,t) = e^{-\frac{t}{\epsilon}\mathcal{P}} \psi_0 \tag{2.57}$$

and write the equation (2.54) in Duhamel form:

$$u(x,t) = e^{-\frac{t}{\epsilon}\mathcal{P}}u_0(x,t) + \int_0^t e^{-\frac{(t-s)}{\epsilon}\mathcal{P}}g(x,s)\,ds, \qquad (2.58)$$

$$g(x,s) = u \wedge \Big(\frac{\Omega}{\epsilon} - \nabla \wedge u\Big). \qquad (2.59)$$

One can observe that the Poincaré operator do have through averaging process regularizing effect and one obtain (cf Babin Mahalov and Nicolaenko [4], [5] who, motivated by the geophysical applications, consider the Navier Stokes equation with a large Coriolis force) that, for ϵ small with respect to ν, (with no hypothesis on the size of $\tilde{\omega}_0$) the Navier Stokes equation, the solution is smooth for all times. In the case of the Euler equation ($\nu = 0$) one obtains that the solution remains smooth on a time interval $]0, T_\epsilon[$ with

$$\lim_{\epsilon \to 0} T_\epsilon = \infty\,. \qquad (2.60)$$

This would lead to two type of considerations

1) A scenario of proof of regularity for the solution of the Navier Stokes equation: If the solution has some tendency to become singular then oscillation in every direction of the velocity and or of the vorticity would by an averaging process stabilize the phenomena.
2) An approach for turbulent regime. Fluids with large Reynolds numbers should produce some type of averaging validating turbulence models as described below.

3. Turbulence, weak convergence and Wigner measures

It is a common fact that definitions of turbulence and intermittency are extremely diverse. At the level of the present contribution I would like to support the following point view concerning a "mathematical definition of turbulence". *The question is the prediction, computation and representation of phenomena which contain no "mystery" in their generation but which are to complex in their realization to be accessible to classical computations.* With this idea in mind one should observe that two approaches would be competing:

A statistical approach in which the velocity of the fluid is a random variable this is *the statistical theory of turbulence*

Situation where a family of solutions does not converges strongly but is L^2 bounded and where nevertheless one can find an equation (or a system of equations) that would be adequate to describe the weak limit of this family.

The randomness in the solution may be generated by randomness in the initial data or in a forcing term (f in the right hand side of the Navier Stokes equation).

The weak convergence may be generated by letting the viscosity ν go to zero (or taken equal to zero) and in the mean time having initial data, forcing term or boundary conditions (even in two space dimension) that prevent strong convergence. In particular the conjunction of the zero viscosity limit and and of viscous boundary condition $u_\nu = 0$ on the boundary of the domain may generate (cf section 4) some genuine weak convergence. The two approaches share in common the following facts

The same type of difficulty appears in the Reynolds stress tensor:

$$\nabla(u \otimes u)$$

In the statistical theory the average $< (u \otimes u) >$ is different from $< u > \otimes < u >$ while when weak convergence is involved with $\overline{u}$ denoting the limit of a sequence u_ν of solutions one has in general $\overline{u \otimes u} \neq \overline{u} \otimes \overline{u}$. In both cases computation of $< u \otimes u >$ (resp. $\overline{u \otimes u}$) involves $< u \otimes u \otimes u >$ (resp. $\overline{u \otimes u \otimes u}$) which in turn involve $< u \otimes u \otimes u \otimes u \otimes u) >$ (resp. $\overline{u \otimes u \otimes u \otimes u}$) and so on ... leading to an infinite hierarchy of equations as in the previous closures. However at variance with the previous derivation the problem of closing the hierarchy is at the mathematical level completely open.

A refined description of the state of the art can be found in ([24]) where in particular it is shown that a finite (up to moments of order N) hierarchy will produce a good approximation of the infinite hierarchy when $N \to \infty$.

In both cases at least for computational and predictability reasons under reasonable hypothesis one should obtain

$$< u \otimes u > - < u > \otimes < u > \;=\; -\nu_{turb}(x,t)\,\{\nabla < u > +^t\nabla < u >\} \tag{3.61}$$

$$\overline{u \otimes u} - \overline{u} \otimes \overline{u} \;=\; -\nu_{turb}(x,t)\{\nabla\overline{u} +^t \nabla\overline{u}\} \tag{3.62}$$

leading to an averaged equation of the form:

$$\partial_t U + \nabla(U \otimes U) - \nu\Delta U - \nabla\big(\nu_{turb}(x,t)\,\{\nabla U +^t \nabla U\}\big) + \nabla p = 0, \nabla u = 0 \tag{3.63}$$

with $\nu_{turb}(x,t)$ being a local (x,t) dependent quantity. For the equation (3.63) to be well posed and "stable for numerical computation" one should assume that $\nu_{turb}(x,t)$ is positive. One suppose that this *turbulent viscosity* ν_{turb} will often be much bigger than the original viscosity ν.

The form of the right hand side of (3.61) or (3.62) is a direct consequence of an hypothesis (made by Taylor and Kolmogorv) of *isotropy, or frame invariance* (cf. [36] page 34). This hypothesis has never been proven but it is very natural since it involves small scales this will also be discussed below. The positivity and more precisely the value of the turbulent diffusion is a more subtle question. It has no reason to be a "markovian" quantity and should depend on the "history" of the flow as indicated by one of the most commonly used model the ϵ–k model (cf also [36] page 51).

In the absence of proof (even under convenient hypothesis) for the derivation of models of turbulence (say like the ϵ k model or others...) information concerning the statistical properties of turbulent flow become valuable at (least to check the validity of numerical simulations). As it is well known this approach mostly due to Kolmogorov ([27]) concerns the turbulence spectra defined through space (one could also use time correlation) according to the formula

$$\langle \hat{R}(x,k,t)\rangle = \int_{\mathbb{R}^3} e^{-ir\cdot k}\Big\langle u\Big(x+\frac{r}{2},t\Big)\otimes u\Big(x-\frac{r}{2},t\Big)\Big\rangle\, dr\,. \tag{3.64}$$

Assuming frame invariance (isotropy) and homogeneity (for the average) ie that the quantity

$$\Big\langle u\Big(x+\frac{r}{2},t\Big)\otimes u\Big(x-\frac{r}{2},t\Big)\Big\rangle$$

is (x,t) independent one obtains the standard formula:

$$\hat{R}(x,k,t) = \frac{E(|k|)}{4\pi|k|^2}\Big(I-\frac{k\otimes k}{|k|^2}\Big)\,. \tag{3.65}$$

$E(|k|)$ is the turbulent spectra for which qualitative properties like the famous 5/3 law are predicted.

In the spirit of the present contribution one should observe that the formula (3.64) makes sense without isotropy and homogeneity hypothesis and introduction of randomness.

$$\hat{R}(x,k,t) = \int_{\mathbb{R}^3} e^{-ir\cdot k}\, u\Big(x+\frac{r}{2},t\Big)\otimes u\Big(x-\frac{r}{2},t\Big)\, dr \tag{3.66}$$

is nothing else than the Wigner transform. To the best of my knowledge this point of view which seems very natural appears in the turbulence literature only in the contribution of D.C. Besnard, F.H. Harlow, R.M. Rauenzahn and C. Zemach [14]. It provides a local and non isotropic definition of the spectra and with the inverse Fourier transform one deduces the relation:

$$u(x,t)\otimes u(x,t) = \Big(\frac{1}{2\pi}\Big)^3\int_{\mathbb{R}^3}\hat{R}(x,k,t)\,dk\,. \tag{3.67}$$

The formula (3.66) leads to a natural connection between the notion of spectra at the level of statistical theory of turbulence and the issues of weak convergence through the H measures (cf. [51], [25]) which underweak convergence evaluate the following type of expression.

$$\overline{u\otimes u} - \overline{u}\otimes\overline{u}$$

In particular for a family of solutions u_ν of the Navier Stokes equation with viscosity ν going to zero, initial data and the forcing term (right hand side) uniformly bounded respectively in L^2 and in $L^\infty(0,T,L^2)$. one has:

$$\tfrac{1}{2}\|u(t,.)\|^2 + \nu\int_0^t |\nabla u(x,s)|^2\,dx\,ds \le C \tag{3.68}$$

With the estimate (3.68) it turns out that the H measure is related to Wigner at the scale ν:

$$\begin{aligned} dW(x,k) &= \lim_{\nu\to 0}\int_{\mathbb{R}^3} e^{-ir\cdot k}\Big[u_\nu\Big(x+\frac{\sqrt{\nu}r}{2},t\Big)-\overline{u}\Big(x+\frac{\sqrt{\nu}r}{2},t\Big)\Big] \\ &\otimes\Big[u_\nu\Big(x-\frac{\sqrt{\nu}r}{2},t\Big)-\overline{u}\Big(x+\frac{\sqrt{\nu}r}{2},t\Big)\Big]\,dr. \end{aligned} \tag{3.69}$$

By the formula (cf [25])

$$\lim_{\nu\to 0}(u_\nu \otimes u_\nu) = \overline{u}\otimes\overline{u} + \lim_{\nu\to 0}(u_\nu - \overline{u})\otimes(u_\nu - \overline{u}) = \overline{u}\otimes\overline{u} + \int dW(x,k)\,. \quad (3.70)$$

Therefore one could at least formulate and may be in some case prove at the level of the Wigner measure the classical issues of turbulence theory: isotropy and universality.

Observe that the Wigner measure

$$\int dW(x,k)$$

is a "microlocal " object which naturally should not depend on the macroscopic characteric of the fluid. The isotropy would be in agreement with the argument of Kolmogorov which says that it is the smallness of the scales which generates isotropy.

In the mean time in term of correlation frame invariance (or isotropy) is currently investigated and discussed and it seems (cf. for instance [19]*) that this property valid for two points correlations is no more valid for higher order correlation. This does not seems to matter too much for the present analysis because the Reynolds stress tensor is computed with the Wigner measure which is a two points correlation.*

4. Some special properties of the dimension 2

In two dimensions the things are much simpler mostly because the vorticity, being perpendicular to plane where the fluid evolves, is conserved along the trajectories:

$$\partial_t\omega + u\nabla\omega = 0\,. \quad (4.71)$$

The mechanism analyzed in the section 2 that would relate the appearance of large vortices to the fact that their direction oscillate in a non control fashion, and which therefore may also lead to some type of average is not present here. At variance one speaks of "inverse" cascade.

However there are in two space variables open problems where the question of turbulence as introduced above do appear . These problems may have their counterpart in 3 space variables but they seems either more natural or slightly more tractable in dimension two. I intend to describe briefly two questions, *The appearance of large scales and the effect of the boundary layer.*

In a bounded domain $\Omega \subset I\!R^2$ one considers the equations

$$\partial_t u + \nabla(u\otimes u) + \nabla p = 0, \quad \nabla\cdot u = 0, \quad (4.72)$$

$$u(x,t)\cdot n = 0 \text{ on } \partial\Omega, \quad u(x,0) = u_0(x) \quad (4.73)$$

and defines a weak solution as a function:

$$u(x,t) \in L^\infty(0,T;(L^2(\Omega))^2,\ \nabla u(x,t) \equiv 0,\ u(x)\cdot n(x) = 0 \text{ on } \partial\Omega$$

which for any smooth test function $\phi \in \mathcal{D}(\Omega \times I\!R_t)$ satisfies the relation

$$\iint (\phi_t u + \nabla\phi : u \otimes u) dx dt = 0 \tag{4.74}$$

and the initial value condition

$$u(x,0) = u_0(x) \,. \tag{4.75}$$

Observe that, with (4.74), (4.75) make sense.

Theorem 4.1. *Assume that the initial vorticity is of the following form*

$$\omega_0 = \omega_0' + \omega_0" \tag{4.76}$$

with ω_0' a compactly positive Radon measure and $\omega"$ a function in L^p $(p > 1)$ then there exists a weak solution of the corresponding 2 Euler equation.

The proof uses the conservation of the L^p norm of the vorticity plus a cleaver argument of Delort [21] (cf. also for a follow up [53]) to prove with the positivity of the non L^p part the absence of concentration.

The general form of the initial data (4.76) is justified by the need to treat initial data with vorticity concentrated on a curve. This is the famous Kelvin Helmholtz problem known to exhibit instabilities.

To the best of my knowledge the uniqueness of the weak solution has only been established (Yudovich ([55]) with the assumption $\omega_0 \in L^\infty$ and is known not to be always true under the only assumption $u_0 \in (L^(\omega))^2$. In fact Scheffer and Shnirelman (cf.[48]) did construct with solutions $u(x,t)$ with compact support in space time and initial data $u(x,0)$ in $L^2(\Omega)$.

Even if it is well known the regularity of the solution (for smooth initial data say $u_0 \in C^{1,\sigma}(\Omega)$) is a subtle question. At present with the conservation of the vorticity this would be a simple adaptation of the theorem 2.1 (Beale Kato Majda). However the original proof given in by Wolibner [54] with the analysis of pair dispersion problem remains interesting. The question comes from the following fact: The relation $\omega(x,t)$ bounded in L^∞ does not implies that the gradient $\nabla u(x,t)$ is also bounded in L^∞ which would be the cornerstone of a simple boot strap argument for the regularity of the solution. At variance one can show the following log lipschitz estimate:

$$|u(x,t) - u(y,t)| \le C\|\omega(.\,,t)\|_{L^\infty(\Omega)}|x-y| \log\Big(\frac{D}{x-y}\Big) \text{ with } D \text{ the diameter of } \Omega \,. \tag{4.77}$$

From (4.77) one deduces for a pair of the trajectories of particles

$$\dot{x}(t) = u(x(t),t),\; \dot{y}(t) = u(y(t),t)\,, \tag{4.78}$$

by a Gronwall type argument, the relation:

$$\Big(\frac{|x(0)-y(0)|}{D}\Big)^{e^{-Ct|\nabla\times u|_{L^\infty(\Omega)}}} \ge \frac{|x(t)-y(t)|}{D} \ge \Big(\frac{|x(0)-y(0)|}{D}\Big)^{e^{Ct|\nabla\times u|_{L^\infty(\Omega)}}} \tag{4.79}$$

which is sufficient to show that the velocity and the vorticity satisfy the following uniform bound:

$$|\nabla u(.\,,t)|_{L^\infty(\Omega)} \leq \frac{C}{\alpha\, e^{-Ct|\nabla\times u|_{L^\infty(\Omega)}}} \, \|\omega(.\,,t)\|_{C^{0,\alpha\, e^{-Ct|\nabla\times u|_{L^\infty(\Omega)^s}}}} \tag{4.80}$$

$$\leq \frac{C}{\alpha}\, e^{Ct|\nabla\times u|_{L^\infty(\Omega)}} \, \|\omega(.\,,0)\|_{C^{0,\alpha}} \tag{4.81}$$

Then by a standard Gronwall argument it follows that the solution will be as regular as the initial data for any finite time, but the "measure" of this regularity may blow up with $t \to \infty$. In fact an example with an initial vorticity in L^∞ but not in $C^{0,\sigma}$ which "saturates" the relation (4.77) has been given by Bahouri and Chemin [6].

The above consideration leads to the idea that for large time and a sequence of initial data ω_ϵ bounded in L^∞ converging in L^∞ weak* but not strongly exotic phenomena may show up. This is a way to access to an explanation of the formation of coherent structures as observed in the Jupiter red spot or in the anticyclone of the Acores. In both cases the problems are $3d$ but due to the smallness of the thickness of the atmosphere they are mostly driven by two dimensional dynamic.

Consider a sequence of solutions of the $2d$ Euler equation written in vorticity form:

$$\epsilon\, \partial_t \omega_\epsilon + u_\epsilon \nabla \omega_\epsilon = 0, \quad \nabla \cdot u_\epsilon = 0, \quad \nabla \wedge u_\epsilon = \omega_\epsilon, \text{ in } \Omega, \tag{4.82}$$

$$u_\epsilon \cdot n = 0 \text{ on } \partial\Omega, \quad \lim_{\epsilon\to 0} \omega_\epsilon(x,0) = \bar{\omega}_o \text{ in weak} L^{\infty *}\,. \tag{4.83}$$

which converges say in $L^\infty(]0,T[\times\Omega)$ weak star to a stationary solution $\overline{u}, \overline{\omega}$

$$\overline{u}\, \nabla \overline{\omega} = 0. \tag{4.84}$$

Assuming for sake of simplicity that this solution is smooth (4.84) implies that the level surfaces of the current function $\overline{\psi}$ ($\overline{u} = \nabla\overline{\psi}^\perp$) and of the vorticity $\overline{\omega} = -\Delta\overline{\psi}$ coincide. Therefore exists a (may be multivalued) function F such that:

$$-\Delta\overline{\psi} = \overline{F}(\overline{\psi}). \tag{4.85}$$

The determination of this function F is the main issue of the rest of the program (cf. [44] and [35]). In the absence of any "dynamical" proof the present discussion relies on three points:

1) The introduction of Liouville (or Young) measures . This is fully consistent with the weak convergence of the vorticity both for $t = 0$ and for $0 \leq t \leq T$, which may be a mathematical formulation of turbulence.
2) The emphasis on the quantities which are conserved and which pass to the limit.
3) An argument borrowed from statistical mechanic which does not seem to have a dynamical counterpart and which is a follow up of an idea of Onsager [38].

Denote by $\mu_0(x,\lambda)$ the Liouville measure associated to the weak convergence of the initial data and by $\overline{\mu(x,\lambda)}$ the Liouville measure of to the convergence to $\overline{\omega}$. The following relations are direct consequences of :

i) The fact that $\overline{\mu(x,\lambda)}$ is a Liouville measure.

ii) The fact that the vorticity is constant along the trajectories of the flow which is measure preserving.

And

iii) The fact that the solution is uniformly bounded in $L^\infty(I\!R_t; H^1(\Omega))$.

$$\int_{I\!R} \overline{\mu}(x,\lambda)d\lambda = 1, \tag{4.86}$$

$$\int_{\Omega} \overline{\mu}(x,\lambda)dx = \sigma(\lambda) \quad \text{prescribed by the initial data,} \tag{4.87}$$

$$\tfrac{1}{2}\int_{\Omega}\big((-\Delta)^{-1}\int_{I\!R}\lambda\overline{\mu}(x,\lambda)d\lambda\big)\big(\int_{I\!R}\lambda\overline{\mu}(x,\lambda)d\lambda\big)dx = E \quad \text{prescribed by the initial data.} \tag{4.88}$$

Then comes the argument borrowed from statistical mechanic which says that, with the above constrains, the "most possible" Liouville measure is the one which minimize the "Kullback" entropy with respect to $\mu_0(x,\lambda)$. It is given by the formula

$$\overline{\mu}(x,\lambda) = \rho(x,\lambda)\,\mu_0(x,\lambda) \tag{4.89}$$

and minimize the quantity

$$\iint_{I\!R\times\Omega} \rho(x;\lambda)\log\rho(x,\lambda)\mu_0(x,\lambda)\;dx\,d\lambda$$

under the constrains (4.86),(4.87),(4.88). Then a variational computation gives:

$$\begin{aligned}
&\iint_{I\!R\times\Omega}(1+\log\rho(x,\lambda))\,\delta\,\rho(x,\lambda)\,\mu_0(x,\lambda)\;dx\,d\lambda\\
&= \beta\iint_{I\!R\times\Omega}\overline{\psi(x)}\,\delta\,\rho(x,\lambda)\,\mu_0(x,\lambda)\;dx\,d\lambda\\
&\quad+\iint_{I\!R\times\Omega}\big(\alpha(\lambda)+\gamma(x)\big)\,\delta\,\rho(x,\lambda)\,\mu_0(x,\lambda)\;dx\,d\lambda.
\end{aligned} \tag{4.90}$$

In (4.90) $\gamma(x)$ is the Lagrange multiplier of the constraint (4.86), $\alpha(\lambda)$ the Lagrange multiplier of the constraint (4.87) and β the Lagrange multiplier of the energy constraint (4.88). Therefore one obtain eventually the formula:

$$\rho(x,\lambda) = \frac{e^{-\alpha(\lambda)-\beta\lambda\overline{\psi(x)}}}{\int e^{-\alpha(\lambda)-\beta\lambda\overline{\psi(x)}}\,\mu_0(x,\lambda)\;d\lambda} \tag{4.91}$$

and with

$$\overline{\omega} = \int \lambda\, \rho(x,\lambda)\, \mu_0(x,\lambda)\, d\lambda \tag{4.92}$$

the "mean field equation":

$$-\Delta\overline{\psi}(x) = \frac{1}{\int e^{-\alpha(\lambda)-\beta\lambda\overline{\psi}(x)}\, \mu_0(x,\lambda)\, d\lambda} \int \lambda e^{-\alpha(\lambda)-\beta\lambda\overline{\psi}(x)}\, \mu_0(x,\lambda)\, d\lambda. \tag{4.93}$$

The above analysis gives an example of derivation of averaged model of the type {IV} from equations of the type {III}. Following this analysis and considering not only the minimization of entropy but the maximization of dissipation of entropy one can construct ([18]) type dependent turbulent model with a "turbulent viscosity".

An other unsolved problem both in 2 and 3 space variables is the inviscid limit of the solutions of the incompressible Navier Stokes when the viscosity goes to 0 and when a viscous boundary layer is prescribed.

The fact that the problem is still open is even more striking in dimension 2 because both for the Navier-Stokes equations and for the Euler equations in two space existence of regular solution is known. The problem is challenging both for evident application (estimation of the drag on a wing and of the vortices in the wake of a plane at takeoff) and for fundamental question in turbulence because it is the action of the boundary which turns out to be the most natural physical way to generate turbulence that will spread in the media.

Let u_ν denote the solution of the incompressible Navier Stokes equation with viscous boundary condition

$$\partial_t u_\nu + \nabla(u_\nu \otimes u_\nu) - \nu\Delta u_\nu + \nabla p_\nu = 0,\ \nabla u_\nu = 0,\ u_\nu(x,t) = 0 \text{ on } \partial\Omega\times]0,T[\tag{4.94}$$

then the question is: Does the solution u_ν converges (with $\nu \to 0$) to the solution of the Euler equation with the natural impermeability boundary condition

$$\partial_t u + \nabla(u\otimes u) + \nabla p = 0,\ \nabla u = 0,\ n\cdot u(x,t) = 0 \text{ on } \partial\Omega\times]0,T[\,? \tag{4.95}$$

Since the type of the boundary condition change this generate a boundary layer which in some cases is analyzed with the Prandtl equation. In fact one observes that it is the tangential variation of the pressure on the boundary which generates vorticity according to the formula:

$$\frac{\partial\omega_\nu}{\partial n} = -\frac{\partial p_\nu}{\partial\tau}\,. \tag{4.96}$$

However since the problem is non linear nothing prevent this boundary layer to move into the media and to introduce oscillations or turbulent behaviour of the solution that may cause

$$\lim_{\nu\to 0}(u_\nu\otimes u_\nu) - (\lim_{\nu\to 0} u_\nu)\otimes(\lim_{\nu\to 0} u_\nu) \neq 0\,. \tag{4.97}$$

In fact such a phenomena is supposed to happen when there is a detachment of the boundary layer and will produce a non trivial turbulent viscosity.

Once again the only estimate available comes from the the energy

$$\tfrac{1}{2}\int_\Omega |u_\nu(x,t)|^2\,dx + \nu\int_0^t\int_\Omega |\nabla\wedge u_\nu(x,s)|^2\,dx\,ds = \tfrac{1}{2}\int_\Omega |u_\nu(x,0)|^2\,dx. \qquad (4.98)$$

From (4.98), with a convenient localizing argument one can deduce:

Theorem 4.2. *(Kato 76)* [26] *The following facts are equivalent:*

1) $u_\nu(x,t)$ *solution of the Navier Stokes equation (4.94) converge strongly (in* $L^2(0,T:(L^2(\Omega))^2)$ *to the solution of the Euler equation (4.95).*
2) *One has*

$$\lim_{\nu\to 0}\ \nu\int_0^T\int_\Omega |\nabla\wedge u_\nu(x,s)|^2dxds = 0\,. \qquad (4.99)$$

3) *One has*

$$\lim_{\nu\to 0}\nu\int_0^T\int_{\Omega\cap d(x,\partial\Omega)\le c\nu} |\nabla\wedge u_\nu(x,s)|^2dxds = 0. \qquad (4.100)$$

Observe that the region

$$\Omega\cap d(x,\partial\Omega)\le c\,\nu$$

is much smaller than the standard region for boundary layer analysis in "parabolic equation" which is of the order of $\sqrt{\nu}$. The above theorem give an indication on the amount of vorticity needed in a very small neighborhood of the boundary to create a detachment. On the other hand it is also important to observe that this statement is not in contradiction with the analysis made under the assumption of the existence of a boundary layer well localized near the boundary. In this case one should have

$$\nu\int_0^T\int_{\Omega\cap d(x,\partial\Omega)\le c\nu} |\nabla\wedge u_\nu(x,s)|^2\,dx\,ds \le C\nu \qquad (4.101)$$

which implies (4.99). This shows that the validity of the boundary layer analysis needs stronger conditions than the *simple convergence to the Euler flow which one could qualify as the absence of turbulence.*

Furthermore in the left hand side of (4.97) the difference come from small scales which are "far away" (with respect to the size of the boundary layer) from the region where the vorticity is generated and it would be natural to assume (this has not yet been proven) that in this configuration the Wigner measure (cf section 3) is frame invariant even if the whole problem due to the presence of the boundary is not. *As a result one should prove, it is not yet done that the limit of the solution of Navier Stokes equation satisfy the following equation:*

$$\partial_t\overline{u} + \nabla(\overline{u}\otimes\overline{u}) - \nu_{turb}(x,t)\Delta\overline{u} + \nabla\overline{p} = 0, \qquad (4.102)$$

$$\nabla\cdot\overline{u}\equiv 0, \qquad (4.103)$$

$$u\cdot n = 0 \text{ on } \partial\Omega,\ \nu_{turb}(x,t)\ge 0,\ u\wedge n(x,t) = 0 \text{ for } x\in\partial\Omega \text{ and } \nu(x,t) > 0. \qquad (4.104)$$

In these equations $\nu_{turb}(x,t)$ represent the *turbulent viscosity* which should be active in some part of the wake of the obstacle.

Eventually the fact that detachment may appear and that the boundary layer analysis may fail is in full agreement with the results which are at present at our disposal. For instance in the half space the Prandtl equation is derived from the $2d$ Navier Stokes equation as follow : In the Navier-Stokes equation (with viscosity $\nu = \epsilon^2$)

$$\partial_t u_1^\epsilon - \epsilon^2 \Delta u_1^\epsilon + u_1^\epsilon \partial_{x_1} u_1^\epsilon + u_2^\epsilon \partial_{x_2} u_1^\epsilon + \partial_{x_1} p^\epsilon = 0\,, \tag{4.105}$$

$$\partial_t u_2^\epsilon - \epsilon^2 \Delta u_2^\epsilon + u_1^\epsilon \partial_{x_1} u_2^\epsilon + u_2^\epsilon \partial_{x_2} u_2^\epsilon + \partial_{x_2} p^\epsilon = 0\,, \tag{4.106}$$

$$\partial_{x_1} u_1^\epsilon + \partial_{x_2} u_2^\epsilon = 0\,, \quad u_1^\epsilon(x_1,0) = u_2^\epsilon(x_1,0) = 0\,, \quad \forall x_1 \in I\!R \tag{4.107}$$

the following scalings, consistent with the persistence of the boundary condition $u_2 \equiv 0$, on the boundary are made:

$$X_1 = x_1,\ X_2 = \frac{x_2}{\epsilon},\ \widetilde{u}_1(x_1,X_2) = u_1(x_1,X_2),\ \widetilde{u}_2(x_1,X_2) = \epsilon u_2(x_1,X_2) \tag{4.108}$$

and the problem becomes the "Prandtl boundary layer " equation:

$$\partial_t \widetilde{u}_1 - \nu\, \partial_{x_2}^2 \widetilde{u}_1 + \widetilde{u}_1\, \partial_{x_1} u_1 + \widetilde{u}_2\, \partial_{x_2} \widetilde{u}_1 + \partial_{x_1} \widetilde{p} = 0 \tag{4.109}$$

$$\partial_{x_2} \widetilde{p} = 0, \quad \widetilde{p}(x_1,x_2,t) = \widetilde{p}(x_1,t), \quad \partial_{x_1} \widetilde{u}_1 + \partial_{x_2} \widetilde{u}_2 = 0 \tag{4.110}$$

$$\widetilde{u}_1(x_1,0) = \widetilde{u}_2(x_1,0) = 0 \text{ for } x_1 \in I\!R, \tag{4.111}$$

$$\lim_{x_2 \to \infty} \widetilde{u}_2(x_1,x_2) = 0, \qquad \lim_{x_2 \to \infty} \widetilde{u}_1(x_1,x_2) = U(x_1,t). \tag{4.112}$$

One can prove that the existence for all $\epsilon > 0$ with uniform estimates of a solution of the Prandtl equation (with convenient data) implies the convergence of the solution of the Navier Stokes equation to the solution of the Euler equation. In would be interesting to explore if the condition described above is sufficient for the validity of a "Prandtl type boundary layer" a positive answer would mean that the question of the absence of "turbulence" in this problem is related to the analysis of the solutions of the "Prandtl " equations and the following facts have been proven

1) For a finite time with convenient analytic initial data the problem has a well defined solution [3], [16] and [17]. The proof given in ([17]) uses only the analyticity with respect to the tangent variable. *This may be related with the formula (4.96)*
2) For special type of profile which would exclude the detachment global existence is proven (cf [41])
3) For some special initial data that do correspond to a strong recirculation a blow up of the solution in finite time is proven ([22])
4) There are configurations (cf. [12]) where the solutions of the NS equations are shown not to converge to the solutions of the Prandtl equation in the boundary layer.

References

[1] K. Aoki and Y. Sone, *Steady gas flows past bodies at small Knudsen numbers,* Transport Theory and Stat. Phys. **16** (1987), 189–199.

[2] V.I. Arnold and B.A. Khesin, *Topological Methods in Hydrodynamics,* Applied Mathematical Sciences **125**, Springer (1997).

[3] K. Asano, *Zero-viscosity limit of the incompressible Navier-Stokes equations,* Conference at the IVth Workshop on Mathematical Aspects of Fluid and Plasma dynamics, Kyoto (1991).

[4] A. Babin, A. Mahalov, and B. Nicolaenko, *Global regularity and integrability of 3D Euler and Navier-Stokes equations for uniformly rotating fluids,* Asymptotic Analysis **15** (1997), 103–150.

[5] A. Babin, A. Mahalov, and B. Nicolaenko, *3DNavier-Stokes and Euler equations with initial data characterized by uniformly large vorticity,* preprint Arizona State University (2001).

[6] H. Bahouri and J.Y. Chemin, *Equations de transport relatives à des champs de vecteurs non-lipschitziens et mécanique des fluides,* Archive for Rat. Mech. and Anal. **127** (1994), 159–182.

[7] C. Bardos, F. Golse and D. Levermore, *Sur les limites asymptotiques de la théorie cinétique conduisant à la dynamique des fluides incompressibles,* C. R. Acad. Sci. **309** (1989), 727–732.

[8] C. Bardos, F. Golse and D. Levermore, *Fluid dynamics limits of kinetic equations I, Formal derivations,* J. of Stat. Phys. **63** (1991), 323–344.

[9] C. Bardos, F. Golse and D. Levermore, *Fluid dynamics limits of kinetic equations II, Convergence proofs for the Boltzmann equation,* Comm. on Pure and Appl. Math. **46** (1993), 667–753.

[10] C. Bardos, F. Golse and D. Levermore, *Acoustic and Stokes limits for the Boltzmann equation,* C.R. Acad. Sci. Paris **327 I** (1998), 323–328.

[11] C. Bardos and S. Ukai, *The Classical incompressible Navier-Stokes limit of the Boltzmann equation,* Math. Models and Methods in the Appl. Sciences **1** (1991), 235–257.

[12] E. Grenier, *On the nonlinear instability of Euler and Prandtl equations,* Comm. Pure and Appl. Math. **53** (2000), 733–744.

[13] J.T. Beale, T. Kato and A. Majda, *Remarks on the breakdown of smooth solutions for the 3D Euler equations,* Commun. Math. Phys. **94** (1984), 61–66.

[14] D.C. Besnard, F.H. Harlow, R.M. Rauenzahn and C. Zemach, *Spectral transport model for turbulence,* Los Alamos Technical Report LA-11821-MS.

[15] L. Caffarelli, R. Kohn and L. Nirenberg, *Partial regularity of suitable weak solutions of the Navier-Stokes equations,* Comm. Pure Appl. Math. **35** (1982), 771–831.

[16] R. Caflisch and M. Sammartino, *Existence and sigularities for the Prandtl boundary layer equations,* Z. Angew. math. Mech. **80** (2000), 733–744.

[17] P.H. M. Cannone, M. C. Lombardo and M. Sammartino, *Existence and uniqueness for the Prandtl equations,* preprint (2000).

[18] P.H. Chavanis and J. Sommeria, *Thermodynamical approach for small scale parametrization in 2D turbulence,* Phys. Rev. Lett. **78**, No. 17, (1997), 3302–3305.

[19] M. Cherkov A. Pumir and B. Shraiman, *Lagrangian tetrad and the phenomenology of turbulence,* The International Conference on Turbulence, Phys. Fluids **11**, No. 8 (1999), 2394–2410.

[20] P. Constantin and Ch. Fefferman, *Direction of vorticity and the problem of global regularity for the Navier-Stokes equations,* Indiana Univ. Math. Journal **42**, No. 3 (1993), 775–789.

[21] J. M. Delort, *Existence de Nappes de Tourbillon en dimension deux,* Journ. A.M.S. **4** (1991).

[22] W. E and B. Engquist, *Blow up of solutions of the unsteady Prandtl Equation,* Comm. Pure Appl. Math. **50** (1997), 1287–1293.

[23] U. Frisch, *Turbulence,* Cambridge University Press 1995.

[24] A. Fursikov and O. Ehmanuilov, *The rate of convergence for the closure of the Friedman Keller chain in the case of large Reynolds numbers,* Russ. Acad. Sci. Sb. Mat. **81** (1995), 235–259.

[25] P. Gérard, *Microlocal defect measures,* Comm. in Partial Differential Equations **16** (1991), 1761–1794.

[26] T. Kato, *Remarks on zero viscosity limit for nonstationart Navier Stokes flows with boundary,* Seminar on nonlinear partial differential equations, Berkley, Calif. 1983, Math. Sci. Res. Inst. Publ. 2, Springer New York–Berlin (1984), 85–98.

[27] A.N. Kolmogorov, *The local structure of turbulence in incompressible viscous fluid for very large Reynolds,* C.R. Acad. Sci. URSS **30** (1941), 301. 2nd ed., Gordon and Breach, New York.

[28] O. Lanford, *Time evolution of large classical systems , Dynamical systems, theory and applications, Recontres,* Battelle Res. Inst., Seattle, Wash., 1974. Lecture Notes in Phys. 38, Springer, Berlin (1975), 1–111.

[29] J. Leray, *Essai sur les mouvements plans d'un liquide visqueux que limitent des parois,* J. Math. Pures Appl., Série 9, **13** (1934), 331–418.

[30] J. Leray, *Sur le mouvement d'un liquide visqueux emplissant l'espace,* Acta Math. **63** (1934), 193–248.

[31] L. Lichtenstein, *Über einige Existenzproblem der Hydrodynamik homogener unzusammendrückbarer, reibunglosser Flüssikeiten und die Helmholtzschen Wirbelsalitze,* Mat. Zeit. Phys. **23** (1925), 89–154; **26** (1927), 193–323; **32** (1930), 608.

[32] F. Lin, *A new proof of the Caffarelli–Kohn–Nirenberg Theorem,* Comm. Pure App. Math **51** (1998), 241–257.

[33] P.L. Lions, *Mathematical Topics in Fluid Mechanics, Volume 1, Incompressible models,* Oxford Lecture Series in Mathematics and its Applications, Oxford (1966).

[34] P.L. Lions and N. Masmoudi, *From Boltzmann equations to incompressible fluid mechanics equations, I and II,* preprints CIMS, April (2000).

[35] J. Miller, P.B. Weichman and C. Cross, *Statistical mechanics, Euler's equation and Jupiter's Red spot,* Phys. Rev. A **45**, No. 4 (1992), 2328–2358.

[36] B. Mohammadi and O. Pironneau, *Analysis of the K-ϵ turbulence model,* Research in Applied Math., No. 31, J.L.Lions and P. Ciarlet (eds.), Masson-Wiley, Paris (1994).

[37] T. Nishida, *Fluid dynamic limit of the nonlinear Boltzmann equation at the level of the compressible Euler equation,* Comm. in Math. Phys. **61** (1978), 119–168.

[38] L. Onsager, *Statistical hydrodynamics,* Supplemento al Nuovo Cimento **6** (1949), 279–287.

[39] J. Neustupa and P. Penel, *Regularity of a suitable weak solution to Navier-Stokes equations as a consequence of one regularity of one velocity component,* Applied Nonlinear Analysis, Kluwer New York 1999, 391–402.

[40] J. Neustupa, A. Novotný and P. Penel, *An interior regularity of a weak solution to the Navier–Stokes equations in dependence on one component of velocity,* to appear in *Topics in Mathematical Fluid Mechanics,* a special issue of Quaderni di Matematica (2001).

[41] O. Oleinik, *On the mathematical theory of boundary layer for unsteady flow of incompressible fluids,* J. Appl. Math. Mech. **30** (1966), 951–974.

[42] H. Poincaré, *Sur la précession des corps déformables,* Bull. Astronomique **27** (1910), 321–356.

[43] J. Rauch, *BV estimates fail for most quasilinear hyperbolic systems in dimension greater than one,* Comm. Math. Phys. **106** (1986), 481–484.

[44] R. Robert and J. Sommeria, *Statistical equilibrium states for two dimensional flows,* J. Fluid Mech. **229** (1991), 291–310.

[45] L. Saint Raymond, *From BGK Boltzmann model to the Navier Stokes equations,* preprint LAN Université de Paris 6 (2000).

[46] V. Scheffer, *An inviscid flow with compact support in space-time,* J. Geom. Anal. **3** (1993), No. 4, 343–401.

[47] J. Serrin, *On the interior regularity of weak solutions of the Navier-Stokes equations,* Arch. Rat. Mech. Anal. **9** (1962), 187–195.

[48] A. Shnirleman, *On the nonuniqueness of weak solution of the Euler equation,* Comm. Pure Appl. Math. **50** (1997), 1261–1286.

[49] T. Sideris, *Formation of singularities in compressible fluids,* Comm. in Math. Phys. **101** (1985), 475–485.

[50] H. Sohr and W.von Wahl, *On the regularity of the pressure of weak solutions of Navier Stokes equations,* Arch. Math. (Basel) **46** (1986), 428–439.

[51] L. Tartar, *t H-measures, a new approach for studying homogeneization, oscillations and concentration effects in partial differential equations,* Proceedings of the Royal Society of Edinburg **115 -A** (1990), 193–230.

[52] S. Ukai, *Les solutions globales de l'équation de Boltzmann dans l'espace tout entier et dans le demi espace,* C.R. Acad. Sci., Paris **282 A** (1976), 317–320.

[53] I. Vecchi and S. Wu, *On L^1 vorticity for 2D incolpressible flows,* Manuscripta Math. **78**, 403–412.

[54] W. Wolibner, *Un théorème sur l'existence du mouvement plan d'un fluide parfait, homogène et incompressible, pendant un temps infiniment long,* Mat. Z. **37** (1933), 698–726.

[55] V.I. Yudovich, *Non stationary flow of an ideal incompressible liquid,* Zh. Vych. Mat. **3** (1963), 1032–1066.

University of Paris 7 and LAN (Univ. Paris 6)
France
E-mail address: `bardos@math.jussieu.fr`

An Iterative Scheme for Steady Compressible Viscous Flow, Modified to Treat Large Potential Forces

Markus Bause, John G. Heywood, Antonín Novotný and Mariarosaria Padula

Abstract. A relatively simple iterative scheme for steady compressible viscous flow was given in a recent paper of Heywood and Padula [5]. It was intended as a basis for both the existence theory and for numerical methods. Among the assumptions they made in introducing the scheme, one was that the force should be small. On the other hand, by another method, Novotný and Pileckas [7] recently extended the existence theory for such flow to the case of forces that are small perturbations of large potential forces. The purpose of the present paper is to incorporate the ideas of Novotný and Pileckas, for treating such forces, into the simpler and more constructive scheme of Heywood and Padula. This is one step in our larger objective, which is to base the existence theory for compressible viscous flow on constructive methods that may be suggestive of new numerical methods.

1. Introduction

Recently Heywood and Padula [5] introduced an iterative scheme for solving the steady Poisson–Stokes equations, which iterates simply between a Stokes problem and a transport equation. Because it consists of a single iteration between standard problems for which there are already well developed numerical methods, it seems to provide a straight forward approach to the computation of compressible viscous flow. Indeed, our main objective in proposing it was to base the existence theory for such flow on a constructive procedure that might suggest new computational methods. However, before that objective can be met, the scheme will require some modifications, and will need to be adapted to more general situations than that in which it was introduced.

This paper addresses one of the adaptations to be made, namely to the case of forces which are not small as assumed in [5], but which are small perturbations of large potential forces. In finding the appropriate modification of the iterative scheme, we have been guided by a recent existence theorem of Novotny and Pileckas [7], where they treat the case of large potential forces by another method. The proof of the convergence of the new scheme, once it has been formulated, follows lines similar to those used for the original scheme in [5], incorporating a number of results from [7], especially in Lemma 4.1.

The scheme given here is equivalent to that given in [5] when the potential force is equal to zero and the problem is normalized so that the density equals one. Nevertheless, even in that case, some terms are grouped differently here, and defined differently than before, yielding a better insight into the problem that may be useful in making further generalizations. In particular, the pressure π_{n+1} in the momentum equation is defined differently. However, it is an important point that the effective pressure H_{n+1} on the right side of the transport equation is defined similarly as in [7].

Let us mention that we have also studied several other modifications in the scheme, intended to enhance its convergence properties for larger data. The first of these is to replace the Stokes problem by an Oseen problem, linearizing about the previous iterate rather than about zero. The second is to modify the transport equation in a manner suggested by a pseudo-transient interpretation of the equation of continuity. When these changes are made, the theorem that follows from the convergence of the resulting scheme still requires an assumption that the data is sufficiently small. But the scheme should perform much better in numerical computations and may actually converge for relatively large data. We originally intended to include all of these modifications in the present paper. However, the inclusion of so many unrelated ideas into one paper seemed to overburden the presentation. We will present these other improvements of the scheme in [2].

2. Equations and notation

The Poisson–Stokes equations describing the steady isothermal motion of a viscous compressible fluid in a bounded domain $\Omega \subset \mathbb{R}^d$, with $d = 2, 3$, are

$$\begin{gathered} \nabla \cdot (\rho v) = 0, \\ \rho v \cdot \nabla v - (\lambda + \mu)\nabla\nabla \cdot v - \mu\Delta v = -\nabla p + \rho b, \\ p = k\rho, \\ v|_{\partial\Omega} = 0, \quad \int_\Omega \rho \, dx = M. \end{gathered} \tag{1}$$

Here, ρ denotes the density of the fluid, v the velocity, p the pressure, b the external force density, and M the total mass of the fluid in the domain Ω. The viscosity coefficients μ and λ will be assumed to be constants, and to satisfy the conditions $\mu > 0$ and $3\lambda + 2\mu \geq 0$. The assumption that the flow is isothermal means that k is a positive constant.

The point of this paper is to consider forces of the form

$$b = \nabla U + f, \tag{2}$$

where f is a small perturbation of a potential force ∇U, that can be arbitrarily large. The exact solution of (1), (2) corresponding to the potential force ∇U is the rest state $v = 0$ and $\rho = \rho_0$, where ρ_0 satisfies the equation

$$\nabla \rho_0 = k^{-1} \rho_0 \nabla U. \tag{3}$$

Thus

$$\rho_0(x) = \rho_M \exp\left(k^{-1} U(x)\right), \tag{4}$$

where the positive constant ρ_M is determined by the equation

$$\int_\Omega \rho_0(x)\, dx = M.$$

We shall assume that

$$U \in C^3(\overline{\Omega}), \tag{5}$$

which clearly implies that $\rho_0 \in C^3(\overline{\Omega})$ and $\rho_0(x) \geq \alpha > 0$. Throughout, we will use these properties of ρ_0 without explicitly mentioning them. Let $\sigma = \rho - \rho_0$ denote the perturbation in the density ρ from the rest state ρ_0. Then problem (1), (2) can be rewritten as

$$\begin{gathered} \nabla \cdot (\rho_0 v) = -\nabla \cdot (\sigma v), \\ (\rho_0 + \sigma) v \cdot \nabla v - (\lambda + \mu) \nabla \nabla \cdot v - \mu \Delta v = -\rho_0 \nabla\left(k \sigma \rho_0^{-1}\right) + (\rho_0 + \sigma) f, \\ p = k(\rho_0 + \sigma), \\ v|_{\partial\Omega} = 0, \quad \int_\Omega \sigma\, dx = 0. \end{gathered} \tag{6}$$

We shall use mostly standard notation. The spatial domain is denoted by $\Omega \subset \mathbb{R}^d$, with $d = 2, 3$. The space of functions that are m-times continuously differentiable in $\overline{\Omega}$ is denoted by $C^m(\overline{\Omega})$. The function spaces $L^p(\Omega)$, $W^{m,p}(\Omega)$, $W_0^{m,p}(\Omega)$, with m an integer, are the standard Lebesgue and Soblev spaces. Further, $W^{-m,p'}(\Omega)$ is the space that is dual to $W_0^{m,p}(\Omega)$, with $1/p + 1/p' = 1$. These spaces are endowed with the standard norms. L^2-norms are denoted simply by $\|\cdot\|$, while all other norms are distinguished by subscripts. For example, $\|\cdot\|_p$ denotes the norm for $L^p(\Omega)$, and $\|\cdot\|_{m,p}$ the norm for $W^{m,p}(\Omega)$. We employ the notation $\langle\cdot,\cdot\rangle$ for the inner product in $L^2(\Omega)$. We do not distinguish through the notation between scalar and vector valued functions, or function spaces, or norms. The precise meaning is always clear from the context.

We also use some special notation. $L_0^2(\Omega)$ will denote the subspace of $L^2(\Omega)$ consisting of functions with vanishing spatial averages. Analogously, for ρ_0 given by (4)–(5), we let

$$L^2_{\rho_0}(\Omega) = \left\{\Pi \in L^2(\Omega) \mid \textstyle\int_\Omega \Pi \rho_0^{-1}\, dx = 0\right\}.$$

For such ρ_0, we also introduce a subspace of $W_0^{1,2}(\Omega)$ defined by

$$J_{1,\rho_0}(\Omega) = \{v \in W_0^{1,2}(\Omega) \mid \nabla \cdot (\rho_0 v) = 0\}.$$

3. Iteration scheme and main results

Our aim is to solve the reformulated problem (6) by a constructive iteration. Using ideas from the existence theorem given in [7], for the case of large potential forces, we propose to modify the iterative scheme given in [5], to read as follows.

Set $v_0 = 0$, $\sigma_0 = 0$. Then, successively, for given v_n and σ_n, $n \geq 0$, compute v_{n+1} and σ_{n+1} by carrying out the following steps:

(i) *Set*

$$g_n = -\nabla \cdot (\sigma_n v_n), \qquad F_n = (\rho_0 + \sigma_n)(f - v_n \cdot \nabla v_n). \tag{7}$$

(ii) *Find v_{n+1}, π_{n+1} as the solution of the Stokes-type problem*

$$\begin{gathered} \nabla \cdot (\rho_0 v_{n+1}) = g_n, \\ -(\lambda + \mu)\nabla\nabla \cdot v_{n+1} - \mu \Delta v_{n+1} = -\rho_0 \nabla (\pi_{n+1}\rho_0^{-1}) + F_n, \\ v_{n+1}|_{\partial\Omega} = 0, \qquad \int_\Omega \pi_{n+1}\, dx = 0. \end{gathered} \tag{8}$$

(iii) *Set*

$$H_{n+1} = \rho_0 \pi_{n+1} - (\lambda + 2\mu)\nabla \cdot (\rho_0 v_{n+1}). \tag{9}$$

(iv) *Find σ_{n+1} as the solution of the transport equation*

$$k\rho_0\sigma_{n+1} + (\lambda + 2\mu)\nabla \cdot (\sigma_{n+1} v_{n+1}) = H_{n+1}. \tag{10}$$

Our main result is the proof of the convergence of the scheme (7)–(10), which is given in Lemma 6.1. By related methods, we also prove the local uniqueness of the resulting solution in Lemma 7.1. Together, these lemmas provide a new proof of existence and local uniqueness for the problem (1), in the case of forces that are small perturbations of arbitrarily large potential forces. It may be summarized as follows.

Theorem 3.1. *Let Ω be a bounded subdomain of $\mathbb{R}^d$, $d = 2, 3$, with boundary of class $C^{2,1}$. Suppose that $f \in W^{1,2}(\Omega)$, and that $U \in C^3(\overline{\Omega})$ and $\rho_0 \in C^3(\overline{\Omega})$ satisfy the conditions (3)–(5). Then, if $\|f\|_{1,2}$ is sufficiently small depending on Ω, k, μ, λ and ρ_0, there exists a locally unique solution v, ρ of problem (1) with $v \in W^{3,2}(\Omega) \cap W_0^{1,2}(\Omega)$, $\rho \in W^{2,2}(\Omega)$ and $\inf_\Omega \rho > 0$. The sequence of the iterates $\{v_n, \sigma_n\}$ generated by the scheme (7)–(10) converges in $W^{2,2}(\Omega) \times W^{1,2}(\Omega)$ to v, σ where $\sigma = \rho - \rho_0$.*

The proof of this theorem is structured like the corresponding theorem in [5]. We believe that it can be easily modified by substituing other choices of the basic function spaces. The weakest spaces we have in mind using would result in the requirement that $f \in L^p(\Omega)$, for some $p > d$, where d is the dimension of Ω, and would yield a solution with $v \in W^{2,p}(\Omega)$ and $\sigma, \rho \in W^{1,p}(\Omega)$. Similar substitutions of stronger function spaces would yield more regularity of the solution, provided the force is correspondingly more regular. However, we do not know how to show, for instance, that a solution $v \in W^{2,p}(\Omega)$, obtained under the assumption that $f \in L^p(\Omega)$, actually belongs to $W^{m+2,p}(\Omega)$, if $f \in W^{m,p}(\Omega)$, for $m > 0$. Similar and further remarks were made in [5].

The assumption about the boundary that is made in Theorem 3.1 can be weakened. What we need is that the Stokes problem (26) should admit a solution $(v_{n+1}, \pi_{n+1}) \in W^{3,2}(\Omega) \times W^{2,2}(\Omega)$, if the right side $(\widetilde{F}, \widetilde{G})$ belongs to $W^{1,2}(\Omega) \times W^{2,2}(\Omega)$, and that the homogeneous Dirichlet and Neumann problem for the Laplacian admit solutions in $W^{2,2}(\Omega)$ for right sides in $L^2(\Omega)$, all with appropriate estimates. These conditions are satisfied if $\partial\Omega \in C^{2,1}$, but are also satisfied by certain other classes of domains; see, e.g., [4].

4. Regularity lemmas

To prove Theorem 3.1, we will show that the iteration $v_n, \sigma_n \to v_{n+1}, \sigma_{n+1}$ remains bounded in $W^{3,2}(\Omega) \times W^{2,2}(\Omega)$, and is a contraction in $W^{2,2}(\Omega) \times W^{1,2}(\Omega)$. This will be based on some estimates for subsidiary problems that are provided below in Lemmas 4.1 to 4.8. It is tacitly assumed in these lemmas that Ω, U and ρ_0 satisfy the assumptions of Theorem 3.1. The numbered constants $c_1, \dots, c_{10}$ and the generic constant c which are introduced in these lemmas depend at most on Ω, k, λ, μ and ρ_0. The reason for numbering some of the constants is for clarity in the proof of Lemma 5.1, to dispel any question of circular reasoning. Only the estimates with numbered constants are needed in §5. In §6 and §7, all constants are generic.

Lemmas 4.1 to 4.8 will be stated in the course of tracing through the regularity obtained at each intermediate step of the iterative procedure, to be sure that the next step is well defined. To that end, we will take the conditions

$$v_n \in W^{3,2}(\Omega) \cap W_0^{1,2}(\Omega), \quad \sigma_n \in W^{2,2}(\Omega), \quad \nabla \cdot (\sigma_n v_n) \in W^{2,2}(\Omega) \cap W_0^{1,2}(\Omega) \quad (11)$$

as induction hypotheses, and show that they are preserved by our iterative procedure, provided that the norms $\|v_n\|_{3,2}$ remain bounded by a constant c_6 introduced in Lemma 4.6. In the next section we will add several estimates to the forgoing induction hypotheses and prove the boundedness of the iterates, including the bound $\|v_n\|_{3,2} \leq c_6$ that is assumed in this section.

Beginning the inductive proof now, we note that the conditions (11) are certainly satisfied by $v_0 = 0$ and $\sigma_0 = 0$. Taking the conditions (11) as induction hypotheses, they are easily seen to guarantee that the functions g_n and F_n defined by (7) satisfy

$$g_n \in W^{2,2}(\Omega) \cap W_0^{1,2}(\Omega), \quad F_n \in W^{1,2}(\Omega).$$

Here, in verifying that $F_n \in W^{1,2}(\Omega)$, we use the inequalities

$$\|\sigma f\|_{1,2} \leq c\|\sigma\|_{2,2}\|f\|_{1,2},$$

$$\|(\rho_0 + \sigma)v \cdot \nabla v\|_{1,2} \leq c\|\rho_0 + \sigma\|_{2,2}\|v\|_{2,2}^2$$

which hold in virtue of the Sobolev embedding inequalities with some constant c that depends only on Ω, and get

$$\|F_n\|_{1,2} \leq c_1(1 + \|\sigma_n\|_{2,2})(\|f_n\|_{1,2} + \|v_n\|_{2,2}^2). \tag{12}$$

Introducing the bilinear form

$$\mathcal{A}(\phi, \psi) = (\lambda + \mu)\langle \nabla \cdot \phi, \nabla \cdot \psi \rangle + \mu\langle \nabla\phi, \nabla\psi \rangle$$

for $\phi, \psi \in W_0^{1,2}(\Omega)$, it is easily seen that

$$|\mathcal{A}(\phi, \psi)| \leq c\|\phi\|_{1,2}\|\psi\|_{1,2}$$

and that

$$\mathcal{A}(\phi, \phi) = (\lambda + \mu)\|\nabla \cdot \phi\|^2 + \mu\|\nabla\phi\|^2 \geq c\|\nabla\phi\|^2 \tag{13}$$

with c a positive constant depending only on μ, λ and d. The assumptions $\mu > 0$ and $3\lambda + 2\mu \geq 0$ are used in proving (13).

Lemma 4.1. *Let $g_n \in L_0^2(\Omega)$ and $F_n \in W^{-1,2}(\Omega)$ be given. Then, there exists a unique solution $v_{n+1} \in W_0^{1,2}(\Omega)$, $\pi_{n+1} \in L_0^2(\Omega)$ of the weak formulation of problem (8):*

$$\begin{aligned} &\langle \nabla \cdot (\rho_0 v_{n+1}), \chi \rangle = \langle g_n, \chi \rangle \quad \text{for all } \chi \in L^2(\Omega), \\ &\mathcal{A}(v_{n+1}, \phi) = \langle \pi_{n+1}\rho_0^{-1}, \nabla \cdot (\rho_0\phi) \rangle + \langle F_n, \phi \rangle \quad \text{for all } \phi \in W_0^{1,2}(\Omega). \end{aligned} \tag{14}$$

Moreover, v_{n+1} and π_{n+1} satisfy the estimates

$$\|v_{n+1}\|_{1,2} \leq c(\|F_n\|_{-1,2} + \|g_n\|), \tag{15}$$

$$\|\pi_{n+1}\| \leq c(\|F_n\|_{-1,2} + \|g_n\|). \tag{16}$$

Proof. We look for a solution v_{n+1} of the form

$$v_{n+1} = w_{n+1} + V_{n+1},$$

where $w_{n+1} \in J_{1,\rho_0}(\Omega)$ and $V_{n+1} \in W_0^{1,2}(\Omega)$ are functions satisfying

$$\nabla \cdot (\rho_0 V_{n+1}) = g_n, \qquad \|\nabla V_{n+1}\| \leq c\|g_n\|. \tag{17}$$

The solvability of (17) is ensured by [7, Lemma 5.3]. Clearly, the function w_{n+1} must satisfy

$$\mathcal{A}(w_{n+1}, \phi) = G(\phi) \tag{18}$$

for all $\phi \in J_{1,\rho_0}(\Omega)$, where

$$G(\phi) = \langle F_n, \phi \rangle - \mathcal{A}(V_{n+1}, \phi).$$

Furthermore,

$$|G(\phi)| \le c(\|F_n\|_{-1,2} + \|\nabla V_{n+1}\|)\|\phi\|_{1,2} \tag{19}$$

for all $\phi \in W_0^{1,2}(\Omega)$. Hence, the existence of $w_{n+1} \in J_{1,\rho_0}(\Omega)$ satisfying (18) follows from the theorem of Lax–Milgram. Moreover, recalling (17), we conclude from (13) and (19) that

$$\|w_{n+1}\|_{1,2} \le c\big(\|F_n\|_{-1,2} + \|\nabla V_{n+1}\|\big). \tag{20}$$

Since $v_{n+1} = w_{n+1} + V_{n+1}$, together (17) and (20) imply that $v_{n+1} \in W_0^{1,2}(\Omega)$ and satisfies (15).

We now consider the functional

$$\mathcal{F}(\phi) = \mathcal{A}(w_{n+1}, \phi) - G(\phi) \tag{21}$$

which is linear and bounded on $W_0^{1,2}(\Omega)$. By (18), $\mathcal{F}(\phi)$ vanishes when $\phi \in J_{1,\rho_0}(\Omega)$. Therefore, there exists (see [7, Lemma 5.4]) a function $\Pi_{n+1} \in L^2_{\rho_0}(\Omega)$ satisfying

$$\mathcal{F}(\phi) = \big\langle \Pi_{n+1}\rho_0^{-1}, \nabla\cdot(\rho_0\phi)\big\rangle \tag{22}$$

for all $\phi \in W_0^{1,2}(\Omega)$. Setting

$$\pi_{n+1} = \Pi_{n+1} - \left(\int_\Omega \Pi_{n+1}\,dx\right)\left(\int_\Omega \rho_0\,dx\right)^{-1}\rho_0, \tag{23}$$

we find that $\pi_{n+1} \in L_0^2(\Omega)$ and

$$\mathcal{F}(\phi) = \big\langle \pi_{n+1}\rho_0^{-1}, \nabla\cdot(\rho_0\phi)\big\rangle$$

for all $\phi \in W_0^{1,2}(\Omega)$. The proof of the existence of functions $v_{n+1} \in W_0^{1,2}(\Omega)$ and $\pi_{n+1} \in L_0^2(\Omega)$ satisfying (14) and (15) is thus established. To prove (16), we consider the problem

$$\nabla\cdot(\rho_0\Psi) = \Pi_{n+1}\rho_0^{-1}, \quad \Psi \in W_0^{1,2}(\Omega), \quad \|\Psi\|_{1,2} \le c\|\Pi_{n+1}\rho_0^{-1}\|. \tag{24}$$

Since $\Pi_{n+1} \in L^2_{\rho_0}(\Omega)$ and thus $\Pi_{n+1}\rho_0^{-1} \in L_0^2(\Omega)$, problem (24) is solvable (see [7, Lemma 5.3]). By (24), (22), (21) and (15),

$$\begin{aligned}\|\Pi_{n+1}\rho_0^{-1}\|^2 &= \langle \Pi_{n+1}\rho_0^{-1}, \nabla\cdot(\rho_0\Psi)\rangle = \mathcal{F}(\Psi)\\ &= (\lambda+\mu)\langle \nabla\cdot v_{n+1}, \nabla\cdot\Psi\rangle + \mu\langle \nabla v_{n+1}, \nabla\Psi\rangle - \langle F_n, \Psi\rangle\\ &\le c(\|F_n\|_{-1,2} + \|g_n\|)\|\Psi\|_{1,2}.\end{aligned}$$

This implies that

$$\|\Pi_{n+1}\| \le c\|\Pi_{n+1}\rho_0^{-1}\| \le c(\|F_n\|_{-1,2} + \|g_n\|).$$

From the definition (23) of π_{n+1} we conclude that

$$\|\pi_{n+1}\| \le \|\Pi_{n+1}\| + c\|\Pi_{n+1}\|_1 \le c\|\Pi_{n+1}\|.$$

Finally, combining the last two estimates proves the assertion (16). Since the equations (14) are linear, the uniqueness of v_{n+1} and π_{n+1} follows directly from (15) and (16) by standard arguments. □

Lemma 4.2. *Let $g_n \in W^{1,2}(\Omega)$ and $F_n \in L^2(\Omega)$ be given. Then, there exists a unique solution $v_{n+1} \in W^{2,2}(\Omega) \cap W_0^{1,2}(\Omega)$, $\pi_{n+1} \in W^{1,2}(\Omega) \cap L_0^2(\Omega)$ of problem (8). Moreover, v_{n+1} and π_{n+1} satisfy*

$$\|v_{n+1}\|_{2,2} + \|\pi_{n+1}\|_{1,2} \le c_2(\|F_n\| + \|g_n\|_{1,2}). \tag{25}$$

Proof. First, Lemma 4.1 ensures that the variational equations (14) have a unique solution v_{n+1}, π_{n+1} in $W_0^{1,2}(\Omega) \times L_0^2(\Omega)$. Next, we note that problem (14) is equivalent to the weak form of the Stokes system

$$\begin{gathered} \nabla \cdot v_{n+1} = \widetilde{G}, \quad -\mu \Delta v_{n+1} + \nabla \pi_{n+1} = \widetilde{F} \quad \text{in } \Omega, \\ v_{n+1}|_{\partial\Omega} = 0, \quad \int_\Omega \pi_{n+1}\, dx = 0 \end{gathered} \tag{26}$$

with right sides $\widetilde{G}$ and $\widetilde{F}$ defined by

$$\begin{gathered} \widetilde{G} = \rho_0^{-1} g_n - \rho_0^{-1} \nabla \rho_0 \cdot v_{n+1}, \\ \widetilde{F} = F_n + \rho_0^{-1}(\nabla \rho_0)\pi_{n+1} + (\lambda+\mu)\nabla(\rho_0^{-1} g_n) - (\lambda+\mu)\nabla(\rho_0^{-1}\nabla\rho_0 \cdot v_{n+1}). \end{gathered} \tag{27}$$

Due to the inequalities

$$\begin{gathered} \|\widetilde{G}\|_{1,2} \le c(\|g_n\|_{1,2} + \|v_{n+1}\|_{1,2}), \\ \|\widetilde{F}\| \le c(\|F_n\| + \|\pi_{n+1}\| + \|v_{n+1}\|_{1,2} + \|g_n\|_{1,2}) \end{gathered} \tag{28}$$

and in virtue of Lemma 4.1, $\widetilde{F} \in L^2(\Omega)$ and $\widetilde{G} \in W^{1,2}(\Omega)$. Therefore, the usual regularity results (see [1, Theorem 4]) for the Stokes system (26) imply that $v_{n+1} \in W^{2,2}(\Omega)$, $\pi_{n+1} \in W^{1,2}(\Omega)$ and

$$\|v_{n+1}\|_{2,2} + \|\pi_{n+1}\|_{1,2} \le c(\|\widetilde{F}\| + \|\widetilde{G}\|_{1,2}). \tag{29}$$

Together, the inequalities (29), (28), (15) and (16) imply the assertion (25). This completes the proof of Lemma 4.2. □

Lemma 4.3. *Let $g_n \in W^{2,2}(\Omega)$ and $F_n \in W^{1,2}(\Omega)$ be given. Then, there exists a unique solution $v_{n+1} \in W^{3,2}(\Omega) \cap W_0^{1,2}(\Omega)$, $\pi_{n+1} \in W^{2,2}(\Omega) \cap L_0^2(\Omega)$ of problem (8). Moreover, v_{n+1} and π_{n+1} satisfy*

$$\|v_{n+1}\|_{3,2} + \|\pi_{n+1}\|_{2,2} \le c_3(\|F_n\|_{1,2} + \|g_n\|_{2,2}). \tag{30}$$

Proof. Let $v_{n+1} \in W^{2,2}(\Omega) \cap W_0^{1,2}(\Omega)$, $\pi_{n+1} \in W^{1,2}(\Omega) \cap L_0^2(\Omega)$ be the unique solution of problem (8) given by Lemma 4.2. Let $\widetilde{F}$ and $\widetilde{G}$ be defined as in (27).

By the Hölder inequality and Sobolev embedding inequalities we have

$$\begin{aligned} \|\widetilde{G}\|_{2,2} &\le c(\|g_n\|_{2,2} + \|v_{n+1}\|_{2,2}), \\ \|\widetilde{F}\|_{1,2} &\le c(\|F_n\|_{1,2} + \|\pi_{n+1}\|_{1,2} + \|v_{n+1}\|_{2,2} + \|g_n\|_{2,2}). \end{aligned} \tag{31}$$

This shows that $\widetilde{G} \in W^{2,2}(\Omega)$ and $\widetilde{F} \in W^{1,2}(\Omega)$. Now, by the same argument as in the proof of Lemma 4.2 we conclude that $v_{n+1} \in W^{3,2}(\Omega)$, $\pi_{n+1} \in W^{2,2}(\Omega)$ and

$$\|v_{n+1}\|_{3,2} + \|\pi_{n+1}\|_{2,2} \le c(\|\widetilde{F}\|_{1,2} + \|\widetilde{G}\|_{2,2}). \tag{32}$$

Combining (32) with (31) and (25) proves the assertion (30). □

Lemma 4.4. *Suppose $g_n \in W^{2,2}(\Omega)$ and $F_n \in W^{1,2}(\Omega)$ are given. Let $v_{n+1} \in W^{3,2}(\Omega)$, $\pi_{n+1} \in W^{2,2}(\Omega)$ be the unique solution of problem (8) given by Lemma 4.3. Then, H_{n+1} defined by (9) belongs to $W^{2,2}(\Omega)$ and satisfies*

$$\|H_{n+1}\|_{1,2} \le c(\|F_n\| + \|v_{n+1}\|_{2,2}), \tag{33}$$

and

$$\begin{aligned} \|H_{n+1}\|_{1,2} &\le c(\|F_n\| + \|g_n\|_{1,2}), \\ \|H_{n+1}\|_{2,2} &\le c_4(\|F_n\|_{1,2} + \|g_n\|_{2,2}), \\ \|\triangle H_{n+1}\| &\le c_5(\|F_n\|_{1,2} + \|g_n\|_{1,2}), \\ \|\triangle H_{n+1}\|_{-1,2} &\le c(\|F_n\| + \|g_n\|). \end{aligned} \tag{34}$$

Proof. From the assumptions of Lemma 4.4 and the definition (9) of H_{n+1} it follows that H_{n+1} belongs to $W^{2,2}(\Omega)$. The first and second inequalities in (34) are thus a consequence of (9) and of (25), (30). To prove the remaining three estimates, we first derive an auxiliary equation for H_{n+1}. Since $-\triangle v = \nabla \times \nabla \times v - \nabla\nabla \cdot v$, the second identity in (8) implies that

$$\mu\nabla \times \nabla \times v_{n+1} - (\lambda + 2\mu)\nabla\nabla \cdot v_{n+1} + \rho_0 \nabla(\pi_{n+1}\rho_0^{-1}) = F_n. \tag{35}$$

Next, we note that

$$\nabla\nabla \cdot (\rho_0 v_{n+1}) = (\nabla^2 \rho_0) v_{n+1} + (\nabla v_{n+1})^T \nabla\rho_0 + (\nabla \cdot v_{n+1})\nabla\rho_0 + \rho_0 \nabla\nabla \cdot v_{n+1}$$

and thus

$$\nabla\nabla \cdot v_{n+1} = \rho_0^{-1} \nabla\nabla \cdot (\rho_0 v_{n+1}) + \rho_0^{-1} L_{n+1}, \tag{36}$$

with L_{n+1} defined by

$$L_{n+1}(v_{n+1}, \nabla v_{n+1}) = -(\nabla^2\rho_0)v_{n+1} - (\nabla v_{n+1})^T(\nabla\rho_0) - (\nabla \cdot v_{n+1})(\nabla\rho_0). \tag{37}$$

Further, we have

$$\rho_0^2 \nabla(\pi_{n+1}\rho_0^{-1}) = \rho_0\nabla\pi_{n+1} - \pi_{n+1}\nabla\rho_0 = \nabla(\rho_0\pi_{n+1}) - 2\pi_{n+1}\nabla\rho_0. \tag{38}$$

We now substitute (36) and (38) into (35) to find

$$\rho_0\mu\nabla \times \nabla \times v_{n+1} - (\lambda + 2\mu)\nabla\nabla \cdot (\rho_0 v_{n+1}) + \nabla(\rho_0\pi_{n+1}) = \rho_0 F_n + \widetilde{G}_{n+1}, \tag{39}$$

with $\widetilde{G}_{n+1}$ defined by

$$\widetilde{G}_{n+1}(v_{n+1}, \nabla v_{n+1}, \pi_{n+1}) = (\lambda + 2\mu)L_{n+1} + 2\pi_{n+1}\nabla\rho_0. \tag{40}$$

In virtue of (9), equation (39) can be rewritten as

$$\rho_0\mu\nabla \times \nabla \times v_{n+1} + \nabla H_{n+1} = \rho_0 F_n + \widetilde{G}_{n+1}. \tag{41}$$

Clearly, by (37) and (40),

$$\|\rho_0 F_n + \widetilde{G}_{n+1}\| \leq c(\|F_n\| + \|v_{n+1}\|_{1,2} + \|\pi_{n+1}\|) \tag{42}$$

and

$$\|\nabla \cdot (\rho_0 F_n + \tilde{G}_{n+1})\| \leq c(\|F_n\|_{1,2} + \|v_{n+1}\|_{2,2} + \|\pi_{n+1}\|_{1,2}).$$

Now, from (41) and (42) one gets

$$\|\nabla H_{n+1}\| \leq c(\|F_n\| + \|v_{n+1}\|_{2,2} + \|\pi_{n+1}\|).$$

Also directly from (9),

$$\|H_{n+1}\| \leq c(\|\pi_{n+1}\| + \|v_{n+1}\|_{1,2}).$$

Using (16) to estimate $\|\pi_{n+1}\|$ in the previous two inequalities yields

$$\|H_{n+1}\|_{1,2} \leq c(\|F_n\| + \|v_{n+1}\|_{2,2} + \|g_n\|).$$

In virtue of the identity $g_n = \nabla \cdot (\rho_0 v_{n+1})$, this last estimate implies (33).

Since $\nabla \cdot \nabla \times w = 0$ for any function $w \in W^{2,2}(\Omega)$, equation (41) implies

$$\triangle H_{n+1} = -\mu\nabla\rho_0 \cdot \nabla \times \nabla \times v_{n+1} + \nabla \cdot \big(\rho_0 F_n + \widetilde{G}_{n+1}\big).$$

Therefore,

$$\|\triangle H_{n+1}\| \leq c\big(\|F_n\|_{1,2} + \|v_{n+1}\|_{2,2} + \|\pi_{n+1}\|_{1,2}\big)$$

and

$$\|\triangle H_{n+1}\|_{-1,2} \leq c\big(\|F_n\| + \|v_{n+1}\|_{1,2} + \|\pi_{n+1}\|\big).$$

These two estimates combined with (15), (16) and (25) yield the third and fourth of the inequalities (34). This completes the proof of Lemma 4.4. □

Carefully analysing the constants in the proof of [6, Theorem 9], we obtain the following lemma.

Lemma 4.5. *Consider the transport equation*

$$k\sigma + \gamma\nabla \cdot (\sigma v) = H. \tag{43}$$

Suppose that $v \in W^{3,2}(\Omega)$, $v \cdot \nu|_{\partial\Omega} = 0$ *and* $k > \frac{\gamma}{2}\|\nabla \cdot v\|_\infty + 2\gamma\|Dv\|_\infty$, *where* $Dv = \frac{1}{2}\big(\nabla v + (\nabla v)^T\big)$ *and* $\|Dv\|_\infty = \sup_\Omega \big(\sum_{i,j}(Dv)^2_{i,j}\big)^{1/2}$. *Then, for any prescribed right side* $H \in W^{2,2}(\Omega)$, *there exists a unique solution* $\sigma \in W^{2,2}(\Omega)$ *of equation (43) and it satisfies the estimates*

$$\|\sigma\|_{1+i,2} \leq C\mathcal{D}\|H\|_{1+i,2}, \quad i \in \{0,1\},$$

$$\|\triangle\sigma\|_{-i,2} \leq \mathcal{D}\|\triangle H\|_{-i,2} + C\mathcal{D}^2\|v\|_{3,2}\|H\|_{2-i,2}, \quad i \in \{0,1\},$$

with $\mathcal{D} = \left(k - \frac{\gamma}{2}\|\nabla\cdot v\|_\infty - 2\gamma\|Dv\|_\infty\right)^{-1}$ *and some constants* $\mathcal{C} = \mathcal{C}(\Omega, \gamma, \|v\|_{3,2}, \mathcal{D})$ *that can be bounded above (monotonically) in terms of* $\mathcal{D}$, γ, $\|v\|_{3,2}$ *and Sobolev constants for* Ω.

The reader who checks these constants carefully will find more precisely that $\mathcal{C} = c(1 + \|v\|_{3,2} + \mathcal{D}\|v\|_{3,2})$, where c depends on γ and Ω. However, only the monotonicity of the dependence of $\mathcal{C}$ on $\mathcal{D}$ and $\|v\|_{3,2}$ is needed in what follows.

Applying now Lemma 4.5 to the transport equation (10) yields:

Lemma 4.6. *There exists a positive constant* c_6*, such that if* $v_{n+1} \in W^{3,2}(\Omega)$ *with* $\nu \cdot v_{n+1}|_{\partial\Omega} = 0$*, and if*

$$\|v_{n+1}\|_{3,2} \le c_6, \tag{44}$$

then for any prescribed right side $H_{n+1} \in W^{2,2}(\Omega)$*, there exists a unique solution* $\sigma_{n+1} \in W^{2,2}(\Omega)$ *of the transport equation (10), and it satisfies*

$$\begin{aligned} \|\sigma_{n+1}\|_{1,2} &\le c\|H_{n+1}\|_{1,2}, \\ \|\sigma_{n+1}\|_{2,2} &\le c_7\|H_{n+1}\|_{2,2}, \\ \|\Delta(\rho_0\sigma_{n+1})\| &\le c\|\Delta H_{n+1}\| + c\|v_{n+1}\|_{3,2}\|H_{n+1}\|_{2,2}, \\ \|\Delta(\rho_0\sigma_{n+1})\|_{-1,2} &\le c\|\Delta H_{n+1}\|_{-1,2} + c\|v_{n+1}\|_{3,2}\|H_{n+1}\|_{1,2}, \end{aligned} \tag{45}$$

and also, as a consequence of the third of these inequalities,

$$\|\Delta\nabla\cdot(\sigma_{n+1}v_{n+1})\| \le c_8\|\Delta H_{n+1}\| + c_9\|v_{n+1}\|_{3,2}\|H_{n+1}\|_{2,2}. \tag{46}$$

Proof. In order to prove Lemma 4.6, we rewrite the identity (10) in the form

$$k(\rho_0\sigma_{n+1}) + (\lambda + 2\mu)\nabla\cdot\left((\rho_0\sigma_{n+1})(v_{n+1}\rho_0^{-1})\right) = H_{n+1}. \tag{47}$$

Now, we apply Lemma 4.5 to equation (47) with the identifications $\sigma = \rho_0\sigma_{n+1}$ and $v = v_{n+1}\rho_0^{-1}$. Noting that $\|\sigma_{n+1}\|_{i,2} = \|(\rho_0\sigma_{n+1})\rho_0^{-1}\|_{i,2} \le c\|\rho_0\sigma_{n+1}\|_{i,2}$ for $i \in \{0,1,2\}$ and that $\|v_{n+1}\rho_0^{-1}\|_{3,2} \le c\|v_{n+1}\|_{3,2}$, we conclude the existence of a unique solution $\sigma_{n+1} \in W^{2,2}(\Omega)$ of equation (10) which satisfies the estimates (45). □

From the second of the inequalities (45) and the transport equation (10), it follows that $\nabla\cdot(\sigma_{n+1}v_{n+1}) \in W^{2,2}(\Omega)$. Thus, as long as the values of $\|v_{n+1}\|_{3,2}$ remain less than c_6, the sequence of functions v_{n+1}, σ_{n+1} will continue to be well defined, and will satisfy the first two of the induction hypotheses (11) and the first inclusion of the third hypothesis in (11).

Further, the induction hypothesis $\nabla\cdot(\sigma_n v_n) \in W^{2,2}(\Omega) \cap W_0^{1,2}(\Omega)$ together with the condition $\nabla\cdot(\rho_0 v_{n+1}) = -\nabla\cdot(\sigma_n v_n)$ in the Stokes-type problem (8) implies that $\nabla\cdot(\rho_0 v_{n+1}) \in W^{2,2}(\Omega) \cap W_0^{1,2}(\Omega)$. Due to this fact, the second inclusion of the third hypothesis in (11) is also preserved under iteration as seen from the following lemma.

Lemma 4.7. *The conditions* $v_{n+1} \in W^{3,2}(\Omega) \cap W_0^{1,2}(\Omega)$, $\sigma_{n+1} \in W^{2,2}(\Omega)$ *and* $\nabla \cdot (\rho_0 v_{n+1}) \in W^{2,2}(\Omega) \cap W_0^{1,2}(\Omega)$ *imply that* $\nabla \cdot (\sigma_{n+1} v_{n+1}) \in W_0^{1,2}(\Omega)$.

Proof. This is easily proved writing

$$\begin{aligned}\nabla \cdot (\sigma_{n+1} v_{n+1}) &= (\nabla \sigma_{n+1}) \cdot v_{n+1} + \sigma_{n+1} \nabla \cdot v_{n+1} \\ &= (\nabla \sigma_{n+1}) \cdot v_{n+1} + \sigma_{n+1} \rho_0^{-1} \nabla \cdot (\rho_0 v_{n+1}) - \sigma_{n+1} \rho_0^{-1} (\nabla \rho_0) \cdot v_{n+1},\end{aligned}$$

and noting that the right side is a sum of terms, each of which is the product of a function belonging to $W^{1,2}(\Omega)$ with a function belonging to $W^{2,2}(\Omega) \cap W_0^{1,2}(\Omega)$. The product of such functions must belong to $W_0^{1,2}(\Omega)$. Indeed, given any $\phi \in W^{2,2}(\Omega) \cap W_0^{1,2}(\Omega)$ and any $\psi \in W^{1,2}(\Omega)$, we can approximate ψ in $W^{1,2}(\Omega)$ by functions $\psi_n \in C^1(\overline{\Omega})$. The products $\psi_n \phi$ belong to $W_0^{1,2}(\Omega)$, since they belong to $W^{1,2}(\Omega) \cap C(\overline{\Omega})$ and vanish on $\partial\Omega$. Since $\psi_n \phi \to \psi\phi$ in $W^{1,2}(\Omega)$, it follows that $\psi\phi \in W_0^{1,2}(\Omega)$. □

This completes the induction argument for the preservation of the conditions (11), modulo the hypothesis $\|v_{n+1}\|_{3,2} \leq c_6$ of Lemma 4.6, which will be established in the next section along with other estimates. There we will also need the following lemma.

Lemma 4.8. *The solutions* v_{n+1}, σ_{n+1} *of Lemmas 4.3 and 4.6 satisfy*

$$\|\nabla \cdot (\sigma_{n+1} v_{n+1})\|_{2,2} \leq c_{10} \|\Delta \nabla \cdot (\sigma_{n+1} v_{n+1})\|.$$

Proof. This is a standard elliptic regularity estimate for $\varphi = \nabla \cdot (\sigma_{n+1} v_{n+1})$ considered as the unique solution of a Dirichlet problem $\Delta\varphi = \Delta(\nabla \cdot (\sigma_{n+1} v_{n+1}))$, $\varphi|_{\partial\Omega} = 0$. □

5. Bounds for the iterates

The following lemma completes the proof that the iterates of scheme (7)–(10) are well defined and bounded. In the course of proving it we introduce further constants $b_1, \ldots, b_{11}$, which, like the constants $c_1, \ldots, c_{10}$ of the last section, depend only on Ω, k, λ, μ and ρ_0.

Lemma 5.1. *If* R, L, B *are chosen to satisfy*

$$\begin{aligned} R \leq R^* = \min\{1, c_6/(c_3 b_1 + c_3), 1/(2b_4), b_{10}, b_{11}\}, \\ L = 2b_6 R^2, \qquad B = R^2, \end{aligned} \tag{48}$$

and if $\|f\|_{1,2} \leq B$, *then the iterates* v_n, σ_n *of scheme (7)–(10) defined in §3 satisfy the hypothesis (44) of Lemma 4.6 and the estimates*

$$\begin{aligned} \|v_n\|_{3,2} + \|\sigma_n\|_{2,2} \leq R, \\ \|\nabla \cdot (\sigma_n v_n)\|_{2,2} \leq L, \end{aligned} \tag{49}$$

as well as the previously considered induction hypotheses (11).

Proof. We take (49) along with (11) as induction hypotheses. During the course of the proof, we will assume a number of restrictions on the size of L, namely that

$$L \leq \min\{1, R, R/(2b_5), 1/(2b_7)\}.$$

At the end of the proof, it will be shown that these restrictions are satisfied by L as defined in (48).

By (12), (48), (49) and under the restriction $R \leq 1$, we have

$$\|F_n\|_{1,2} \leq 2c_1(1+R)R^2 \leq b_1 R^2$$

where $b_1 = 4c_1$. Therefore, the estimate (30) yields

$$\|v_{n+1}\|_{3,2} \leq c_3 b_1 R^2 + c_3 L. \tag{50}$$

Using the second of the estimates (34), together with (48) and (49), gives

$$\|H_{n+1}\|_{2,2} \leq b_2 R^2 + c_4 L \tag{51}$$

where $b_2 = c_4 b_1$. Recalling (7), we find by Hölder's inequality and the Sobolev embedding theorems that

$$\|g_n\|_{1,2} = \|\nabla \cdot (\sigma_n v_n)\|_{1,2} \leq b_3 \|\sigma_n\|_{2,2} \|v_n\|_{2,2}.$$

Therefore, by the third inequality of (34) we obtain

$$\|\triangle H_{n+1}\| \leq c_5(b_1 + b_3)R^2. \tag{52}$$

The hypothesis $\|v_{n+1}\|_{3,2} \leq c_6$ of Lemma 4.6 is satisfied in virtue of (50) and the restrictions $L \leq R$, $R \leq 1$, and $R \leq c_6/(c_3 b_1 + c_3)$. Therefore, the second of the estimates (45) yields

$$\|\sigma_{n+1}\|_{2,2} \leq c_7 b_2 R^2 + c_7 c_4 L. \tag{53}$$

Combining (50) and (53), we have

$$\|v_{n+1}\|_{3,2} + \|\sigma_{n+1}\|_{2,2} \leq b_4 R^2 + b_5 L,$$

where $b_4 = c_3 b_1 + c_7 b_2$ and $b_5 = c_3 + c_7 c_4$. Therefore, v_{n+1}, σ_{n+1} satisfy the first of the induction estimates (49) in virtue of the restrictions $L \leq R/(2b_5)$, $R \leq 1/(2b_4)$ and $R \leq 1$.

To prove the second of the estimates (49) we use first Lemma 4.8 and then (46), to obtain

$$\begin{aligned} \|\nabla \cdot (\sigma_{n+1} v_{n+1})\|_{2,2} &\leq c_{10} \|\triangle \nabla \cdot (\sigma_{n+1} v_{n+1})\| \\ &\leq c_8 c_{10} \|\triangle H_{n+1}\| + c_9 c_{10} \|v_{n+1}\|_{3,2} \|H_{n+1}\|_{2,2}. \end{aligned}$$

Next, by (52), (50), (51) and the restriction $L \leq R \leq 1$,

$$\begin{aligned} \|\nabla \cdot (\sigma_{n+1} v_{n+1})\|_{2,2} &\leq c_5 c_8 c_{10}(b_1 + b_3)R^2 + c_9 c_{10}(c_3 b_1 R^2 + c_3 L)(b_2 R^2 + c_4 L) \\ &\leq b_6 R^2 + b_7 L^2 \end{aligned}$$

with suitably defined b_6 and b_7. In virtue of the restriction $L \leq 1/(2b_7)$, we have $b_6R^2 + b_7L^2 \leq L$, confirming that v_{n+1}, σ_{n+1} satisfy (49), provided that

$$2b_6R^2 \leq L. \tag{54}$$

This last requirement is consistent with our preceding restrictions. Indeed, the restrictions used so far are $R \leq R_1 \equiv \min\{1, c_6/(c_3b_1 + c_3), 1/(2b_4)\}$ and $L \leq b_8$ and $L \leq b_9R$, where $b_8 = \min\{1, 1/(2b_7)\}$ and $b_9 = \min\{1, 1/(2b_5)\}$. The curve $2b_6R^2 = L$, bounding the region (54) in the R, L-plane, intersects $L = b_8$ at $R = b_{10} \equiv \sqrt{b_8/2b_6}$, and intersects $L = b_9R$ at $R = b_{11} \equiv b_9/2b_6$. These then present two additional restrictions, namely $R \leq b_{10}$ and $R \leq b_{11}$. If $R \leq R^* = \min\{1, c_6/(c_3b_1 + c_3), 1/(2b_4), b_{10}, b_{11}\}$ and $L = 2b_6R^2$, then L will satisfy all of the restrictions that have been put upon it.

Therefore, we have reached the desired conclusion that every iterate will satisfy (49), provided $\|f\|_{1,2} \leq B \equiv R^2$. This completes the proof. □

6. Convergence of the iterates

It remains to prove the convergence of the iterative scheme. We denote by v'_{n+1}, π'_{n+1}, g'_{n+1}, F'_{n+1}, H'_{n+1} and σ'_{n+1} the differences $v'_{n+1} \equiv v_{n+1} - v_n$ etc.. For $n \geq 1$, they satisfy the Stokes-type problem

$$\nabla \cdot (\rho_0 v'_{n+1}) = g'_n,$$

$$-(\lambda + \mu)\nabla\nabla \cdot v'_{n+1} - \mu\Delta v'_{n+1} = -\rho_0\nabla\left(\pi'_{n+1}\rho_0^{-1}\right) + F'_n,$$

$$v'_{n+1}|_{\partial\Omega} = 0, \qquad \int_\Omega \pi'_{n+1}\,dx = 0.$$

The right sides g'_n and F'_n can be rewritten as

$$\begin{aligned} g'_n &= -\nabla \cdot (\sigma_n v_n) + \nabla \cdot (\sigma_{n-1}v_{n-1}) = -\nabla \cdot (\sigma_n v'_n) - \nabla \cdot (\sigma'_n v_{n-1}), \\ F'_n &= \sigma'_n f - (\rho_0 + \sigma_n)v_n \cdot \nabla v'_n - (\rho_0 + \sigma_n)v'_n \cdot \nabla v_{n-1} - \sigma'_n v_{n-1} \cdot \nabla v_{n-1}. \end{aligned} \tag{55}$$

Further, they satisfy the linear equation (9) as well as the transport equation

$$k\rho_0\sigma'_{n+1} + (\lambda + 2\mu)\nabla \cdot (\sigma'_{n+1}v_{n+1}) = \mathcal{H}'_{n+1},$$

with right side $\mathcal{H}'_{n+1}$ defined by

$$\mathcal{H}'_{n+1} = H'_{n+1} - (\lambda + 2\mu)\nabla \cdot (\sigma_n v'_{n+1}). \tag{56}$$

Hence, $v'_{n+1}, \pi'_{n+1}, H'_{n+1}$ and σ'_{n+1} satisfy the estimates of Lemmas 4.2, 4.4 and 4.6, with F'_n, g'_n replacing F_n, g_n in Lemmas 4.2 and 4.4, and $\mathcal{H}'_{n+1}$ replacing H_{n+1} in Lemma 4.6.

For convenience, let us also recall from Lemma 5.1, that if $R \leq R^*$, and if

$$\|f\|_{1,2} \leq B = R^2, \tag{57}$$

then, for all $n \geq 0$ we have

$$\|v_n\|_{3,2} + \|\sigma_n\|_{2,2} \leq R. \tag{58}$$

Lemma 6.1. *If R and f satisfy the hypotheses of Lemma 5.1, then*

$$\|v'_{n+1}\|_{2,2} + \|\sigma'_{n+1}\|_{1,2} \leq cR\big(\|v'_n\|_{2,2} + \|\sigma'_n\|_{1,2} + \|v'_{n-1}\|_{2,2} + \|\sigma'_{n-1}\|_{1,2}\big), \tag{59}$$

for all $n \geq 1$, where c is a constant that depends only on Ω, k, λ, μ and ρ_0, and where we have set $v'_0 \equiv 0$ and $\sigma'_0 \equiv 0$. Under the additional assumption that $cR < 1/2$, the inequality (59) implies the geometric convergence of the iterates v_n, σ_n of the scheme (7)–(10) in $W^{2,2}(\Omega) \times W^{1,2}(\Omega)$. The limit v, σ belongs to $W^{3,2}(\Omega) \times W^{2,2}(\Omega)$ and satisfies all of the equations and conditions of problem (1).

Proof. From Lemmas 4.2 and 4.6, we have

$$\begin{aligned} \|v'_{n+1}\|_{2,2} &\leq c\|F'_n\| + c\|g'_n\|_{1,2}, \\ \|\sigma'_{n+1}\|_{1,2} &\leq c\|\mathcal{H}'_{n+1}\|_{1,2}. \end{aligned} \tag{60}$$

To prove (59) we need to bound the terms on the right sides of (60) by expressions of the form $cR\big(\|v'_n\|_{2,2} + \|\sigma'_n\|_{1,2} + \|v'_{n-1}\|_{2,2} + \|\sigma'_{n-1}\|_{1,2}\big)$.

From (55), using (57) and (58), we obtain

$$\begin{aligned} \|F'_n\| &\leq cR(1+R)\|v'_n\|_{1,2} + cR^2\|\sigma'_n\|_{1,2}, \\ \|g'_n\| &\leq cR\|v'_n\|_{1,2} + cR\|\sigma'_n\|_{1,2}. \end{aligned} \tag{61}$$

When estimating g'_n in $W^{1,2}(\Omega)$, we need to treat separately the case $n = 1$. Since $v_0 = 0$, and $v'_1 = v_1$, it follows from (55) and (58) that

$$\|g'_1\|_{1,2} = \|\nabla \cdot (\sigma_1 v_1)\|_{1,2} \leq c\|\sigma_1\|_{2,2}\|v_1\|_{2,2} \leq cR\|v'_1\|_{2,2}, \tag{62}$$

which provides a suitable estimate for g'_1 on the right side of (60).

To estimate the $W^{1,2}$-norm of g'_n for $n \geq 2$, we take the difference between the transport equation (10) for σ_n and for σ_{n-1} to get

$$g'_n = \frac{k}{\lambda + 2\mu}\rho_0\sigma'_n - \frac{1}{\lambda + 2\mu}H'_n. \tag{63}$$

Arguing as in the proof of Lemma 4.8, and then using (63) and finally the fourth of the estimates (34) along with (61) and the restriction $R \leq 1$, we get

$$\begin{aligned} \|g'_n\|_{1,2} &\leq c\|\Delta g'_n\|_{-1,2} \leq c\|\Delta(\rho_0\sigma'_n)\|_{-1,2} + c\|\Delta H'_n\|_{-1,2} \\ &\leq c\|\Delta(\rho_0\sigma'_n)\|_{-1,2} + c\|F'_{n-1}\| + c\|g'_{n-1}\|. \end{aligned} \tag{64}$$

Since $\|\Delta(\rho_0\sigma'_n)\|_{-1,2} \leq c\|\Delta\mathcal{H}'_n\|_{-1,2} + c\|v_n\|_{3,2}\|\mathcal{H}'_n\|_{1,2}$ by Lemma 4.6, we obtain when substituting for $\mathcal{H}'_n$ from (56),

$$\begin{aligned} \|\Delta(\rho_0\sigma'_n)\|_{-1,2} &\leq c\|\Delta H'_n\|_{-1,2} + c\|\Delta\nabla \cdot (\sigma_{n-1}v'_n)\|_{-1,2} \\ &\quad + c\|v_n\|_{3,2}\|H'_n\|_{1,2} + c\|v_n\|_{3,2}\|\nabla \cdot (\sigma_{n-1}v'_n)\|_{1,2}. \end{aligned} \tag{65}$$

Now, we use in (65) the fourth estimate of (34) to deal with $\triangle H_n'$, the inequality (33) to deal with H_n', the Hölder inequality, Sobolev imbeddings and $\|\nabla\cdot\varphi\|_{-1,2} \le \|\varphi\|$ to estimate the nonlinear terms, and get

$$\begin{aligned}\|\triangle(\rho_0\sigma_n')\|_{-1,2} &\le c\|F_{n-1}'\| + c\|g_{n-1}'\| + c\|\sigma_{n-1}\|_{2,2}\|v_n'\|_{2,2} + c\|v_n\|_{3,2}\|F_{n-1}'\| \\ &\quad + c\|v_n\|_{3,2}\|v_n'\|_{2,2} + c\|v_n\|_{3,2}\|\sigma_{n-1}\|_{2,2}\|v_n'\|_{2,2}.\end{aligned} \tag{66}$$

Combining (66) with (61), taking into account estimates (57), (58) and the restriction $R \le 1$, one gets

$$\|\triangle(\rho_0\sigma_n')\|_{-1,2} \le cR(\|v_n'\|_{2,2} + \|v_{n-1}'\|_{1,2} + \|\sigma_{n-1}'\|_{1,2}). \tag{67}$$

Substituting (61) and (67) into (64) gives the desired estimate of g_n',

$$\|g_n'\|_{1,2} \le cR(\|v_n'\|_{2,2} + \|v_{n-1}'\|_{1,2} + \|\sigma_{n-1}'\|_{1,2}). \tag{68}$$

Now, from the first inequality of (60) when using (61) and (68), one obtains

$$\|v_{n+1}'\|_{2,2} \le cR\big(\|v_n'\|_{2,2} + \|\sigma_n'\|_{1,2} + \|v_{n-1}'\|_{1,2} + \|\sigma_{n-1}'\|_{1,2}\big). \tag{69}$$

Finally, we estimate $\mathcal{H}_{n+1}'$ on the right side of (60). Recalling its definition (56), we easily find, $\|\mathcal{H}_{n+1}'\|_{1,2} \le \|H_{n+1}'\|_{1,2} + c\|\nabla\cdot(\sigma_n v_{n+1}')\|_{1,2}$. After applying the first estimate of (34) to the term $\|H_{n+1}'\|_{1,2}$ and the Hölder inequality and Sobolev imbeddings to the second term, we get

$$\|\mathcal{H}_{n+1}'\|_{1,2} \le c(\|F_n'\| + \|g_n'\|_{1,2} + \|\sigma_n\|_{2,2}\|v_{n+1}'\|_{2,2}),$$

which in turn with (61), (68) and (69) yields

$$\|\mathcal{H}_{n+1}'\|_{1,2} \le cR\big(\|v_n'\|_{2,2} + \|\sigma_n'\|_{1,2} + \|v_{n-1}'\|_{1,2} + \|\sigma_{n-1}'\|_{1,2}\big). \tag{70}$$

Estimate (59) now follows from (60), (69) and (70).

To prove that (59) implies the geometric convergence of the iterates v_n, σ_n in $W^{2,2}(\Omega)\times W^{1,2}(\Omega)$, let $\alpha_n \equiv \|v_n'\|_{2,2} + \|\sigma_n'\|_{1,2}$ for $n \ge 0$ and let $\gamma \equiv cR$. Then,

$$\alpha_0 = 0, \qquad \alpha_1 = \|v_1\|_{2,2} + \|\sigma_1\|_{1,2}, \qquad \alpha_{n+1} \le \gamma(\alpha_n + \alpha_{n-1}) \quad \text{for } n \ge 1.$$

The terms of this sequence are majorized, after a normalization of α_1, by those of the sequence

$$a_0 = 0, \qquad a_1 = 1, \qquad a_{n+1} = \gamma(a_n + a_{n-1}) \quad \text{for } n \ge 1.$$

We will now show this sequence to be geometrically decreasing if $\gamma < 1/2$. Letting

$$r_{n+1} = \frac{a_{n+1}}{a_n} = \gamma\Big(1 + \frac{1}{r_n}\Big), \tag{71}$$

we note that

$$|r_{n+2} - r_{n+1}| = \frac{\gamma}{r_{n+1}r_n}|r_{n+1} - r_n|.$$

Since $r_{n+1} = \gamma(1 + r_n^{-1}) > \gamma$, there holds

$$\frac{\gamma}{r_{n+1}r_n} = \frac{1}{1+r_n} < \frac{1}{1+\gamma}.$$

Hence, the sequence of ratios r_{n+1} are geometrically convergent to some limit r. By passing to the limit in (71), we find that

$$r = \frac{\gamma + \sqrt{\gamma^2 + 4\gamma}}{2},$$

which is less than 1 for $\gamma < 1/2$. Thus, for $\gamma < 1/2$, the sequence a_n is also geometrically convergent, and, by comparison, so is the sequence α_n. This completes the proof of the convergence of v_n, σ_n in $W^{2,2}(\Omega) \times W^{1,2}(\Omega)$ to a limit v, σ. Since the iterates are bounded in $W^{3,2}(\Omega) \times W^{2,2}(\Omega)$, their limit must also belong to this space, by a standard argument. If R is sufficiently small that

$$\|\sigma\|_{2,2} \leq R \quad \text{implies} \quad \sup_{\Omega} |\sigma| < \alpha \leq \rho_0(x), \tag{72}$$

the density will be positive. Passing to the limit in the equations (7), (8), (9) and (10), setting $\rho = \rho_0 + \sigma$ and $p = k\rho$ and recalling (3), it is easily seen that (v, ρ) solves the problem (1). This completes the proof of Lemma 6.1. □

7. Uniqueness in the ball of existence

The proof of convergence given in Lemma 6.1 suggests an analogous argument for the local uniqueness of the solution that was obtained. The resulting uniqueness theorem is independent of the existence theorem, in that it applies to any solutions satisfying the stated hypotheses. The condition (72) can be dropped by simply assuming that the solutions under consideration have positive densities.

Lemma 7.1. *If R and f satisfy the hypotheses of Lemmas 5.1 and 6.1, then there can be at most one solution of the problem (1) satisfying* $\inf_\Omega(\rho_0 + \sigma) > 0$ *and*

$$\|v\|_{3,2} + \|\sigma\|_{2,2} \leq R.$$

Proof. Given one solution v, σ such as that obtained in Lemma 6.1 when (72) is assumed, we set

$$\begin{aligned} g &= -\nabla \cdot (\sigma v), \qquad & F &= (\rho_0 + \sigma)(f - v \cdot \nabla v), \\ p &= k(\rho_0 + \sigma), & \pi &= k\sigma. \end{aligned} \tag{73}$$

Then, we have the identity

$$\nabla \cdot (\rho_0 v) = g,$$

$$-(\lambda + \mu)\nabla\nabla \cdot v - \mu\Delta v = -\rho_0 \nabla(\pi \rho_0^{-1}) + F,$$

$$v|_{\partial\Omega} = 0, \qquad \int_\Omega \pi \, dx = 0.$$

If we set

$$H = \rho_0 \pi - (\lambda + 2\mu)\nabla \cdot (\rho_0 v), \tag{74}$$

the identity $k\rho_0\sigma + (\lambda + 2\mu)\nabla \cdot (\sigma v) = H$ also holds.

If $\overline{v}, \overline{\sigma}$ is a second solution satisfying the hypotheses of the lemma, we define $\overline{g}, \overline{F}, \overline{p}, \overline{\pi}, \overline{H}$ similarly as in (73) and (74), and then define $v', \sigma', g', F', p', \pi', H'$ by setting $v' = \overline{v} - v$ etc..

Analogously to (59), we will now show that

$$\|v'\|_{2,2} + \|\sigma'\|_{1,2} \leq 2cR(\|v'\|_{2,2} + \|\sigma'\|_{1,2}), \tag{75}$$

if $R \leq R^*$, and hence that $v' = 0$, $\sigma' = 0$ if R also satisfies $cR < 1/2$. The constant c here is the same as in (59), because the proofs of (75) and (59) are based, step by step, on the same estimates.

The differences v', g', F', π' satisfy

$$\nabla \cdot (\rho_0 v') = g',$$

$$-(\lambda + \mu)\nabla\nabla \cdot v' - \mu\Delta v' = -\rho_0 \nabla(\pi' \rho_0^{-1}) + F',$$

$$v'|_{\partial\Omega} = 0, \qquad \int_\Omega \pi' \, dx = 0.$$

Also, we can write

$$g' = -\nabla \cdot (\overline{\sigma}\,\overline{v}) + \nabla \cdot (\sigma v) = -\nabla \cdot (\overline{\sigma} v') - \nabla \cdot (\sigma' v),$$

$$F' = \sigma' f - (\rho_0 + \bar{\sigma})\bar{v} \cdot \nabla v' - (\rho_0 + \bar{\sigma})v' \cdot \nabla v - \sigma' v \cdot \nabla v.$$

For H' an analogue of equation (74) holds. Also, σ' satisfies the transport equation

$$k\rho_0\sigma' + (\lambda + 2\mu)\nabla \cdot (\sigma'\overline{v}) = \mathcal{H}',$$

with right side

$$\mathcal{H}' = H' - (\lambda + 2\mu)\nabla \cdot (\sigma v').$$

From Lemmas 4.2 and 4.6 we have

$$\begin{aligned} \|v'\|_{2,2} &\leq c\,\|F'\| + c\|g'\|_{1,2}, \\ \|\sigma'\|_{1,2} &\leq c\|\mathcal{H}'\|_{1,2}. \end{aligned} \tag{76}$$

For F', we have directly

$$\|F'\| \leq cR(1 + R)\|v'\|_{1,2} + cR^2\|\sigma'\|_{1,2}. \tag{77}$$

The quantity g' satisfies

$$\|g'\|_{1,2} \leq 2cR(\|v'\|_{2,2} + \|\sigma'\|_{1,2}), \tag{78}$$

as one easily verifies repeating the arguments (63)–(68) in this slightly different situation.

Finally, similarly to (70), we get

$$\|\mathcal{H}'\|_{1,2} \leq 2cR(\|v'\|_{2,2} + \|\sigma'\|_{1,2}). \tag{79}$$

Estimates (76) together with (77), (78) and (79) yield under the condition $R \leq 1$, the inequality (75). This completes the proof. □

Acknowledgments

This research was supported by the Natural Sciences and Engineering Research Council of Canada under grant A4150, and by the Deutsche Forschungsgemeinschaft.

References

[1] C. Amrouche and V. Girault, *On the existence and regularity of the solutions of Stokes problem in arbitrary dimension*, Proc. Japan Acad., **67** (1991), pp. 171–175.

[2] M. Bause, J. G. Heywood, A. Novotny and M. Padula, *Numerical approximation schemes for the Poisson–Stokes equations of steady compressible viscous flow*, to appear in: JMFM.

[3] F. Golay and P. Helluy, *Numerical simulation of a viscous compressible fluid based on a splitting method*, Preprint 2000/08, Université de Toulon et du Var, 2000.

[4] P. Grisvard *Elliptic problems in nonsmooth domains*, Pitman, London, 1985.

[5] J. G. Heywood and M. Padula, *On the existence and uniqueness theory for steady compressible viscous flow*, in G. Galdi, J. Heywood, R. Rannacher (eds.): Fundamental Directions in Mathematical Fluid Mechanics, Birkhäuser, Basel, 2000.

[6] J. G. Heywood and M. Padula, *On the steady transport equation*, in G. Galdi, J. Heywood, R. Rannacher (eds.): Fundamental Directions in Mathematical Fluid Mechanics, Birkhäuser, Basel, 2000.

[7] A. Novotný and K. Pileckas, *Steady compressible Navier–Stokes equations with large potential forces via a method of decomposition*, Math. Meth. Appl. Sci., **21** (1998), pp. 665–684.

Markus Bause
Institut für Angewandte Mathematik, Universität Erlangen-Nürnberg
Martensstraße 3, 91058 Erlangen, Germany
E-mail address: `bause@am.uni-erlangen.de`

John G. Heywood
Department of Mathematics, University of British Columbia
1984 Mathematics Road, Vancouver, B.C. V6T 1Z2, Canada
E-mail address: `heywood@math.ubc.ca`

Antonín Novotný
Mathematiques, Université de Toulon et du Var
BP 132, 83957 La Garde, France
E-mail address: `novotny@univ-tln.fr`

Mariarosaria Padula
Dipartimento di Matematica, Università di Ferrara
Via Machiavelli 35, 44100 Ferrara, Italy
E-mail address: `pad@dns.unife.it`

Asymptotic Results for the Linear Stage of the Rayleigh–Taylor Instability

Catherine Cherfils–Clérouin, Olivier Lafitte and Pierre–Arnaud Raviart

Abstract. We present a mathematical rigorous analysis of the linear Rayleigh–Taylor instability for some simple model flows which occur in connection with inertial confinement fusion. We consider the effect of a smooth density profile and of a convection velocity on the growth rate of the instability. We use an asymptotic analysis for studying the growth of large wavelength perturbations.

1. Introduction

The study of the stability of fluid flows is a classical and fundamental subject in Fluid Mechanics and, since the pioneering work of Lord Rayleigh [10], an extensive litterature has been devoted to the field. This is true in particular of the Rayleigh-Taylor (RT) instability [10, 11] which occurs at the interface of two fluids of different densities when the system experiments a constant acceleration field directed toward the heavier fluid. Recently, one have observed a renewal of interest in the RT instability for inviscid fluid flows in the context of Astrophysics and Inertial Confinement Fusion (ICF). In ICF, a spherical DT capsule is imploded by laser beams which create a low density plasma; the surface between the dense and cold part of the target and the plasma becomes RT unstable during a stage of the implosion, and the development of this instability may inhibit the ignition of thermonuclear fusion. It is thus crucial in practice to be able to predict the growth rate of this instability. We refer to Chandrasekhar's book [3] for an overview of the linear theory, and to Kull's article [7] and Goncharov's thesis [5] for a general survey of the RT instability in the scope of ICF.

Quite recently, Goncharov *and al.* [5, 6] have developed an analysis of the linear Rayleigh-Taylor instability for an isobaric model flow introduced by Kull and Anisimov [8]. This theory, based on a sophisticated formal boundary layer analysis, enables them to predict growth rates of linear stability which agree fairly well with numerical experiments. The purpose of this paper is to present a mathematical rigorous analysis of the linear RT instability for some simple flows connected to the work of Goncharov *and al.*. This analysis may be viewed as a first step for deriving a general mathematical theory of the linear stage of the RT instability in the scope of ICF.

The plan of the paper is as follows. In Section 2, we define the problem of the linear stability of incompressible transverse perturbations for a fairly general inviscid compressible stationary plane flow. In Section 3, we consider the case where the basic flow is an hydrostatic equilibrium and presents a smooth density profile. Assuming that the wavelength of perturbations is large compared with the characteristic length of this profile, we are able to give an asymptotic explicit expression of the largest growth rate of the instability. We also show that this growth rate depends strongly on the behavior of the density at infinity. Next, in Section 4, we add the convection mechanism to the basic flow. We analyze a very simple model where heat conduction is not taken into account, but which nevertheless leads to a mathematical problem of higher complexity. Selecting the relevant parameters of the problem enables us to carry out an asymptotic analysis of the largest growth rate of the instability and to predict the stabilizing or destabilizing effects of the convection velocity.

2. The linearized stability problem

Consider a compressible inviscid fluid in an external force field $\mathbf{g}$ per unit mass. For simplicity, we assume that $\mathbf{g}$ is a constant vector and we chose the x-axis in such a way that

$$\mathbf{g} = g\mathbf{e_x}$$

where $\mathbf{e_x}$ is the corresponding unit vector. We are given a plane stationary flow of the fluid along the x-axis, called the basic flow. We denote by ρ^0, u^0 and p^0 respectively the density, the velocity and the pressure of the flow. We assume that the triple (ρ^0, u^0, p^0) is defined on the whole real line $\mathbf{R}$ and satisfies the mass and momentum conservation equations

$$\frac{d}{dx}\left(\rho^0 u^0\right) = 0\,, \tag{1}$$

$$\frac{d}{dx}\left(\rho^0 (u^0)^2 + p^0\right) = \rho^0 g\,. \tag{2}$$

Note that, using (1), (2) reads equivalently

$$\frac{dp^0}{dx} = \rho^0 g \left(\rho^0 - u^0 \frac{du^0}{dx}\right). \tag{3}$$

We consider the case where ρ^0 is a decreasing function such that the limits

$$\rho_L^0 = \lim_{x\to-\infty} \rho^0(x) \;\;,\;\; \rho_R^0 = \lim_{x\to+\infty} \rho^0(x).$$

exist. Since $\mathbf{g}.\nabla\rho^0$ is negative, the basic flow will suffer in general from the Rayleigh-Taylor (RT) instability that we will now analyze.

We then consider *small incompressible* multidimensional perturbations of the basic flow. Just for the sake of simplicity, we restrict ourselves to 2D perturbations but the same techniques apply as well to the analysis of 3D perturbations. The perturbed density ρ^1, velocities (u^1, v^1) and pressure p^1 satisfy the *linearized* mass and momentum equations together with the incompressibility condition

$$\left\{\begin{array}{l} \dfrac{\partial \rho^1}{\partial t} + u^0 \dfrac{\partial \rho^1}{\partial x} + \dfrac{\partial u^0}{\partial x}\rho^1 + \dfrac{\partial \rho^0}{\partial x}u^1 = 0 \\ \dfrac{\partial u^1}{\partial t} + u^0 \dfrac{\partial u^1}{\partial x} + \dfrac{1}{\rho_0}\dfrac{\partial p^1}{\partial x} - \dfrac{1}{\rho_0{}^2}\dfrac{\partial p^0}{\partial x}\rho^1 + \dfrac{\partial u^0}{\partial x}u^1 = 0 \\ \dfrac{\partial v^1}{\partial t} + u^0 \dfrac{\partial v^1}{\partial x} + \dfrac{1}{\rho_0}\dfrac{\partial p^1}{\partial y} = 0 \\ \dfrac{\partial u^1}{\partial x} + \dfrac{\partial v^1}{\partial y} = 0. \end{array}\right. \tag{4}$$

Let us look for *normal mode* solutions of (4), i.e. for solutions of the form

$$\varphi^1(x,y,t) = \varphi \exp(iky + \gamma t), \quad \varphi = \rho, u, v, p, \tag{5}$$

where the wave number k is to be seen as a parameter. Such a normal mode is unstable if $\mathcal{R}e\gamma$ is strictly positive. We are in fact interested in determining the largest growth rate $\gamma = \gamma(k)$ of perturbations corresponding to the wavenumber k. Using (3) and (4), the normal modes (5) are characterized as the solutions of the differential system

$$\left\{\begin{array}{l} \gamma\rho + u^0 \dfrac{\partial \rho}{\partial x} + \dfrac{\partial u^0}{\partial x}\rho + \dfrac{\partial \rho^0}{\partial x}u = 0 \\ \gamma u + u^0 \dfrac{\partial u}{\partial x} + \dfrac{1}{\rho^0}\dfrac{\partial p}{\partial x} - \dfrac{1}{\rho^0}(g - u^0\dfrac{\partial u^0}{\partial x})\rho + \dfrac{\partial u^0}{\partial x}u = 0 \\ \gamma v + u^0 \dfrac{\partial v}{\partial x} + \dfrac{ik}{\rho^0}p = 0 \\ \dfrac{\partial u}{\partial x} + ikv = 0. \end{array}\right. \tag{6}$$

In addition, we require for the perturbations to vanish at infinity, i.e.,

$$\varphi(x) \to 0 \text{ as } x \to \pm\infty \ , \ \varphi = \rho, u, v, p. \tag{7}$$

The linearized problem (6), (7) involves several dimensionless parameters. Indeed, let us scale the equations (6). We introduce characteristic quantities : a length $\overline{x}$, a density $\overline{\rho}$, a velocity $\overline{u}$, a sound speed $\overline{c}$, a pressure $\overline{p} = \overline{\rho c}^2$ and a growth rate $\overline{\gamma}$. We then define the dimensionless numbers:

$$\begin{aligned} \epsilon &= k\overline{x} , \\ M &= \frac{\overline{u}}{\overline{c}} = \text{characteristic Mach number} , \\ F_r &= \frac{\overline{u}^2}{g\overline{x}} = \text{characteristic Froude number.} \end{aligned}$$

On the other hand, it appears natural to compare the growth rate γ with the one of the classical RT instability $\sqrt{Agk}$ where A denotes the *Atwood number*

$$A = \frac{\rho_L^0 - \rho_R^0}{\rho_L^0 + \rho_R^0}.$$

We thus choose

$$\overline{\gamma} = \sqrt{gk}.$$

Then, denoting by φ' a dimensionless quantity ($\varphi = \overline{\varphi}\varphi'$), we obtain the dimensionless form of system (6)

$$\left\{ \begin{array}{l} \sqrt{\frac{\epsilon}{Fr}}\gamma'\rho' + u'^0\frac{\partial \rho'}{\partial x'} + \frac{\partial u'^0}{\partial x'}\rho' + \frac{\partial \rho'^0}{\partial x'}u' = 0 \\ \sqrt{\frac{\epsilon}{Fr}}\gamma' u' + u'^0\frac{\partial u'}{\partial x'} + \frac{1}{M^2\rho'^0}\frac{\partial p'}{\partial x'} - \frac{1}{\rho'^0}(\frac{1}{Fr} - u'^0\frac{\partial u'^0}{\partial x'})\rho' + \frac{\partial u'^0}{\partial x'}u' = 0 \\ \sqrt{\frac{\epsilon}{Fr}}\gamma' v' + u'^0\frac{\partial v'}{\partial x'} + \frac{i\epsilon}{M^2\rho'^0}p' = 0 \\ \frac{\partial u'}{\partial x'} + i\epsilon v' = 0 \end{array} \right. \tag{8}$$

Observe that, at the expense of replacing $\frac{p'}{M^2}$ by p', we may as well suppose $M = 1$ so that (8) depends in fact on the two dimensionless parameters ϵ and Fr. In all the sequel, we will take the density profile ρ^0 of the basic flow of the form

$$\rho^0(x) = \widehat{\rho^0}\left(\frac{x}{L}\right),$$

where L is a characteristic length of the profile. For instance, we can take for L the minimum density gradient length

$$L_{\min} = \min_{x \in R}\left(-\frac{1}{\rho^0}\frac{d\rho^0}{dx}\right)^{-1}.$$

It is then natural to choose $\overline{x} = L$ so that

$$\rho'^0(x') = \frac{1}{\overline{\rho}}\widehat{\rho^0}(x').$$

Our goal is now to study the dimensionless rate of growth γ' of the instability in terms of the two parameters $\epsilon = kL$ and Fr.

3. The case of an hydrostatic equilibrium

In this section, we assume $u^0 \equiv 0$, i.e., the basic flow corresponds to an hydrostatic equilibrium. Then the first order differential system (8) is degenerate, and it is a simple matter to check that, dropping the primes for the ease of notations, solving system (8) amounts to find a pair (γ, u) satisfying the second order differential equation, called the *Rayleigh equation*

$$-\frac{d}{dx}\left(\rho^0\,\frac{du}{dx}\right) + \epsilon^2\,\rho^0\,u = -\frac{\epsilon}{\gamma^2}\,\frac{d\rho^0}{dx}\,u \tag{9}$$

supplemented by the conditions at infinity

$$u(x) \to 0 \quad \text{as} \quad x \to \pm\infty. \tag{10}$$

Then, the functions ρ, v and p are given by

$$v = \frac{i}{\epsilon}\,\frac{du}{dx}, \qquad \rho = -\sqrt{\frac{Fr}{\epsilon}}\,\frac{1}{\gamma}\,\frac{\partial\rho^0}{\partial x}\,u, \qquad p = -M^2\,\frac{\rho^0}{\epsilon^2}\,\sqrt{\frac{\epsilon}{Fr}}\,\gamma\,v.$$

Problem (9),(10) looks like a classical symmetric spectral problem. Indeed, setting

$$\lambda = \frac{1}{\gamma^2},$$

a variational formulation of problem (9),(10) amounts to find a pair $(\lambda, u) \in \mathbf{R} \times H^1(\mathbf{R})$ solution of

$$\int_{-\infty}^{\infty} \rho^0\left(\frac{du}{dx}\,\frac{dv}{dx} + \epsilon^2 u\,v\right)\,dx = \epsilon\lambda \int_{-\infty}^{\infty} \frac{d\rho^0}{dx}\,u\,v\,dx, \qquad \forall v \in \mathrm{H}^1(\mathrm{R}). \tag{11}$$

We are interested in the smallest "generalized" eigenvalue of problem (11) which corresponds to the largest growth rate of the normal modes associated with the wave number k.

We first check

Lemma 3.1. *Any non trivial solution $(\lambda, u) \in \mathbf{R} \times H^1(\mathbf{R})$ of (9) satisfies $\lambda \geq 1$.*

Proof. Choosing $v = u$ in (11), we obtain

$$\int_{-\infty}^{\infty} \rho^0\left(\left|\frac{du}{dx}\right|^2 + \epsilon^2\,|u|^2\right)\,dx = -\epsilon\,\lambda \int_{-\infty}^{\infty} \frac{d\rho^0}{dx}\,|u|^2\,dx$$

$$= \epsilon\,\lambda \int_{-\infty}^{\infty} \rho^0\,\frac{d}{dx}|u|^2\,dx = 2\,\epsilon\,\lambda \int_{-\infty}^{\infty} \rho^0\,u\,\frac{du}{dx}\,dx$$

and by the triangular inequality

$$\int_{-\infty}^{\infty} (1-\lambda)\,\rho^0 \left(\left|\frac{du}{dx}\right|^2 + \epsilon^2 \, |u|^2 \right) dx \,\le\, 0$$

which proves our assertion. □

Next, for going further, we need to make an assumption on the behavior of ρ^0 at $\pm\infty$.

Lemma 3.2. *Assume the hypothesis*

$$M^2 = \int_{-\infty}^{0} |\rho^0(x) - \rho_L^0|^2 \, dx + \int_{0}^{+\infty} |\rho^0(x) - \rho_R^0|^2 \, dx \,<\, +\infty. \tag{12}$$

Let $(\lambda_\epsilon, u_\epsilon)$ be a sequence of nontrivial solutions of (11) such that λ_ϵ remains bounded as ϵ tends towards zero. Then

$$\lambda_\epsilon \to A^{-1} \quad as \quad \epsilon \to 0.$$

Proof. We perform the change of variable $y = \epsilon x$ and we set

$$\tilde{\rho}^0_\epsilon(y) = \rho^0\left(\frac{y}{\epsilon}\right), \quad \tilde{u}_\epsilon(y) = u\left(\frac{y}{\epsilon}\right).$$

Then the pair $(\lambda_\epsilon, u_\epsilon) \in \mathbf{R} \times H^1(\mathbf{R})$ is a solution of (11) if and only if the pair $(\lambda_\epsilon, \tilde{u}_\epsilon) \in \mathbf{R} \times H^1(\mathbf{R})$ satisfies

$$\int_{-\infty}^{\infty} \tilde{\rho}^0_\epsilon \left(\frac{d\tilde{u}_\epsilon}{dy}\,\frac{dv}{dy} + \tilde{u}_\epsilon \, v \right) dy \;=\; -\lambda_\epsilon \int_{-\infty}^{\infty} \frac{d\tilde{\rho}^0_\epsilon}{dy}\,\tilde{u}_\epsilon \, v \, dy \quad \forall v \in H^1(\mathbf{R}). \tag{13}$$

We normalize the function $\tilde{u}_\epsilon$ in such a way that $\|\tilde{u}_\epsilon\|_{H^1(\mathbf{R})} = 1$.
Hence, using the hypothesis on λ_ϵ, we can extract from the sequence $(\lambda_\epsilon, \tilde{u}_\epsilon)$ a subsequence, still denoted by $(\lambda_\epsilon, \tilde{u}_\epsilon)$, such that

$$\lambda_\epsilon \to \lambda, \quad \tilde{u}_\epsilon \to \tilde{u} \text{ weakly in } H^1(\mathbf{R}),$$

$$\tilde{u}_\epsilon \to \tilde{u} \text{ uniformly on any compact set of } \mathbf{R} \text{ as } \epsilon \to 0.$$

Next, setting

$$\tilde{\rho}^0(y) = \begin{cases} \rho_L^0, & y < 0 \\ \rho_R^0, & y > 0 \end{cases}$$

we observe that

$$\int_{-\infty}^{+\infty} |\tilde{\rho}^0_\epsilon - \tilde{\rho}^0|^2 \, dy = \epsilon \, M^2 \tag{14}$$

and therefore by (12)

$$\tilde{\rho}^0_\epsilon \to \tilde{\rho}^0 \text{ strongly in } L^2(\mathbf{R}) \text{ as } \epsilon \to 0.$$

On the other hand, it is a simple matter to check that

$$\frac{d\rho_\epsilon}{dy} = (\rho^0_R - \rho^0_L)\,\delta + e_\epsilon, \quad \|e_\epsilon \tilde{u}_\epsilon\|_{H^{-1}(\mathbf{R})} \le M\sqrt{\epsilon}.$$

Then, replacing v by $\tilde{u}_\epsilon$ in (13), we obtain the estimate

$$\rho^0_L \le \lambda_\epsilon \left((\rho^0_R - \rho^0_L)\, |\tilde{u}_\epsilon(0)|^2 + M\sqrt{\epsilon} \right),$$

which proves that $|\tilde{u}_\epsilon(0)|$ remains bounded above by a constant independent on ϵ, and therefore $\tilde{u}$ is not identically 0.

Now, let $v \in C^\infty_c(\mathbf{R})$; we have

$$\int_{-\infty}^{\infty} \tilde{\rho}^0_\epsilon \left(\frac{d\tilde{u}_\epsilon}{dy}\frac{dv}{dy} + \tilde{u}_\epsilon v \right) dy \;=\; \lambda_\epsilon \int_{-\infty}^{\infty} \tilde{\rho}^0_\epsilon \, \frac{d}{dy}(\tilde{u}_\epsilon v)\, dy.$$

By passing to the limit, we obtain

$$\int_{-\infty}^{\infty} \tilde{\rho}^0 \left(\frac{d\tilde{u}}{dy}\frac{dv}{dy} + \tilde{u}v - \lambda \frac{d}{dy}(\tilde{u}v) \right) dy = 0$$

which yields

$$-\frac{d}{dy}\left(\tilde{\rho}^0 \frac{d\tilde{u}}{dy} \right) + \tilde{\rho}^0 \tilde{u} + \lambda \frac{d\tilde{\rho}^0}{dy}\, \tilde{u} \;=\; 0.$$

We obtain on one hand

$$-\frac{d^2\tilde{u}}{dy^2} + \tilde{u} = 0 \quad \text{in } (-\infty, 0) \cup (0, +\infty)$$

which gives

$$\tilde{u}(y) = \tilde{u}(0)\, e^{-|y|}$$

and on the other hand

$$-\left(\rho^0_R \frac{d\tilde{u}}{dy}(0_+) - \rho^0_L \frac{d\tilde{u}}{dy}(0_-) \right) + \lambda\, (\rho^0_R - \rho^0_L)\, \tilde{u}(0) \;=\; 0$$

which yields

$$\lambda = \frac{\rho^0_L + \rho^0_R}{\rho^0_L - \rho^0_R} = A^{-1}.$$

In addition, by the uniqueness of the limit λ, we find that the whole sequence (λ_ϵ) converges. □

In fact, we want to prove that problem (11) has a smallest eigenvalue λ_ϵ which is simple and converges toward A^{-1} as $\epsilon \to 0$, while the other eigenvalues behave as $O(\frac{1}{\epsilon})$. The behavior of λ_ϵ in ϵ will depend on the behavior of ρ^0 at infinity. Since in the limit $\epsilon \to 0$, $u_\epsilon(x)$ behaves as $\tilde{u}(\epsilon x) = C\exp(-\epsilon|x|)$, it is natural to look for a solution $u = u_\epsilon$ of (11) in the form

$$u_\epsilon(x) = v_\epsilon(x)\exp(-\epsilon|x|) \ .$$

Then, we observe that the pair $(\lambda_\epsilon, v_\epsilon)$ satisfies for $x < 0$

$$\frac{d}{dx}\left(\rho^0\frac{dv_\epsilon}{dx}\right) + 2\epsilon\frac{d}{dx}(\rho^0 v_\epsilon) = \epsilon(\lambda_\epsilon + 1)\frac{d\rho^0}{dx}v_\epsilon \tag{15}$$

and for $x > 0$

$$\frac{d}{dx}\left(\rho^0\frac{dv_\epsilon}{dx}\right) - 2\epsilon\frac{d}{dx}(\rho^0 v_\epsilon) = \epsilon\,(\lambda_\epsilon - 1)\frac{d\rho^0}{dx}v_\epsilon. \tag{16}$$

In addition, assuming $\rho^0 \in \mathrm{W}^{1,\infty}(\mathbf{R})$ so that u_ϵ belongs to $H^2(\mathbf{R})$, we have to prescribe the continuity of u_ϵ and $\dfrac{du_\epsilon}{dx}$ at $x = 0$ which yields

$$\left\{\begin{array}{c} v_\epsilon(0_+) = v_\epsilon(0_-) = v_\epsilon(0) \\ \dfrac{dv_\epsilon}{dx}(0_+) - \dfrac{dv_\epsilon}{dx}(0_-) = 2\,\epsilon\,v_\epsilon(0). \end{array}\right. \tag{17}$$

Let us suppose that v_ϵ has the following behavior at $\pm\infty$:

$$\left\{\begin{array}{c} \lim\limits_{x\to-\infty} v_\epsilon(x) = v_{\epsilon,L}\,, \quad \lim\limits_{x\to+\infty} v_\epsilon(x) = v_{\epsilon,R} \\ \lim\limits_{x\to\pm\infty} \dfrac{dv_\epsilon}{dx}(x) = 0. \end{array}\right. \tag{18}$$

We then look for a *formal* asymptotic expansion of the pair $(\lambda_\epsilon, v_\epsilon)$ in the form

$$\left\{\begin{array}{l} \lambda_\epsilon = \lambda_0 + \epsilon\lambda_1 + \epsilon^2\lambda_2 + \dots \\ \\ v_\epsilon = v_0 + \epsilon v_1 + \epsilon^2 v_2 + \dots \end{array}\right. \tag{19}$$

Proposition 3.1. *Under the assumption (18), we obtain*

$$\lambda_0 = A^{-1}, \quad \lambda_1 = \frac{4\rho_L^0\rho_R^0}{(\rho_L^0 - \rho_R^0)^3}\int_{-\infty}^{+\infty}\frac{(\rho_L^0 - \rho^0)(\rho^0 - \rho_R^0)}{\rho^0}\,dx. \tag{20}$$

Proof. For $x < 0$, (15) yields at the orders $0, 1, 2, \ldots$ in ϵ

$$\begin{aligned} \frac{d}{dx}\left(\rho^0 \frac{dv_0}{dx}\right) &= 0, && (21)\\ \frac{d}{dx}\left(\rho^0 \frac{dv_1}{dx}\right) + 2\frac{d}{dx}\left(\rho^0 v_0\right) &= (\lambda_0 + 1)\frac{d\rho^0}{dx} v_0\,, && (22)\\ \frac{d}{dx}\left(\rho^0 \frac{dv_2}{dx}\right) + 2\frac{d}{dx}\left(\rho^0 v_1\right) &= (\lambda_0 + 1)\frac{d\rho^0}{dx} v_1 + \lambda_1 \frac{d\rho^0}{dx} v_0\,, \cdots && (23) \end{aligned}$$

It then follows from (21) and (18) that

$$\rho^0 \frac{dv_0}{dx} = \lim_{x \to -\infty} \left(\rho^0 \frac{dv_0}{dx}\right)(x) = 0$$

and therefore

$$v_0 = v_{0,L}. \tag{24}$$

Next, integrating (22) from $-\infty$ to x and using again (18) we find

$$\rho^0 \frac{dv_1}{dx} + 2(\rho^0 - \rho^0_L)\, v_{0,L} = (\lambda_0 + 1)\,(\rho^0 - \rho^0_L)\, v_{0,L}$$

or

$$\rho^0 \frac{dv_1}{dx} = (\lambda_0 - 1)\,(\rho^0 - \rho^0_L)\, v_{0,L} \tag{25}$$

which yields

$$v_1 = v_{1,L} + (\lambda_0 - 1)\, v_{0,L} \int_{-\infty}^{x} \frac{(\rho^0 - \rho^0_L)}{\rho^0}(y)\, dy. \tag{26}$$

Similarly, integrating (23) and using (18) and (25) gives

$$\rho^0 \frac{dv_2}{dx} = (\lambda_0 - 1)\,(\rho^0 v_1 - \rho^0_L v_{1,L}) - (\lambda_0^2 - 1)\, v_{0,L} \int_{-\infty}^{x} (\rho^0 - \rho^0_L)(y)\, dy + \lambda_1\, v_{0,L}\,(\rho^0 - \rho^0_L). \tag{27}$$

In exactly the same way, we obtain for $x > 0$

$$v_0 = v_{0,R}, \tag{28}$$

$$v_1 = v_{1,R} - (\lambda_0 + 1)\, v_{0,R} \int_{x}^{+\infty} \frac{(\rho^0 - \rho^0_R)}{\rho^0}(y)\, dy, \tag{29}$$

$$\rho^0 \frac{dv_2}{dx} = (\lambda_0 + 1)\,(\rho^0 v_1 - \rho^0_R v_{1,R}) + (\lambda_0^2 - 1)\, v_{0,R} \int_{x}^{+\infty} (\rho^0 - \rho^0_R)(y)\, dy + \lambda_1\, v_{0,R}\,(\rho^0 - \rho^0_R). \tag{30}$$

It remains to take into account the continuity conditions (17) which yield at the orders 0, 1, 2, ... in ϵ

$$v_0(0_+) = v_0(0_-) = v_0(0) \quad , \quad \frac{dv_0}{dx}(0_+) = \frac{dv_0}{dx}(0_-) \quad , \tag{31}$$

$$v_1(0_+) = v_1(0_-) = v_1(0) \quad , \quad \frac{dv_1}{dx}(0_+) - \frac{dv_1}{dx}(0_-) = 2v_0(0), \tag{32}$$

$$v_2(0_+) = v_2(0_-) = v_2(0) \quad , \quad \frac{dv_2}{dx}(0_+) - \frac{dv_2}{dx}(0_-) = 2v_1(0). \tag{33}$$

On the one hand, we have by (24), (28) and (31)

$$v_0 = v_{0,L} = v_{0,R}. \tag{34}$$

On the other hand, we have by (26), (29) and (32)

$$\begin{aligned} &\rho^0(0)\left(\frac{dv_1}{dx}(0_+) - \frac{dv_1}{dx}(0_-)\right) \\ &\qquad = v_0\left\{(\lambda_0+1)(\rho^0(0)-\rho^0_R(0)) - (\lambda_0-1)(\rho^0(0)-\rho^0_L(0))\right\} \\ &\qquad = 2\rho^0(0)v_0 \end{aligned}$$

which gives

$$\lambda_0 = \frac{\rho^0_L + \rho^0_R}{\rho^0_L - \rho^0_R} = \frac{1}{A}.$$

We find also

$$\begin{aligned} v_1(0_+) &= v_{1,R} - (\lambda_0+1)v_0 \int_0^{+\infty} \frac{\rho^0 - \rho^0_R}{\rho_0}\,dx \\ &= v_1(0_-) = v_{1,L} + (\lambda_0-1)v_0 \int_{-\infty}^{0} \frac{\rho^0 - \rho^0_L}{\rho_0}\,dx \end{aligned}$$

and thus

$$v_{1,R} - v_{1,L} = v_0\left\{(\lambda_0+1)\int_0^{+\infty} \frac{\rho^0-\rho^0_R}{\rho^0}\,dx + (\lambda_0-1)\int_{-\infty}^{0} \frac{\rho^0-\rho^0_L}{\rho^0}\,dx\right\}.$$

Finally, using (27), (30) and (33), we obtain

$$\rho^0(0)\left(\frac{dv_2}{dx}(0_+)-\frac{dv_2}{dx}(0_-)\right)$$
$$=(\lambda_0+1)(\rho^0(0)v_1(0)-\rho^0_R v_{1,R})+(\lambda_0^2-1)v_0\int_0^{+\infty}(\rho^0-\rho^0_R)\,dx+\lambda_1\,v_0\,(\rho^0-\rho^0_R)$$
$$-(\lambda_0-1)(\rho^0(0)v_1(0)-\rho^0_L v_{1,L})+(\lambda_0^2-1)v_0\int_{-\infty}^{0}(\rho^0-\rho^0_L)\,dx-\lambda_1\,v_0\,(\rho^0-\rho^0_L)$$
$$=2\,\rho^0(0)\,v_1(0).$$

Hence

$$-(\lambda_0+1)\rho^0_R v_{1,R}+(\lambda_0-1)\rho^0_L v_{1,L}+(\lambda_0^2-1)v_0\Big(\int_0^{+\infty}(\rho^0-\rho^0_R)dx+\int_{-\infty}^{0}(\rho^0-\rho^0_L)dx\Big)$$
$$+\lambda_1 v_0(\rho^0_L-\rho^0_R)=0.$$

Since

$$\lambda_0+1=\frac{2\rho^0_L}{\rho^0_L-\rho^0_R}\;,\quad \lambda_0-1=\frac{2\rho^0_R}{\rho^0_L-\rho^0_R}\;,$$

we find by using (31)

$$-\quad\frac{4\rho^0_L\rho^0_R}{(\rho^0_L-\rho^0_R)^2}\left(\rho^0_L\int_O^{+\infty}\frac{(\rho^0-\rho^0_R)}{\rho^0}\,dx+\rho^0_R\int_{-\infty}^{0}\frac{(\rho^0-\rho^0_L)}{\rho^0}\,dx\right)$$
$$+\quad\frac{4\rho^0_L\rho^0_R}{(\rho^0_L-\rho^0_R)^2}\left(\int_O^{+\infty}(\rho^0-\rho^0_R)\,dx+\int_{-\infty}^{0}(\rho^0-\rho^0_L)\,dx\right)+\lambda_1\,(\rho^0_L-\rho^0_R)=0\,,$$

and (20) follows. □

Clearly, one can extend the above method of proof in order to determine recursively and in an unique way the coefficients (λ_j,v_j) of the asymptotic expansions (19). It remains to investigate under what hypothesis on the density profile ρ^0 these asymptotic expansions hold. Assuming that ρ^0 is a C^1 strictly decreasing function, it is convenient to perform the change of variable

$$y=\rho^0(x) \tag{35}$$

which maps $\mathbf{R}$ onto the interval (ρ^0_R,ρ^0_L).

Setting

$$q(y)=\frac{d\rho^0}{dx}\le 0\;,\;w_\epsilon(y)=v_\epsilon(x), \tag{36}$$

Eq. (15) becomes

$$\frac{d}{dy}(yq\frac{dw_\epsilon}{dy}+2\epsilon yw^\epsilon)=\epsilon(\lambda_\epsilon+1)w_\epsilon\ ,\ y\in(\rho^0(0),\rho_L^0) \tag{37}$$

while Eq. (16) reads

$$\frac{d}{dy}(yq\frac{dw_\epsilon}{dy}-2\epsilon yw^\epsilon)=\epsilon(\lambda_\epsilon-1)w_\epsilon\ ,\ y\in(\rho_R^0,\rho^0(0)). \tag{38}$$

Let us focus for instance on Eq.(37). Since $q(\rho_L^0)=0$, we observe that (37) degenerates at ρ_L^0. We thus consider the boundary value problem

$$\begin{cases} \dfrac{d}{dy}\left(yq\dfrac{dw_\epsilon^-}{dy}+2\epsilon yw_\epsilon^-\right)=\epsilon(\lambda_\epsilon+1)w_\epsilon^-\ ,\ y\in(\rho^0(0),\rho_L^0) \\ w_\epsilon^-(\rho_L^0)=1. \end{cases} \tag{39}$$

Lemma 3.3. *Assume that there exists a constant $C>0$ such that*

$$|q(y)|\geq C\left(\rho_L^0-y\right),\ y\in[\rho^0(0),\rho_L^0]. \tag{40}$$

Then, for $|\lambda|\leq\Lambda$, there exists $\epsilon_0=\epsilon_0(\Lambda)$ such that, for $\epsilon\leq\epsilon_0$, problem (39) has an unique solution $w_\epsilon^-(.;\lambda)\in C^0([\rho^0(0),\rho_L^0])\cup C^1([\rho^0(0),\rho_L^0])$. Moreover, w_ϵ^- depends analytically on (ϵ,λ) and may be written as the sum of a series

$$w_\epsilon^-(.;\lambda)=1+\sum_{p\geq1}\epsilon^p w_p^-(.;\lambda) \tag{41}$$

which converges in $C^0([\rho^0(0),\rho_L^0])$ and in $C^1([\rho^0(0),\rho_L^0-\alpha])$ for all $\alpha>0$.

Proof. We only sketch the proof. Suppressing the upper index 0 for simplicity, we set

$$w_\epsilon^-(y)=1+\epsilon(y-\rho_L)s_\epsilon^-. \tag{42}$$

Then it is an easy matter to check that w_ϵ^- is solution of (39) if and only if s_ϵ^- satisfies the equation

$$s_\epsilon^-=s_0^-+\epsilon K^-s_\epsilon^-$$

where

$$s_0^-(y,\lambda)=\frac{\lambda-1}{y-\rho_L}\int_{\rho_L}^{y}\frac{z-\rho_L}{zq(z)}dz$$

and $K^-=K^-(\lambda)$ is the integral operator

$$\begin{aligned}(K^-s)(y)= & -\frac{2}{y-\rho_L}\int_{\rho_L}^{y}\frac{z-\rho_L}{q(z)}s(z)\,dz \\ & +\frac{\lambda+1}{y-\rho_L}\int_{\rho_L}^{y}\left(\frac{1}{q(z)}\int_{\rho_L}^{z}(t-\rho_L)s(t)dt\right)dz.\end{aligned} \tag{43}$$

Using the assumption (40), one can check that ϵK^- is a strict contraction mapping in $C^0([\rho^0(0), \rho_L^0])$ for $\epsilon C(\lambda) < 1$, where $C(\lambda)$ is the norm of $K^-(\lambda)$ as an operator of $C^0([\rho^0(0), \rho_L^0])$ into itself . Hence, we obtain the existence and uniqueness of the function

$$s_\epsilon^- = \left(I - \epsilon K^-\right)^{-1} s_0^- = \sum_{p \geq 0} \epsilon^p (K^-)^p s_0^-$$

for ϵ small enough. The above series converges in $C^0([\rho^0(0), \rho_L^0])$. On the other hand, one can prove that the series

$$\sum_{p \geq 0} \epsilon^p \frac{d}{dy}\left((K^-)^p s_0^-\right)$$

converges in $C^1([\rho^0(0), \rho_L^0 - \alpha])$ for all $\alpha > 0$. Hence (41) holds with

$$w_p^-(.;\lambda) = (y - \rho_L)(K^-)^{p-1} s_0^- .$$

In particular, we have

$$w_1^-(y;\lambda) = (y - \rho_L) s_0^-(y;\lambda) = (\lambda - 1)\int_{\rho_L}^{y} \frac{z - \rho_L}{z q(z)}\, dz.$$

More generally, $w_\epsilon^-(.;\lambda)$ is easily seen to be a polynomial of degree p in λ.

Finally, the analyticity of $w_\epsilon^-(.,\lambda)$ in (ϵ, λ) is a consequence of the analytic dependence of (39) in (ϵ, λ) and of the uniqueness of the solution on the domain $\epsilon C(\lambda) < 1$. □

Clearly the analogue of Lemma 3.2 holds for the solution w_ϵ^+ of the boundary value problem

$$\left\{ \begin{aligned} & \frac{d}{dy}\left(yq\frac{dw_\epsilon^+}{dy} - 2\epsilon y w_\epsilon^+\right) = \epsilon(\lambda_\epsilon - 1) w_\epsilon{}^+ \ , \ y \in (\rho_R^0, \rho^0(0)) \\ & w_\epsilon^+(\rho_R^0) = 1. \end{aligned} \right. \tag{44}$$

Theorem 3.1. *Assume that there exists a constant $C > 0$ such that*

$$\left\{ \begin{aligned} & |q(y)| \geq C|y - \rho_L^0| \ , \ \ y \in (\rho^0(0), \rho_L^0) \\ & |q(y)| \geq C|y - \rho_R^0| \ , \ \ y \in (\rho_R^0, \rho^0(0)). \end{aligned} \right. \tag{45}$$

Then, given a bounded neighborhood $\mathcal{N}$ of A^{-1}, one can find $\epsilon_0 \geq 0$ such that, for $\epsilon \leq \epsilon_0$, there exists a unique eigenvalue $\lambda_\epsilon \in \mathcal{N}$. Moreover λ_ϵ is an analytic function of ϵ and we have

$$\lambda_\epsilon = A^{-1} + \epsilon\lambda_1 + \dots \tag{46}$$

where λ_1 is given by (20).

Proof. We first note that the continuity constraints (17) read on the function w_ϵ (dropping again the superscript 0)

$$\left\{\begin{array}{c} w_\epsilon(\rho(0)_+) - w_\epsilon(\rho(0)_-) = 0 \\ \dfrac{d\rho}{dx}(0)\left(\dfrac{dw_\epsilon}{dy}(\rho(0)_+) - \dfrac{dw_\epsilon}{dy}(\rho(0)_-)\right) - \epsilon(w_\epsilon(\rho(0)_+) - w_\epsilon(\rho(0)_-)) = 0. \end{array}\right. \tag{47}$$

Thus, finding $(\lambda_\epsilon, v_\epsilon)$ solution of (15)-(17) amounts to find $(\lambda_\epsilon, w_\epsilon)$ solution of (37), (38) and (47). Moreover, it follows from Lemmas 3.1 and 3.2 that we can look for λ_ϵ in any bounded neighborhood $\mathcal{N}$ of A^{-1} ($\mathcal{N} = [1, 2A^{-1}]$, say). Now, for a given $\lambda \in \mathcal{N}$, any solution w_ϵ of (37),(38), with λ_ϵ replaced by λ, is of the form

$$w_\epsilon = \left\{\begin{array}{l} Bw_\epsilon^- \,,\ y \in [\rho^0(0), \rho_L^0] \\ Cw_\epsilon^+ \,,\ y \in [\rho_R^0, \rho^0(0)] \end{array}\right.$$

with

$$w_\epsilon^-(y) = 1 + \epsilon(\lambda - 1)\int_{\rho_L}^{y} \frac{z - \rho_L}{zq(z)}\, dz + O(\epsilon^2)$$

and

$$w_\epsilon^+(y) = 1 + \epsilon(\lambda + 1)\int_{\rho_R}^{y} \frac{z - \rho_R}{zq(z)}\, dz + O(\epsilon^2).$$

where the remainder terms are $O(\epsilon^2)$ uniformly in $\lambda \in \mathcal{N}$. The conditions (47) then give

$$\left\{\begin{array}{c} (Cw_\epsilon^+ - Bw_\epsilon^-)(\rho(0)) = 0 \\ \dfrac{d\rho}{dx}(0)\left(C\dfrac{dw_\epsilon^+}{dy} - B\dfrac{dw_\epsilon^-}{dy}\right)(\rho(0)) - \epsilon\left(Cw_\epsilon^+ + Bw_\epsilon^-\right)(\rho(0)) = 0. \end{array}\right.$$

Since $\dfrac{d\rho}{dx} = q(\rho(0))$, we observe that the above linear system in (B, C) has a nontrivial solution provided that

$$\left(q\left(w_\epsilon^+\frac{dw_\epsilon^-}{dy} - w_\epsilon^-\frac{dw_\epsilon^+}{dy}\right) + 2\epsilon w_\epsilon^+ w_\epsilon^-\right)(\rho(0)) = 0.$$

On the other hand, we have

$$\left(q\left(w_\epsilon^+\frac{dw_\epsilon^-}{dy} - w_\epsilon^-\frac{dw_\epsilon^+}{dy}\right) + 2\epsilon w_\epsilon^+ w_\epsilon^-\right)(y) = \epsilon\,\frac{\rho_L + \rho_R - \lambda(\rho_L - \rho_R)}{y} + O(\epsilon).$$

Next, we set

$$\phi(\epsilon, \lambda) = \frac{\rho(0)}{\epsilon}\left(q\left(w_\epsilon^+\frac{dw_\epsilon^-}{dy} - w_\epsilon^-\frac{dw_\epsilon^+}{dy}\right) + 2\epsilon w_\epsilon^+ w_\epsilon^-\right)(\rho(0)).$$

Clearly, ϕ is an analytic function of (ϵ, λ). Moreover

$$\phi(0, \lambda) = \rho_L + \rho_R - \lambda(\rho_L - \rho_R)$$

vanishes at $\lambda_0 = A^{-1}$. In addition

$$\frac{\partial \phi}{\partial \lambda} = -(\rho_L - \rho_R) \neq 0.$$

Applying the implicit function theorem, we obtain that the dispersion relation

$$\phi(\epsilon, \lambda_\epsilon) = 0$$

has an unique solution $\lambda_\epsilon \in \mathcal{N}$ for $\epsilon < \epsilon_0$ small enough. Moreover λ_ϵ is an analytic function of ϵ and (46) follows from Proposition 1. □

In fact, the assumption (45) can be somewhat relaxed; it is enough to suppose that, for some constant $\sigma > 0$, we have

$$\begin{cases} |q(y)| \geq C|y - \rho_L^0|\,, & y \in [\rho_L^0 - \sigma, \rho_L^0] \\ |q(y)| \geq C|y - \rho_R^0|\,, & y \in [\rho_R^0, \rho_R^0 + \sigma]. \end{cases} \tag{48}$$

We refer to [9] for this more general result and a detailed proof. The condition (48) is still a strong hypothesis on the behavior of $\frac{d\rho^0}{dx}$ at infinity: it means that $-\frac{d\rho^0}{dx}$ behaves as $\exp(Cx)$ (resp. as $\exp(-Cx)$) as x tends to $-\infty$ (resp. $+\infty$). In such a case, using the asymptotic expansion (46) of λ_ϵ, we have

$$\frac{1}{\lambda_\epsilon} = A(1 - \delta\epsilon + ...) \tag{49}$$

where

$$\delta = \frac{4\rho_L^0 \rho_R^0 (\rho_L^0 + \rho_R^0)}{(\rho_L^0 - \rho_R^0)^2} \int_{-\infty}^{+\infty} \frac{(\rho_L^0 - \rho^0)(\rho^0 - \rho_R^0)}{\rho^0}\, dx. \tag{50}$$

We thus obtain in the limit $\epsilon = kL$ tending to zero

$$\gamma = \sqrt{A(1 - \delta\epsilon + \dots)}. \tag{51}$$

Indeed, at the zero order in ϵ, we find the classical (dimensionless) growth rate of the Rayleigh-Taylor instability but, since δ is > 0, the first order term in ϵ is a stabilizing one which depends on the form of the density profile [4].

Remark 3.1. *If we go back to the physical quantities, we find the more familiar expression of the growth rate*

$$\gamma = \sqrt{Agk(1 - \delta kL + ...)} \tag{52}$$

where δ is still given by (50) at the only difference that now ρ^0 stands for the physical density.

Let us notice that the existence of an asymptotic expansion (19) in whole powers of ϵ is no longer true if the density profile ρ^0 does not behave in such a way that (48) holds. In fact, in problems encountered in Inertial Confinement Fusion, the density profile behaves like $x^{-\mu}$, $\mu > 0$, as x tends to $+\infty$ [5, 9]. Let us check that in this case we cannot expect such a simple asymptotic expansion. Consider the simple model case where

$$\rho^0(x) = \frac{\rho^0(0)}{(1+x)^\mu}\ , \ x > 0\ ,\ \mu > 0. \tag{53}$$

Then, replacing ρ^0 by its expression in (16), we obtain that the pair $(\lambda_\epsilon, v_\epsilon)$ satisfies for $x > 0$

$$(1+x)\frac{d^2 v_\epsilon}{dx^2} - (\mu + 2\epsilon(1+x))\frac{dv_\epsilon}{dx} + \epsilon(\lambda_\epsilon + 1)\mu v_\epsilon = 0. \tag{54}$$

Hence, setting

$$v_\epsilon(x) = \varphi_\epsilon(t)\ ,\ \ t = 2\epsilon(1+x),$$

the function φ_ϵ is solution of the second order differential equation of Kummer's type

$$t\frac{d^2\varphi_\epsilon}{dt^2} - (\mu + t)\frac{d\varphi_\epsilon}{dt} + \frac{1}{2}(\lambda_\epsilon + 1)\mu\varphi_\epsilon = 0\ \ , t > 2\epsilon, \tag{55}$$

and is therefore a linear combination of confluent hypergeometric functions.

Let us quote, for reader's convenience, some classical results on confluent hypergeometric functions (see [1] for instance). When b is not an integer, the power series equation of Kummer's type

$$M(a,b,t) = \sum_{n=0}^{\infty} \frac{(a)_n}{(b)_n}\frac{x^n}{n!},\ \ (a)_n = a(a+1)...(a+n-1),\ (a)_0 = 1 \tag{56}$$

defines an analytic function of t solution of the second order differential equation

$$t\frac{d^2\varphi_\epsilon}{dt^2}(t) + (b-t)\frac{d\varphi_\epsilon}{dt}(t) - a\varphi_\epsilon(t) = 0. \tag{57}$$

Moreover, any solution of (57) is of the form

$$AM(a,b,t) + Bt^{1-b}M(1+a-b, 2-b, t)$$

and, for $a \geq 0$, the solutions of (57) which remain bounded as t tends to $+\infty$ are proportional to the function

$$M(a,b,t) - \frac{\Gamma(1+a-b)\Gamma(b)}{\Gamma(2-b)\Gamma(a)} t^{1-b} M(1+a-b, 2-b, t)$$

where Γ denotes the familiar Gamma function. Indeed the above function behaves like $\frac{1}{\pi}\sin(\pi b)\Gamma(1+a-b)\Gamma(b)t^{-a}$ as $t \to +\infty$.

As a consequence of the above results, we find

$$v_\epsilon(x) = B_\epsilon^+ v_\epsilon^+(x)\ ,\ \ x > 0$$

where

$$v_\epsilon^+(x) = M\left(-\frac{\mu(\lambda_\epsilon+1)}{2}, -\mu, 2\epsilon(1+x)\right)$$

$$-C(\mu,\lambda_\epsilon)\left(2\epsilon(1+x)^{1+\mu}M(1+\mu-\frac{\mu(\lambda_\epsilon+1)}{2}, 2+\mu, 2\epsilon(1+x)\right)$$

with

$$C(\mu,\lambda_\epsilon) = \frac{\Gamma\left(1+\mu-\frac{\mu(\lambda_\epsilon+1)}{2}\right)\Gamma(-\mu)}{\Gamma(2+\mu)\Gamma\left(-\frac{\mu(\lambda_\epsilon+1)}{2}\right)}.$$

Now, using (56), we obtain $v_\epsilon^+(x)$ as the sum of two asymptotic expansions in ϵ^n, $n \geq 0$, and $\epsilon^{n+\mu}$, $n \geq 1$, respectively. Thus, we have

$$\begin{aligned} v_\epsilon^+(x) = 1 \quad &+ \quad \epsilon(\lambda_\epsilon+1)(1+x)\,(1+O(\epsilon)) \\ &- \quad \epsilon^{1+\mu}2^{1+\mu}C(\mu,\lambda_\epsilon)(1+x)^{1+\mu}\,(1+O(\epsilon))\,. \end{aligned}$$

On the other hand, assuming that $\rho_0^L - \rho^0(x)$ tends to zero like $\exp(\mu_L x)$, $\mu_L > 0$, as $x \to -\infty$, we may apply the results of Lemma 3.2 which yield

$$v_\epsilon(x) = B_\epsilon^- v_\epsilon^-(x)\,, \quad x < 0$$

where

$$v_\epsilon^-(x) = 1 + \epsilon(\lambda_\epsilon - 1)\int_{-\infty}^{x} \frac{\rho^0(\xi)-\rho_L^0}{\rho^0(\xi)}d\xi + O(\epsilon^2).$$

Next, we observe that the continuity conditions (17) hold if and only if

$$v_\epsilon^+(0)\frac{dv_\epsilon^-(0)}{dx}(0) - v_\epsilon^-(0)\frac{dv_\epsilon^+(0)}{dx}(0) + 2\epsilon v_\epsilon^+(0)v_\epsilon^-(0) = 0.$$

Then, it is an easy matter to check that

$$\lambda_\epsilon = \left(1+\epsilon^\mu\frac{2^{1+\mu}(1+\mu)}{\Gamma(2+\mu)}\frac{\rho^0(0)}{\rho_L^0}\right)(1+O(\epsilon))\,.$$

The asymptotic expansion of λ_ϵ has indeed a more general form than (19). Note that here $A = 1$. Hence, in the case (53), the stabilization of the linear Rayleigh Taylor instability depends in a crucial way on the behavior of ρ^0 at $+\infty$. We refer to [9] for extensions of the previous results to the case of more general density profiles.

Up to now, we have considered the smallest eigenvalue of the spectral problem (11). The other eigenvalues are of order $O(\frac{1}{\epsilon})$. In fact, setting

$$\mu = \epsilon\lambda = \frac{\epsilon}{\gamma^2}$$

(9) becomes

$$-\frac{d}{dx}\left(\rho^0\frac{du}{dx}\right)+\epsilon^2\rho^0 u=-\mu\frac{d\rho^0}{dx}u \tag{58}$$

which appears as a perturbation of the spectral problem

$$-\frac{d}{dx}\left(\rho^0\frac{du}{dx}\right)=-\mu\frac{d\rho^0}{dx}u.$$

If we perform the change of variable (35), we obtain

$$\frac{d}{dy}(yq\frac{dw_0}{dy})=\mu_0 w_0\ ,\ y\in(\rho_R^0,\rho_L^0)$$

which is a standard spectral problem for an elliptic operator which degenerates at the boundary. If we assume the conditions (48), we can apply the results of [2]: there exists a countable infinite sequence of eigenvalues $(\mu_{0,n})_{n\geq 1}$ which tend towards $+\infty$ as n tends to $+\infty$. We then *conjecture* that the perturbed spectral problem (58) has a countable infinite sequence of eigenvalues $(\mu_{\epsilon,n})_{n\geq 1}$ such that

$$\mu_{\epsilon,n}=\mu_{0,n}+O(\epsilon).$$

4. The influence of a convection velocity

We pass to the case $u^0\not\equiv 0$ and we want to analyze the effect of the convection velocity u^0 on the growth rate of the linear Rayleigh Taylor instability. We start again from the dimensionless first order differential system (8) with $M=1$, which does not degenerate in this case. It is convenient to introduce the parameters

$$\alpha=\sqrt{\frac{\epsilon}{Fr}}\ ,\quad \beta=\sqrt{\epsilon Fr}. \tag{59}$$

In addition, we choose $\bar{\rho}$ and $\bar{u}$ such that

$$\rho^0 u^0\equiv 1. \tag{60}$$

Then, it is an easy matter to put the system (8) in the form (where we have dropped the primes)

$$\frac{d\mathbf{U}}{dx}+\mathbf{A}(x;\alpha,\beta,\gamma)\mathbf{U}=\mathbf{0} \tag{61}$$

with

$$\mathbf{U}=(\rho,u,v,p)^T \tag{62}$$

and

$$\mathbf{A}(x;\alpha,\beta,\gamma) = \begin{pmatrix} \rho^0\left(\dfrac{du^0}{dx}+\alpha\gamma\right) & \rho^0\dfrac{d\rho^0}{dx} & 0 & 0 \\ 0 & 0 & i\alpha\beta & 0 \\ 0 & 0 & \alpha\gamma\rho^0 & i\alpha\beta \\ u^0\dfrac{du^0}{dx}-\dfrac{\alpha}{\beta} & \rho^0\left(\dfrac{du^0}{dx}+\alpha\gamma\right) & -i\alpha\beta & 0 \end{pmatrix}. \tag{63}$$

In addition, we require the perturbation $\mathbf{U}$ to vanish at infinity, i.e.,

$$\lim_{x\to\pm\infty}\mathbf{U}(x)=\mathbf{0}. \tag{64}$$

We want to determine γ so that the problem (61), (64) has a nontrivial solution. In full generality, this problem appears to be a fairly difficult one. Hence, we will restrict here for simplicity to the case where ρ^0 is constant for $x < a$ and $x > b$, i.e.,

$$\rho^0(x) = \begin{cases} \rho^0_L\,,\ x<a \\ \\ \rho^0_R\,,\ x>b. \end{cases} \tag{65}$$

However, we have to keep in mind that, as in Section 3, γ can be strongly influenced by the behavior at infinity of the density profile ρ^0. As a consequence of the assumption (65), the differential system (61) has constant coefficients for $x < a$ and $x > b$, so that we can find an explicit form of the solution $\mathbf{U}$ in $(-\infty, a]$ and $[b, +\infty)$.

One can check that, for constant ρ^0 (or u^0), the matrix $\mathbf{A}$ possesses two simple eigenvalues $\pm\epsilon$ and a double eigenvalue $-\alpha\gamma\rho^0$ with a one-dimensional eigenspace. Moreover the corresponding *generalized* eigenvectors can be chosen as

$$\begin{cases} \mathbf{X}_1 = (0,\beta,i\beta,-\gamma\rho^0-\beta)^T, \\ \mathbf{X}_2 = (0,\beta,-i\beta,\gamma\rho^0-\beta)^T, \\ \mathbf{X}_3 = (0,\beta,-i\gamma\rho^0,0)^T, \\ \mathbf{X}_4 = (\beta^2-(\gamma\rho^0)^2,0,i,\dfrac{\gamma\rho^0}{\beta})^T. \end{cases} \tag{66}$$

Hence, we can state

Lemma 4.1. *For constant ρ^0 (or u^0), any solution of the differential system (61) is of the form*

$$\mathbf{U}(x) = c_1\mathbf{X}_1e^{\epsilon x} + c_2\mathbf{X}_2e^{\epsilon x} + \{(c_3+c_4x)\mathbf{X}_3 + c_4\mathbf{X}_4\}\,e^{-\alpha\gamma\rho^0 x}, \tag{67}$$

where the constants c_i, $1 \le i \le 4$, are arbitrary.

We are concerned with the *unstable* normal modes, i.e., with the solutions of (61),(64) corresponding to $\mathcal{R}e\gamma > 0$. Then, we have $\mathcal{R}e(\alpha\gamma\rho^0) > 0$ and, using

(64), we obtain from Lemma 4.1

$$\mathbf{U}(x) = \begin{cases} c_1 \mathbf{X}_1 e^{\epsilon x} \ , \ x < a \\ \\ c_2 \mathbf{X}_2 e^{-\epsilon x} + \{(c_3 + c_4 x)\mathbf{X}_3 + c_4 \mathbf{X}_4\} \, e^{-\alpha\gamma\rho^0 x} \ , \ x > b. \end{cases}$$

Now, it follows from the continuity of $\mathbf{U}$ at $x = a$ that $\mathbf{U}(a)$ is colinear to the vector

$$\mathbf{U}^{(1)} = (0, \beta, i\beta, -\gamma\rho^0(a) - \beta)^T. \tag{68}$$

On the other hand, the continuity of $\mathbf{U}$ at $x = b$ implies that $\mathbf{U}(b)$ must belong to the three-dimension subspace of $\mathbf{R}^4$ spanned by the vectors

$$\begin{cases} \mathbf{U}^{(2)} = (0, \beta, -i\beta, \gamma\rho^0(b) - \beta)^T, \\ \mathbf{U}^{(3)} = (0, \beta, -i\gamma\rho^0(b), 0)^T, \\ \mathbf{U}^{(4)} = (\beta^2 - (\gamma\rho^0(b))^2, 0, i, \dfrac{\gamma\rho^0(b)}{\beta})^T. \end{cases} \tag{69}$$

Now, let $\mathbf{U} = \mathbf{U}(.; \alpha, \beta, \gamma)$ be the solution of the Cauchy problem

$$\begin{cases} \dfrac{d\mathbf{U}}{dx} + \mathbf{AU} = \mathbf{0}, \ x > a \\ \\ \mathbf{U}(a) = \mathbf{U}^{(1)}. \end{cases} \tag{70}$$

Then, we look for $\gamma = \gamma(\alpha, \beta)$ in such a way that $\mathbf{U}(b)$ belongs to the space spanned by the vectors $\mathbf{U}^{(j)}$, $2 \leq j \leq 4$. This amounts to write

$$D(\alpha, \beta, \gamma) \equiv det \left(\mathbf{U}(b; \alpha, \beta, \gamma), \mathbf{U}^{(2)}(\beta, \gamma), \mathbf{U}^{(3)}(\beta, \gamma), \mathbf{U}^{(4)}(\beta, \gamma) \right) = 0. \tag{71}$$

This is the desired *dispersion relation* which connects the growth rate of the perturbations to the parameters α and β. We want to show that (71) enables us to determine γ as an analytic function of (α, β), at least in a neighborhood of $\alpha = 0$.

We begin by checking that the solution $\mathbf{U}$ of (70) depends analytically on the parameters α, β and γ. In fact, we prove a more precise result.

Lemma 4.2. *The solution* $\mathbf{U}$ *of (70) is of the form*

$$\mathbf{U} = \beta\mathbf{V} - \mathbf{Q} \ , \ \mathbf{Q} = (0, 0, 0, q)^T \tag{72}$$

where the functions $\mathbf{V} = \mathbf{V}(x; \alpha, \beta, \gamma)$ *and* $q = q(x; \alpha, \beta, \gamma)$ *depend analytically on* (α, β, γ).

Proof. We set

$$\mathbf{A} = \mathbf{A}_0(x) = \begin{pmatrix} \rho^0 \dfrac{du^0}{dx} & \rho^0 \dfrac{d\rho^0}{dx} & 0 & 0 \\ 0 & 0 & 0 & 0 \\ 0 & 0 & 0 & 0 \\ u^0 \dfrac{du^0}{dx} & \rho^0 \dfrac{du^0}{dx} & 0 & 0 \end{pmatrix}$$

and

$$\mathbf{B}_1 = \begin{pmatrix} 0 & 0 & 0 & 0 \\ 0 & 0 & 0 & 0 \\ 0 & 0 & 0 & 0 \\ 1 & 0 & 0 & 0 \end{pmatrix}, \mathbf{B}_2 = \begin{pmatrix} 1 & 0 & 0 & 0 \\ 0 & 0 & 0 & 0 \\ 0 & 0 & 1 & 0 \\ 0 & 1 & 0 & 0 \end{pmatrix}, \mathbf{B}_3 = \begin{pmatrix} 0 & 0 & 0 & 0 \\ 0 & 0 & 1 & 0 \\ 0 & 0 & 0 & 1 \\ 0 & 0 & -1 & 0 \end{pmatrix}.$$

so that

$$\mathbf{A} = \mathbf{A}_0 - \frac{\alpha}{\beta}\mathbf{B}_1 + \alpha\gamma\rho^0\mathbf{B}_2 + i\alpha\beta\mathbf{B}_3.$$

We then observe that finding $\mathbf{U}$ of the form (72) solution of the Cauchy problem (70) amounts to look for a pair $(\mathbf{V}, q)$ solution of the Cauchy problem

$$\begin{cases} \dfrac{d\mathbf{V}}{dx} + \mathbf{A}_0\mathbf{V} + \alpha\gamma\rho^0\mathbf{B}_2\mathbf{V} + i\alpha\beta\mathbf{B}_3\mathbf{V} - i\alpha\mathbf{B}_3\mathbf{Q} = 0 \\ \dfrac{dq}{dx} + \alpha\mathbf{V}_1 = 0\,, \end{cases} \tag{73}$$

$$\begin{cases} \mathbf{V}(a) = (0, 1, i, -1)^T \\ q(a) = \gamma. \end{cases} \tag{74}$$

At the difference of the original problem (70), the Cauchy problem (73),(74) depends analytically on the parameters α, β, γ. The conclusion follows from the general theory of differential equations. □

Next, we study the dispersion relation (71). By developing the determinant $D(\alpha, \beta, \gamma)$ with respect to its first column, we obtain

Lemma 4.3. *We have*

$$D(\alpha, \beta, \gamma) = -i\beta\left(\gamma\rho^0(b) - \beta\right)^2 \Phi(\alpha, \beta, \gamma) \tag{75}$$

where

$$\Phi(\alpha, \beta, \gamma) = \frac{1}{\beta}\rho(b) + \left(\gamma\rho^0(b) + \beta\right)\left(-\frac{\gamma\rho^0(b)}{\beta}u(b) + iv(b) + p(b)\right). \tag{76}$$

Observe that $\gamma = \dfrac{\beta}{\rho^0(b)} = \beta u^0(b)$ is a double root of the dispersion relation. This is in fact a parasitic root since, for $\gamma = \beta u^0(b)$, $-\epsilon$ becomes a triple eigenvalue of $\mathbf{A}(b; \alpha, \beta, \gamma)$ with an associate eigenspace of dimension 1, and one can check that the dispersion relation (71) is no longer valid in this case. Finally, we are left with the dispersion equation

$$\Phi(\alpha,\beta,\gamma)=0. \tag{77}$$

Now, it follows from Lemma 4.2 that the functions $\frac{1}{\beta}\rho$, $\frac{1}{\beta}u$, v and p depend analytically on (α,β,γ) so that Φ is *an analytic function* of (α,β,γ).

Theorem 4.1. *For α small enough, the dispersion equation (77) has a unique solution $\gamma=\gamma(\alpha,\beta)$ which is an analytic function of (α,β). Moreover, we have*

$$\gamma(0,\beta)=\sqrt{A}-\beta u^0(b). \tag{78}$$

Proof. Let us first check that the equation

$$\Phi(0,\beta,\gamma)=0. \tag{79}$$

has a unique solution $\gamma=\gamma(0,\beta)$ given by (78). Setting

$$\mathbf{V}_0(x;\beta,\gamma)=\mathbf{V}(x;0,\beta,\gamma)\ ,\ q_0(x;\beta,\gamma)=q(x;0,\beta,\gamma)\ ,$$

it follows from (73), (74) that the pair$(\mathbf{V}_0,q_0)$ is solution of the differential system

$$\begin{cases}\dfrac{d\mathbf{V}_0}{dx}+\mathbf{A}_0\mathbf{V}_0=0\\[2mm] \dfrac{dq_0}{dx}=0\end{cases}$$

with the initial conditions

$$\begin{cases}\mathbf{V}_0(a)=(0,1,i,-1)^T\\[2mm] q_0(a)=\gamma.\end{cases}$$

In fact, one can easily solve this initial value problem explicitely. We obtain

$$\mathbf{V}_0=(\rho^0(1-\rho^0),1,i,-u^0)^T\ ,\ q_0=\gamma\ ,$$

and therefore

$$\mathbf{U}_0=\beta\mathbf{V}_0-\mathbf{Q}_0=(\beta\rho^0(1-\rho^0),\beta,i\beta,-\gamma-\beta u^0)^T.$$

Hence, by (76), we have

$$\Phi(0,\beta,\gamma)=\rho^0(b)(1-\rho^0(b))-(\gamma\rho^0(b)+\beta)^2(1+u_0(b))$$

and the unique solution of (79) satisfying $\mathcal{R}e\gamma\geq 0$ is indeed given by (78).

Moreover, since

$$\frac{\partial \Phi}{\partial \gamma}(0, \beta, \gamma) = -2\rho^0(b)(\gamma\rho^0(b) + \beta)$$

and therefore

$$\frac{\partial \Phi}{\partial \gamma}(0, \beta, \gamma(0, \beta)) = -2\rho^0(b)\sqrt{A}u^0(b) = -2\sqrt{A} < 0\,,$$

we can apply the implicit function theorem and our result follows. □

Similarly, one can determine explicitely the first terms of the series expansion of $\gamma(\alpha, \beta)$ at the expense however of lengthy but straightforward calculations. In problems that we have in mind, the small parameter is here $\alpha = \sqrt{\frac{\epsilon}{Fr}}$ rather than ϵ as the Froude number is always large. We will then consider in the following the expansion of $\gamma(\alpha, \beta)$ in a series of powers of α, β being a parameter, given up to the first order by

$$\begin{aligned}
\gamma(\alpha, \beta) &= \sqrt{A} - \beta u^0(b) \\
&+ \alpha\left(\frac{\beta\rho^0(a)}{\sqrt{A}\rho^0(b)(\rho^0(a)+\rho^0(b))^2}\int_a^b (\rho^0(a)-\rho^0)(\rho^0-\rho^0(b))\,dx\right. \\
&- \left.\frac{\beta^2 u^0(b)}{2\rho^0(b)(\rho^0(a)+\rho^0(b))}\int_a^b \frac{(\rho^0(a)-\rho^0)(\rho^0-\rho^0(b))^2}{\rho^0}\,dx\right) + O(\alpha^2). \qquad (80)
\end{aligned}$$

Let us next assume that the Froude number is $O(\frac{1}{\epsilon})$. Setting $Fr = \frac{F}{\epsilon}$ and using (80), we obtain

$$\begin{aligned}
\gamma &= \sqrt{A} - Fu^0(b) \\
&+ \epsilon\left(\frac{\rho^0(a)}{\sqrt{A}\rho^0(b)(\rho^0(a)+\rho^0(b))^2}\int_a^b (\rho^0(a)-\rho^0)(\rho^0-\rho^0(b))\,dx\right. \\
&- \left.\frac{Fu^0(b)}{2\rho^0(b)(\rho^0(a)+\rho^0(b))}\int_a^b \frac{(\rho^0(a)-\rho^0)(\rho^0-\rho^0(b))^2}{\rho^0}\,dx\right) + O(\epsilon^2).
\end{aligned}$$

At the order 0 in ϵ, we do not recover the classical growth rate of the linearized RT instability since the stabilizing effect of the convection velocity occurs at this order. On the other hand, the term in $O(\epsilon)$ which takes into account the density profile is either stabilizing or destabilizing depending how large is the Froude number. In any case, we obtain that, at least for long-wavelength perturbations, a large Froude number leads to a significative stabilization of the RT instability.

Remark 4.1. *Going back to the physical quantities, the dimensionless expression (80) lead to the following growth rate*

$$\begin{aligned}\gamma &= \sqrt{Agk} - ku^0(b) \\ &+ \sqrt{Agk}\frac{k\rho^0(a)}{\sqrt{A}\rho^0(b)(\rho^0(a)-\rho^0(b))^2}\int_a^b (\rho^0(a)-\rho^0)(\rho^0-\rho^0(b))\,dx \\ &- \frac{k^2u^0(b)}{2\rho^0(b)(\rho^0(a)+\rho^0(b))}\int_a^b \frac{(\rho^0(a)-\rho^0)(\rho^0-\rho^0(b))^2}{\rho^0}\,dx + O(\epsilon^2).\end{aligned}$$

where ρ^0 stands for the physical density.

Let us emphasize that the model analyzed in this Section is still far from being physically realistic. We have first assumed that the density is constant at infinity, which is not true in the applications and more importantly, we have neglected the heat diffusion which plays an essential role in the mechanism of ICF. These points have been considered from a physical point of view in [5, 6], and studied successfully by means of a formal boundary layer technique. The rigorous stability analysis of such a complex flow still appears as a challenge for mathematicians.

References

[1] M. Abramovitz and I. Stegun, *Handbook of Mathematical Functions,* (Dover, New York, 1968).

[2] M.S. Baouendi and C. Goulaouic, *Régularité et théorie spectrale pour une classe d'opérateurs elliptiques non dégénérés,* Arch. Rat. Mech. Anal **34** (1969), 361–379.

[3] S. Chandrasekhar, *Hydrodynamic and Hydromagnetic Stability,* (Oxford University Press, London, 1968).

[4] C. Cherfils and O. Lafitte, *Analytic solutions of the Rayleigh equation for linear density profiles,* Phys. Rev. E **62** (2000), 2967–2970.

[5] V.N. Goncharov, *Self consistent stability analysis of ablation fronts in Inertial Confinement Fusion,* Ph.D.Thesis, University of Rochester, 1998.

[6] V.N. Goncharov, R. Betti, R.L. McCrory, P. Sorotokin and C.P. Verdon, *Self-consistent stability analysis of ablation fronts with large Froude numbers,* Phys. Plasmas **3** (1996), 1402–1414;
R. Betti, V.N. Goncharov, R.L. McCrory, P. Sorotokin and C.P. Verdon, *Self-consistent stability analysis of ablation fronts in Inertial Confinement Fusion,* Phys. Plasmas **3** (1996), 2122–2128;
V.N. Goncharov, R. Betti, R.L. McCrory and C.P. Verdon, *Self-consistent stability analysis of ablation fronts with small Froude numbers,* Phys. Plasmas **3** (1996), 4665–4676.

[7] H.J. Kull, *Theory of the Rayleigh-Taylor instability,* Phys. Rep. **206** (1991), 197–235.

[8] H.J. Kull and S.I. Anisimov, *Ablative stabilization in the incompressible Rayleigh-Taylor instability,* Phys. Fluids **29** (1986), 2067–2075.

[9] O. Lafitte, *Analysis of the discrete spectrum of the Rayleigh equation: Application to the linear Rayleigh-Taylor instability,* Preprint, CMAT **24-2000** (Ecole Polytechnique, Paris, 2000).

[10] Lord Rayleigh, *Analytic solutions of the Rayleigh equation for linear density profiles,* Proc. London Math. Soc. **14** (1883), 170–177.

[11] G. Taylor, *The instability of liquid surfaces when accelerated in a direction perpendicular to their planes,* Proc. R. Soc. London, Ser. A **201** (1950), 192–196.

Catherine Cherfils–Clérouin
CEA/DIF
91680 Bruyères le Châtel Cedex
France
E-mail address: `catherine.cherfils@cea.fr`

Olivier Lafitte
CEA/DMT Centre d'Etudes de Saclay
91191 Gif sur Yvette Cedex
France
E-mail address: `olivier.lafitte@cea.fr`

Pierre–Arnaud Raviart
CMAP Ecole Polytechnique
91128 Palaiseau Cedex
France
E-mail address: `raviart@cmapx.polytechnique.fr`

Recent Progress in the Mathematical Theory of Viscous Compressible Fluids

Eduard Feireisl

Abstract. We review some recent results on *a priori* estimates, compactness, global existence, and the long-time behaviour of solutions to the Navier-Stokes equations of a compressible fluid flow.

1. Mathematical models

1.1. Equations

Let $\Omega \subset R^N$, $N = 1, 2, 3$ be a spatial domain filled with a fluid and let $I \subset R$ be a time interval. For each time $t \in I$, the state of the fluid is characterized by its *density* $\varrho(t,x)$, the *velocity* $\vec{u}(t,x)$, and the *temperature* $\theta(t,x)$. The time evolution of the quantities ϱ, $\vec{u}$, and θ is governed by the three fundamental principles of classical physics:

Conservation of Mass:

$$\partial_t \varrho + \operatorname{div}(\varrho \vec{u}) = 0; \tag{1}$$

Balance of Momentum:

$$\partial_t(\varrho \vec{u}) + \operatorname{div}(\varrho \vec{u} \otimes \vec{u}) + \nabla p = \operatorname{div}\ \mathrm{T} + \varrho \vec{f}; \tag{2}$$

Conservation of Energy:

$$\partial_t E + \operatorname{div}((E+p)\vec{u}) = \operatorname{div}\ (\mathrm{T}.\vec{u}) - \operatorname{div}\ \vec{q} + \varrho \vec{f}.\vec{u} \tag{3}$$

Here p denotes the *pressure*, T is the *viscous stress tensor*, E stands for the *total energy*, $\vec{q}$ is the *heat flux*, and $\vec{f}$ denotes a given *body force*.

We restrict ourselves to the *Newtonian fluids*, i.e., the stress tensor T depends only on the first derivatives of the velocity, specifically,

$$T = T(\nabla \vec{u}) = \mu\Big(\nabla \vec{u} + (\nabla \vec{u})^T\Big) + \lambda \operatorname{div} \vec{u}\ \mathrm{Id}$$

where μ and λ are *viscosity coefficients.*

The total energy E can be decomposed as

$$E = \frac{1}{2}\varrho|\vec{u}|^2 + \varrho e$$

Work supported by the Grant A1019002 of GA AV ČR

where $e = e(\varrho, \theta)$ is the *specific internal energy*. In what follows, we shall assume a very simple constitutive relation between the temperature and the specific internal energy, namely,

$$e = c_v \theta, \ c_v > 0,$$

where c_v is the *specific heat* constant.

The heat flux $\vec{q}$ obeys the *Fourier law*:

$$\vec{q} = -\kappa \nabla \theta, \ \kappa \geq 0.$$

Finally, the pressure p will be determined by a general constitutive relation

$$p = p(\varrho, e).$$

To obtain a rigorous mathematical theory, however, we shall consider more specific situations. For instance, the *ideal gas* consitutive relation reads

$$p = (\gamma - 1)\varrho e \text{ where } \gamma > 1 \text{ is the } \textit{adiabatic constant.} \tag{4}$$

Introducing the *entropy*

$$S = S(\varrho, e) = \log(e) + (1 - \gamma)\log(\varrho)$$

we can rewrite the energy equation (3) in the form

$$\partial_t(\varrho S) + \operatorname{div}\Big(\varrho S \vec{u} + \frac{\vec{q}}{e}\Big) = \frac{\mathrm{T} : \nabla \vec{u}}{e} - \frac{\vec{q}.\nabla e}{e^2}. \tag{5}$$

In accordance with the second law of thermodynamics, the right-hand side of the above equation should be nonnegative which yields the restriction

$$\mu \geq 0, \ \lambda + \frac{2}{N}\mu \geq 0. \tag{6}$$

Here, we focus on *viscous fluids*, i.e., we shall always assume $\mu > 0$.

In the case $\mu = \lambda = \kappa = 0$, the equation (5) reduces to a conservation law for the entropy S. Taking a spatially homogeneous initial distribution of the entropy $S(t_0) = S_0$ we easily deduce $S(t) = S_0$ for any $t > t_0$ and, consequently,

$$p(\varrho) = (\gamma - 1)\exp(S_0)\varrho^\gamma.$$

In other words, the pressure p is a function of the density only. In this case, the equations (1), (2) represent a closed system describing the motion of *isentropic* compressible viscous fluids. Similar situation occurs in the *isothermal* case where we suppose e in (4) to be constant. Both situations represent a particular case of a general *barotropic* flow when the pressure p is given by the constitutive relation

$$p = p(\varrho) \tag{7}$$

For a general barotropic flow, the total energy takes the form

$$E_b = \frac{1}{2}\varrho|\vec{u}|^2 + P(\varrho)$$

and obeys (formally) the equation

$$\partial_t E_b + \operatorname{div}\Big((E_b + p)\vec{u}\Big) = \operatorname{div}(\mathrm{T}.\vec{u}) - \mathrm{T} : \nabla\vec{u} + \varrho\vec{f}.\vec{u} \tag{8}$$

where the function P is related to p by the formula

$$P'(z)z - P(z) = p(z).$$

By virtue of (6), the term $\mathrm{T} : \nabla\vec{u}$ is always nonnegative and represents a dissipative mechanism induced by viscosity.

1.2. Boundary conditions

The boundary $\partial\Omega$, if any, will be always assumed rigid (impermeable) and thermally insulated, i.e., $\vec{u}$ will obey the boundary conditions

$$\vec{u}.\vec{n}|_{\partial\Omega} = 0 \tag{9}$$

while $\vec{q}$ will comply with

$$\vec{q}.\vec{n}|_{\partial\Omega} = 0 \tag{10}$$

where $\vec{n}$ denotes the outer normal vector. Moreover, we shall suppose some kind of dissipative boundary conditions for the tangential component of the velocity:

$$\Big(T(\vec{u}).\vec{u}\Big).\vec{n}|_{\partial\Omega} \leq 0. \tag{11}$$

Probably the simplest case when both (9) and (11) hold is the situation when the velocity $\vec{u}$ satisfies

No-slip Boundary Condition:

$$\vec{u}|_{\partial\Omega} = 0 \tag{12}$$

1.3. Total mass conservation and energy inequalities

Integrating (1) over Ω and making use of the boundary conditions (9) we obtain

$$\frac{\mathrm{d}}{\mathrm{d}t}\int_\Omega \varrho(t)\ \mathrm{d}x = 0, \tag{13}$$

i.e., the *total mass*

$$m = \int_\Omega \varrho\ \mathrm{d}x$$

is a constant of motion.

Similarly, using (3) together with (9) - (11) we deduce the *energy inequality*

$$\frac{\mathrm{d}}{\mathrm{d}t}\int_\Omega E(t)\ \mathrm{d}x \leq \int_\Omega \varrho(t)\vec{f}(t).\vec{u}(t)\ \mathrm{d}x \tag{14}$$

.

For barotropic flows we can replace (3) by (8) to get

$$\frac{\mathrm{d}}{\mathrm{d}t}\int_\Omega E_b(t)\ \mathrm{d}x + \int_\Omega \mathrm{T}(\nabla\vec{u}(t)) : \nabla\vec{u}(t)\ \mathrm{d}x \leq \int_\Omega \varrho(t)\vec{f}(t).\vec{u}(t)\ \mathrm{d}x. \tag{15}$$

If, in addition, the velocity $\vec{u}$ satifies the no-slip boundary conditions (12), a simple by parts integration yields

$$\int_\Omega \mathrm{T}(\nabla\vec{u}) : \nabla\vec{u}\ \mathrm{d}x = \mu \int_\Omega |\nabla\vec{u}|^2\ \mathrm{d}x + (\lambda+\mu)\int_\Omega |\mathrm{div}\ \vec{u}|^2\ \mathrm{d}x.$$

The right-hand side in (14), (15) can be estimated by means of the Cauchy-Schwartz inequality and (13) to deduce

$$\frac{\mathrm{d}}{\mathrm{d}t}\int_\Omega E(t)\ \mathrm{d}x \le \sqrt{2m}F\Big(\int_\Omega E(t)\ \mathrm{d}x\Big)^{\frac{1}{2}} \tag{16}$$

,

$$\frac{\mathrm{d}}{\mathrm{d}t}\int_\Omega E_b(t)\ \mathrm{d}x \le \sqrt{2m}F\Big(\int_\Omega E_b(t)\ \mathrm{d}x\Big)^{\frac{1}{2}} \tag{17}$$

respectively, where

$$F = \mathrm{ess\,sup}_{t\in I,\ x\in\Omega}|\vec{f}(t,x)|.$$

2. Variational formulation, weak solutions

2.1. Distributional solutions

We shall say that the equations (1) - (3) are satisfied in the sense of distributions, shortly in $\mathcal{D}'(I\times\Omega)$, if all the quantities

$$\varrho,\ \varrho u^i,\ \varrho u^i u^j,\ p,\ pu^i,\ \mathrm{T},\ \mathrm{T}.u,\ E,\ E.u,\ q,\ i,j=1,...,N \text{ belong to } L^1_{loc}(I\times\Omega)$$

and their generalized derivatives of first order satisfy (1) - (3).

Note that the requirement $\mathrm{T}\in L^1_{loc}$ is equivalent to $\nabla\vec{u}\in L^1_{loc}$, i.e., the first spatial derivative of the velocity is supposed to be a function.

2.2. Renormalized solutions of the continuity equation

Motivated by the work of Kruzkhov DiPERNA and LIONS [4] introduced the concept of *renormalized solutions* of the continuity equation (1). Formally, one can multiply (1) by $b'(\varrho)$ where b is any continuously differentiable function to obtain the identity:

$$\partial_t b(\varrho) + \mathrm{div}(b(\varrho)\vec{u}) + \Big(b'(\varrho)\varrho - b(\varrho)\Big)\ \mathrm{div}\ \vec{u} = 0. \tag{18}$$

Obviously, any strong (classical) solution of (1) satisfies automatically (18). For the weak solutions, however, (18) represents an additional constraint which may not be always satisfied.

Following [4] we shall say that ϱ is a *renormalized solution* of (1) if $\vec{u}$, $\nabla\vec{u}\in L^1_{loc}(I\times\Omega)$ and (18) holds in $\mathcal{D}'(I\times\Omega)$ for any function b such that

$$b\in C^1(R),\ b'(z)=0 \text{ for all } z \text{ large enough, say, } |z|\ge M. \tag{19}$$

An important class of functions satisfying (19) are the *cut-off functions*:

$$\mathcal{T}_k(z) = k\ \mathcal{T}(\frac{z}{k}),\ k\ge 1 \tag{20}$$

where $\mathcal{T} \in C^1(R)$ is an odd function ($\mathcal{T}(-z) = -\mathcal{T}(z)$) such that

$$\mathcal{T}(z) = z \text{ for } 0 \leq z \leq 1, \ \mathcal{T}(z) = 2 \text{ for } z \geq 3, \ \mathcal{T} \text{ concave on } [0, \infty).$$

A remarkable feature of the renormalized solutions is that the compositions $\mathcal{T}_k(\varrho)$ are continuous functions of time with values in the space L^1_{loc} for any $k \geq 1$. To see this, let us review first an auxilliary assertion which is basically due to Friedrichs:

Lemma 2.1. *Let $\vec{u}$, $\nabla\vec{u}$ belong to $L^p_{loc}(\Omega)$ and $\sigma \in L^q_{loc}(\Omega)$, $1 \leq p, q \leq \infty$, $1/r = 1/p + 1/q \leq 1$. Let ϑ_ε be a regularizing sequence, i.e., $\vartheta_\varepsilon \in \mathcal{D}(R^N)$, ϑ_ε radially symmetric and radially decreasing,*

$$\int_{R^N} \vartheta_\varepsilon \, \mathrm{d}x = 1, \ \vartheta_\varepsilon(x) \to 0 \ \textit{as } \varepsilon \to 0 \textit{ for any fixed } x \in R^N \setminus \{0\}.$$

Then

$$\|\vartheta_\varepsilon * [\mathrm{div}(\sigma\vec{u})] - \mathrm{div}([\vartheta_\varepsilon * \sigma]\vec{u})\|_{L^r(K)} \leq c(K)\|\vec{u}\|_{W^{1,p}(K)}\|\sigma\|_{L^q(K)}$$

for any compact $K \subset \Omega$ and

$$r_\varepsilon = \vartheta_\varepsilon * [\mathrm{div}(\sigma\vec{u})] - \mathrm{div}([\vartheta_\varepsilon * \sigma]\vec{u}) \to 0 \ \textit{in } L^r_{loc}(\Omega) \ \textit{as } \varepsilon \to 0.$$

Now we are ready to prove the following

Proposition 2.2. *Let $\vec{u}$, $\nabla\vec{u}$ be locally integrable on $I \times \Omega$. Let ϱ be a renormalized solution of the equation (1).*

Then for any compact subinterval $J \subset I$ and any compact $K \subset \Omega$, the functions $\mathcal{T}_k(\varrho)$ satisfy

$$\mathcal{T}_k(\varrho) \in C(J; L^1(K)), \ k \geq 1 \ \textit{arbitrary}.$$

Proof. Given k, J, K, we take $b = \mathcal{T}_k$ in (18) to deduce that

$$\mathcal{T}_k(\varrho) \in C(J; L^1_{weak}(\Omega))$$

changing $\varrho(t)$ for t belonging to a set of zero measure if necessary.

Making use of Lemma 2.1 we can regularize (18) to obtain

$$\partial_t(\vartheta_\varepsilon * \mathcal{T}_k(\varrho)) + \mathrm{div}([\vartheta_\varepsilon * \mathcal{T}_k(\varrho)]\vec{u}) + \vartheta_\varepsilon * \left[\Big(\mathcal{T}_k'(\varrho)\varrho - \mathcal{T}_k(\varrho)\Big)\mathrm{div}\ \vec{u}\right] = r_\varepsilon.$$

Finally, multiplying the above equality by $b'(\vartheta_\varepsilon * \mathcal{T}_k(\varrho))$ and letting $\varepsilon \to 0$ we conclude

$$b(\vartheta_\varepsilon * \mathcal{T}_k(\varrho)) \to b(\mathcal{T}_k(\varrho)) \text{ in } C(J; L^1_{weak}(K)) \text{ as } \varepsilon \to 0$$

for any function $b \in C^1(R)$. In particular, $b(\mathcal{T}_k(\varrho)) \in C(J; L^1_{weak}(K))$ for any $b \in C^1(R)$ which implies strong continuity of $\mathcal{T}_k(\varrho)$.

□

2.3. Weak limits of renormalized solutions

To investigate weak limits of sequences of renormalized solutions we need a tool to describe possible oscillations in the density field. The following assertion will be useful.

Lemma 2.3. *Let ϱ_n be a sequence of functions on a bounded set $Q \subset I \times \Omega$, $\varrho_n \to \varrho$ weakly in $L^1(Q)$.*

Then

$$\overline{T_k(\varrho)} \to \varrho \text{ in } L^1(Q) \text{ as } k \to \infty.$$

Remark: Here and in what follows, the bar denotes a weak limit.

Proof. We have

$$\|\overline{T_k(\varrho)} - \varrho\|_{L^1(Q)} \leq \liminf_{n\to\infty} \|T_k(\varrho_n) - \varrho_n\|_{L^1(Q)} \leq 2 \sup_n \int_{\varrho_n \geq k} |\varrho_n| \, dx$$

On the other hand,

$$|\{\varrho_n \geq k\}| \leq \frac{1}{k} \sup_n \|\varrho_n\|_{L^1(Q)} \to 0 \text{ as } k \to \infty$$

and the desired result follows from Dunford Pettis theorem, namely, the family $\{\varrho_n\}$ is equi-integrable.

□

Now, let ϱ_n be a locally equi-integrable sequence defined on $I \subset \Omega$ such that

$$\varrho_n \to \varrho \text{ weakly in } L^1_{loc}(I \times \Omega).$$

The amplitude of oscillations of $\varrho_n - \varrho$ on a set $Q \subset I \times \Omega$ measured in the L^p–norm will be desribed by means of the quantity

$$\mathbf{osc}_p[\varrho_n - \varrho](Q) = \sup_{k\geq 1} \limsup_{n\to\infty} \|T_k(\varrho_n) - T_k(\varrho)\|_{L^p(Q)}.$$

Now, we claim the following result.

Proposition 2.4. *Let ϱ_n be a sequence of renormalized solutions of (1) with $\vec{u} = \vec{u}_n$. Assume*

$$\varrho_n \to \varrho \text{ weakly in } L^1_{loc}(I \times \Omega),\ \vec{u}_n \to \vec{u},\ \nabla\vec{u}_n \to \nabla\vec{u} \text{ weakly in } L^q_{loc}(I \times \Omega),$$

and

$$\mathbf{osc}_p[\varrho_n - \varrho](Q) \leq c(Q) \text{ for any compact } Q \subset I \times \Omega$$

where

$$\frac{1}{p} + \frac{1}{q} < 1.$$

Then ϱ (and $\vec{u}$) is a renormalized solution of (1).

Proof. Given an open time interval $J \subset \overline{J} \subset I$ and a subdomain $K \subset \overline{K} \subset \Omega$, it is enough to show that (18) holds in $\mathcal{D}'(J \times K)$.

Taking $b = \mathcal{T}_k$ in (18) and seeing that $q > 1$ we deduce easily

$$\mathcal{T}_k(\varrho_n) \to \overline{\mathcal{T}_k(\varrho)} \text{ in } C(J; L^{\alpha}_{weak}(K)) \text{ for all } 1 \leq \alpha < \infty;$$

and, by means of the Lions-Aubin Lemma,

$$T_k(\varrho_n)\vec{u}_n \to \overline{\mathcal{T}_k(\varrho)}\vec{u} \text{ weakly in } L^q(J \times \Omega).$$

Here and in what follows, bar denotes a weak limit. Consequently,

$$\partial_t \overline{\mathcal{T}_k(\varrho)} + \operatorname{div}(\overline{\mathcal{T}_k(\varrho)}\vec{u}) + \overline{\left(\mathcal{T}_k'(\varrho)\varrho - \mathcal{T}_k(\varrho)\right) \operatorname{div} \vec{u}} = 0 \text{ in } \mathcal{D}'(J \times K) \tag{21}$$

where

$$\left(\mathcal{T}_k'(\varrho_n)\varrho_n - \mathcal{T}_k(\varrho_n)\right) \operatorname{div} \vec{u}_n \to \overline{\left(\mathcal{T}_k'(\varrho)\varrho - \mathcal{T}_k(\varrho)\right) \operatorname{div} \vec{u}} \text{ weakly in } L^q(J \times K).$$

Now, similarly as in the proof of Proposition 2.2, we can regularize (21) convolving it with ϑ_ε, multiply by $b'(\overline{\mathcal{T}_k(\varrho)})$, and let $\varepsilon \to 0$ to conclude

$$\partial_t b(\overline{\mathcal{T}_k(\varrho)}) + \operatorname{div}(b(\overline{\mathcal{T}_k(\varrho)})\vec{u}) + \left(b'(\overline{\mathcal{T}_k(\varrho)})\overline{\mathcal{T}_k(\varrho)} - b(\overline{\mathcal{T}_k(\varrho)})\right)\operatorname{div} \vec{u} = \tag{22}$$

$$b'(\overline{\mathcal{T}_k(\varrho)}) \; \overline{\left(\mathcal{T}_k(\varrho) - \mathcal{T}_k'(\varrho)\varrho\right) \operatorname{div} \vec{u}} = 0 \text{ in } \mathcal{D}'(J \times K)$$

for all b as in (19).

By virtue of Lemma 2.3, the left-hand side of (22) will tend to that of (18) for $k \to \infty$; whence it is enough to show

$$b'(\overline{\mathcal{T}_k(\varrho)}) \; \overline{\left(\mathcal{T}_k(\varrho) - \mathcal{T}_k'(\varrho)\varrho\right) \operatorname{div} \vec{u}} \to 0 \text{ in } L^1(J \times K) \text{ for } k \to \infty. \tag{23}$$

Denoting

$$Q_{k,M} = \{(t,x) \in J \times K \mid |\overline{\mathcal{T}_k(\varrho)}| \leq M\}$$

where M is the constant appearing in (19), we get

$$\left\| b'(\overline{\mathcal{T}_k(\varrho)}) \; \overline{\left(\mathcal{T}_k(\varrho) - \mathcal{T}_k'(\varrho)\varrho\right) \operatorname{div} \vec{u}} \right\|_{L^1(J \times K)} \leq \tag{24}$$

$$\Big(\sup_{|z| \leq M} |b'(z)|\Big)\Big(\sup_n \|\operatorname{div} \vec{u}_n\|_{L^q(J\times K)}\Big) \liminf_{n \to \infty} \|\mathcal{T}_k(\varrho_n) - \mathcal{T}_k'(\varrho_n)\varrho_n\|_{L^{q'}(Q_{k,M})}$$

where $1/q + 1/q' = 1$.

Now, by interpolation,

$$\|\mathcal{T}_k(\varrho_n) - \mathcal{T}_k'(\varrho_n)\varrho_n\|_{L^{q'}(Q_{k,M})} \leq \tag{25}$$

$$\|\mathcal{T}_k(\varrho_n) - \mathcal{T}_k'(\varrho_n)\varrho_n\|^{\beta}_{L^1(J\times K)} \|\mathcal{T}_k(\varrho_n) - \mathcal{T}_k'(\varrho_n)\varrho_n\|^{1-\beta}_{L^p(Q_{k,M})}$$

with $\beta = (p/q' - 1)(p-1)^{-1}$. Here, similarly as in the proof of Lemma 2.3, equi-integrability of $\{\varrho_n\}$ implies

$$\sup_n \left\{ \|\mathcal{T}_k(\varrho_n) - \mathcal{T}_k'(\varrho_n)\varrho_n\|_{L^1(J\times K)} \right\} \to 0 \text{ as } k \to \infty. \tag{26}$$

Finally, seeing that $|\mathcal{T}_k'(z)z| \le |\mathcal{T}_k(z)|$ we get

$$\|\mathcal{T}_k(\varrho_n) - \mathcal{T}_k'(\varrho_n)\varrho_n\|_{L^p(Q_{k,M})} \le$$

$$2\Big(\|\mathcal{T}_k(\varrho_n) - \mathcal{T}_k(\varrho)\|_{L^p(J\times K)} + \|\mathcal{T}_k(\varrho) - \overline{\mathcal{T}_k(\varrho)}\|_{L^p(J\times K)} + \|\overline{\mathcal{T}_k(\varrho)}\|_{L^p(Q_{k,M})} \Big) \le$$

$$2\Big(\|\mathcal{T}_k(\varrho_n) - \mathcal{T}_k(\varrho)\|_{L^p(J\times K)} + \mathbf{osc}_p[\varrho_n - \varrho](J\times K) + |J\times K|^{\frac{1}{p}} M \Big);$$

whence

$$\limsup_{n\to\infty} \|\mathcal{T}_k(\varrho_n) - \mathcal{T}_k'(\varrho_n)\varrho_n\|_{L^p(Q_{k,M})} \le 4\ \mathbf{osc}_p[\varrho_n - \varrho](J\times K) + 2M|J\times K|^{\frac{1}{p}}. \tag{27}$$

Now, (23) follows easily from (24) - (27).

□

2.4. Renormalized solutions on the whole space

We shall assume that $\Omega \subset R^N$ is a Lipschitz domain, the velocity field $\vec{u}$ such that $\nabla\vec{u} \in L^2_{loc}(I\times\overline{\Omega})$, and $\varrho \in L^p_{loc}(I\times\overline{\Omega})$. Moreover, we shall suppose that $\vec{u}$ satisfies the no-slip boundary conditions (12) in the sense of traces. Equivalently, making use of the Hardy inequality, we can say that

$$|\vec{u}|\ \mathrm{dist}[x,\partial\Omega]^{-1} \in L^2_{loc}(I\times\overline{\Omega}). \tag{28}$$

Now, prolonging $\vec{u}$ to be zero outside Ω we still have $\nabla\vec{u} \in L^2_{loc}(I\times R^N)$ and the following assertion holds.

Lemma 2.5. *Let $\varrho \in L^p_{loc}(I\times\overline{\Omega})$, $p\ge 1$ be a renormalized solution of (1) on $I\times\Omega$ where $\nabla\vec{u} \in L^2_{loc}(I\times\overline{\Omega})$ satisfies the no-slip boundary conditions in the sense (28).*
Then, prolonging ϱ, $\vec{u}$ to be zero outside Ω, we have

$$\partial_t b(\varrho) + \mathrm{div}(b(\varrho)\vec{u}) + \Big(b'(\varrho)\varrho - b(\varrho)\Big)\ \mathrm{div}\ \vec{u} = 0 \tag{29}$$

in $\mathcal{D}'(I\times R^N)$ for any $b \in C^1(R\setminus\{0\})$ such that

$$|b'(z)| \le c(|z|^{-\alpha} + |z|^{\frac{p}{2}-1}) \text{ for all } z,\ |z|\ge 1 \tag{30}$$

where $\alpha \in [0,1)$.

Proof. To begin, it is a routine matter to construct a sequence b_n satisfying (19) and such that

$$b_n(z) \to b(z),\ b_n'(z)z \to b'(z)z \text{ for all } z\in R,$$

$$|b_n(z)| + |b_n'(z)z| \le c(1+|z|^{\frac{p}{2}}) \text{ for all } z \in R.$$

Using (18) and the Lebesgue convergence theorem we infer that (29) holds in $\mathcal{D}'(I \times \Omega)$ for any b satisfying (30). Moreover $b(\varrho) \in L^2_{loc}(I \times R^N)$ provided $\varrho \equiv 0$ outside Ω.

Consequently, it remains to show that

$$\int_I \int_{R^N} b(\varrho)\ \partial_t\varphi + b(\varrho)\vec{u}.\nabla\varphi + \Big(b(\varrho) - b'(\varrho)\varrho\Big)\ \text{div}\ \vec{u}\ \varphi\ \text{d}x\ \text{d}t = 0$$

for any test function $\varphi \in \mathcal{D}(I \times \Omega)$. We shall assume the existence of a bounded ball in $B \subset R^N$ such that supp $\varphi \subset I \times B$ and $B \cap \partial\Omega \neq \emptyset$ since otherwise there is nothing to prove.

Now, we can construct a sequence of functions $\phi_n \in \mathcal{D}(\Omega)$ such that

$$0 \leq \phi_n \leq 1,\ |\nabla\phi_n| \leq 2n,\ \phi_n(x) \equiv 1 \text{ for all } x \in B,\ \text{dist}[x, \partial\Omega] \geq \frac{1}{n}. \tag{31}$$

We have

$$\int_I \int_{R^N} b(\varrho)\partial_t\varphi\ \text{d}x\ \text{d}t = \int_I \int_\Omega b(\varrho)\partial_t(\phi_n\varphi) + b(\varrho)(1-\phi_n)\partial_t\varphi\ \text{d}x\ \text{d}t,$$

and, similarly,

$$\int_I \int_{R^N} b(\varrho)\vec{u}.\vec{\varphi}\ \text{d}x\ \text{d}t = \int_I \int_\Omega b(\varrho)\vec{u}.\nabla(\phi_n\varphi) + b(\varrho)(1-\phi_n)\vec{u}.\nabla\varphi\ \text{d}x\ \text{d}t -$$

$$\int_I \int_\Omega b(\varrho)\varphi\ \vec{u}.\nabla\phi_n \text{d}x\ \text{d}t,$$

$$\int_I \int_{R^N} \Big(b(\varrho) - b'(\varrho)\varrho\Big)\text{div}\ \vec{u}\ \varphi\ \text{d}x\ \text{d}t = \int_I \int_\Omega \Big(b(\varrho) - b'(\varrho)\varrho\Big)\text{div}\ \vec{u}\ \varphi\phi_n\ \text{d}x\ \text{d}t +$$

$$\int_I \int_\Omega \Big(b(\varrho) - b'(\varrho)\varrho\Big)\text{div}\ \vec{u}\ \varphi(1-\phi_n)\ \text{d}x\ \text{d}t.$$

Thus, in accordance with our hypotheses, it is enough to show

$$\int_I \int_\Omega b(\varrho)\varphi\ \vec{u}.\nabla\phi_n \text{d}x\ \text{d}t \to 0 \text{ as } n \to \infty. \tag{32}$$

Now, (32) follows from (28) and the Lebesgue convergence theorem as $b(\varrho) \in L^2_{loc}(I \times R^N)$ and

$$|\nabla\phi_n|\ \text{dist}[x, \partial\Omega] \leq 2,\ |\nabla\phi_n(x)| \to 0 \text{ as } n \to \infty \text{ for all } x \in B \cap \Omega.$$

$\square$

Taking $p = 2$ in Lemma 2.5 we get the following

Corollary 2.6. *Let $\Omega \subset R^N$ be a Lipschitz domain, $\varrho \in L^2_{loc}(I \times \overline{\Omega})$ a renormalized solution of (1) with $\nabla \vec{u} \in L^2_{loc}(I \times \overline{\Omega})$.*

Then ϱ is also a weak solution of (1). If, moreover, Ω is bounded and $\vec{u} \in L^2_{loc}(I; W_0^{1,2}(\Omega))$, then the total mass

$$m = \int_\Omega \varrho(t) \,\mathrm{d}x$$

is a constant of motion, i.e., it is independent of t.

3. Pressure estimates for isentropic flows

3.1. Finite energy weak solutions

In this section, we shall assume that the flow is isentropic, i.e., the pressure p is given by the relation

$$p = p(\varrho) = a\varrho^\gamma \text{ where } a > 0,\ \gamma > 1. \tag{33}$$

Consequently, the system (1) - (3) reduces to the $N+1$ equations (1) - (2). For definiteness, we shall also suppose that $\Omega \subset R^N$, $N = 2, 3$ is a bounded Lipschitz domain and that the velocity $\vec{u}$ complies with the no-slip boundary conditions (12). In accordance with Section 1, the energy is given by the formula

$$E = \frac{1}{2}\varrho|\vec{u}|^2 + \frac{a}{\gamma - 1}\varrho^\gamma. \tag{34}$$

Motivated by the results of Section 2 we introduce a concept of *finite energy weak solutions* to the problem (1), (2), (33), and (12):

- The density ϱ is a nonnegative function belonging to the space
$$\varrho \in L^\infty(I; L^\gamma(\Omega)),$$
the velocity
$$\vec{u} \in L^2(I; W_0^{1,2}(\Omega)).$$
- The energy E given by formula (34) belongs to $L^1_{loc}(I; L^1(\Omega))$ and the energy inequality
$$\frac{\mathrm{d}}{\mathrm{d}t}\int_\Omega E \,\mathrm{d}x + \int_\Omega \mu|\nabla\vec{u}|^2 + (\lambda + \mu)|\mathrm{div}\,\vec{u}|^2 \,\mathrm{d}x \le \int_\Omega \varrho\vec{f}.\vec{u} \,\mathrm{d}x \tag{35}$$
holds in $\mathcal{D}'(I)$.
- The functions ϱ, $\vec{u}$ prolonged to be zero outside Ω solve the continuity equation (1) in $\mathcal{D}'(I \times R^N)$. In addition, ϱ solves (1) in the sense of renormalized solutions, specifically, (18) holds in $\mathcal{D}'(I \times R^N)$ for any b satisfying (19).
- The equations of motion (2) are satisfied in $\mathcal{D}'(I \times \Omega)$.

Here and always, the external force density $\vec{f}$ is a bounded measurable function such that

$$\operatorname{ess\,sup}_{t\in I,\ x\in\Omega}|\vec{f}(t,x)| \leq F.$$

3.2. On the inverse of the *div* operator

It is the aim of the present section to derive L^p–estimates of the pressure p in terms of the energy E and the amplitude F of the driving force $\vec{f}$. To this end, some preliminary results concerning solvability of the equation

$$\operatorname{div} \vec{v} = g - \oint_\Omega g \ \mathrm{d}x,\ \ \vec{v}|_{\partial\Omega} = 0 \tag{36}$$

are needed. Here we have denoted

$$\oint g \ \mathrm{d}x = \frac{1}{|\Omega|}\int_\Omega \ \mathrm{d}x.$$

The equation (36) has been studied by many authors. We adopt the approach which is essentially due to BOGOVSKII [1]. It can be shown that the problem (36) admits a solutions operator $\mathcal{B} : g \mapsto \vec{v}$ enjoying the following properties

- $\mathcal{B} = [\mathcal{B}_1, ..., \mathcal{B}_N]$ is a bounded linear operator from $L^p(\Omega)$ into $[W_0^{1,p}(\Omega)]^N$, specifically,

$$\|\mathcal{B}[g]\|_{W_0^{1,p}(\Omega)} \leq c(p)\|g\|_{L^p(\Omega)} \text{ for any } 1 < p < \infty. \tag{37}$$

- The function $\vec{v} = \mathcal{B}[g]$ solves the problem (36).
- If, moreover, $g \in L^p(\Omega)$ can be written in the form $g = \operatorname{div}\vec{h}$ where $\vec{h} \in [L^r(\Omega)]^N$, $\vec{h}.\vec{n} = 0$ on $\partial\Omega$, then

$$\|\mathcal{B}[g]\|_{L^r(\Omega)} \leq c(p,r)\|\vec{h}\|_{L^r(\Omega)}. \tag{38}$$

The proof of the existence of the operator $\mathcal{B}$ as well as the above properties can be found in GALDI [7] or BORCHERS and SOHR [2].

3.3. Higher integrability of the pressure term

The operator $\mathcal{B}$ introduced above will be now used to construct test functions for the moment equations (2). More specifically, we take

$$\phi_i(t,x) = \psi(t)\mathcal{B}_i[\vartheta_\varepsilon * T_k(\varrho^\theta)],\ i = 1,...,N$$

where

$$\psi \in \mathcal{D}(I),\ 0 \leq \psi \leq 1,\ \int_I |\partial_t\psi| \ \mathrm{d}t \leq 2$$

and ϑ_ε is a regularizing sequence as in Lemma 2.1. As always, ϱ is set zero outside Ω.

By virtue of Lemma 2.5, ϱ^θ satisfies (29) whenever $0 < \theta \leq 1/2$. Accordingly, we have

$$\partial_t(\vartheta_\varepsilon * T_k(\varrho^\theta)) + \operatorname{div}([\vartheta_\varepsilon * T_k(\varrho^\theta)]\vec{u}) + \vartheta_\varepsilon * \left[\left(\theta T_k'(\varrho^\theta)\varrho^\theta - T_k(\varrho^\theta)\right)\operatorname{div}\vec{u}\right] = r_\varepsilon \tag{39}$$

where, similarly as in Lemma 2.1,

$$r^{\varepsilon} \to 0 \text{ in } L^2_{loc}(I; L^2(\Omega)) \text{ for } \varepsilon \to 0. \tag{40}$$

Using φ_i, $i = 1, \ldots, N$ as test functions for (2) we arrive at the following expression:

$$\int_I \int_\Omega \psi p(\varrho)[\vartheta_\varepsilon * T_k(\varrho^\theta)] \, \mathrm{d}x \, \mathrm{d}t = \sum_{j=1}^{9} I^j_{\varepsilon,k} \tag{41}$$

where

$$I^1_{\varepsilon,k} = \int_I \psi \Big(\int_\Omega p(\varrho) \, \mathrm{d}x \Big) \Big(\oint_\Omega \vartheta_\varepsilon * T_k(\varrho^\theta) \, \mathrm{d}x \Big) \, \mathrm{d}t,$$

$$I^2_{\varepsilon,k} = - \int_I \int_\Omega \psi \; \varrho u^i u^j \partial_{x_j} \mathcal{B}_i[\vartheta_\varepsilon * T_k(\varrho^\theta)] \, \mathrm{d}x \, \mathrm{d}t,$$

$$I^3_{\varepsilon,k} = \mu \int_I \int_\Omega \psi \; \partial_{x_j} u^i \partial_{x_j} \mathcal{B}_i[\vartheta_\varepsilon * T_k(\varrho^\theta)] \, \mathrm{d}x \, \mathrm{d}t,$$

$$I^4_{\varepsilon,k} = (\lambda + \mu) \int_I \int_\Omega \psi \; \mathrm{div} \, \vec{u} \; [\vartheta_\varepsilon * T_k(\varrho^\theta)] \, \mathrm{d}x \, \mathrm{d}t,$$

$$I^5_{\varepsilon,k} = - \int_I \int_\Omega \psi \; \varrho f^i \mathcal{B}_i[\vartheta_\varepsilon * T_k(\varrho^\theta)] \, \mathrm{d}x \, \mathrm{d}t,$$

$$I^6_{\varepsilon,k} = - \int_I \int_\Omega \partial_t \psi \; \varrho u^i \mathcal{B}_i[\vartheta_\varepsilon * T_k(\varrho^\theta)] \, \mathrm{d}x \, \mathrm{d}t,$$

$$I^7_{\varepsilon,k} = \int_I \int_\Omega \psi \; \varrho u^i \mathcal{B}_i \Big[\mathrm{div}([\vartheta_\varepsilon * T_k(\varrho^\theta)] \vec{u}) \Big] \, \mathrm{d}x \, \mathrm{d}t,$$

$$I^8_{\varepsilon,k} = \int_I \int_\Omega \psi \; \varrho u^i \mathcal{B}_i \Big[\vartheta_\varepsilon * \Big((\theta T_k'(\varrho^\theta) \varrho^\theta - T_k(\varrho^\theta)) \; \mathrm{div} \, \vec{u} \Big) \Big] \, \mathrm{d}x \, \mathrm{d}t,$$

$$I^9_{\varepsilon,k} = - \int_I \int_\Omega \psi \; \varrho u^i \mathcal{B}_i[r_\varepsilon] \, \mathrm{dx} \, \mathrm{dt}.$$

Denote $J \subset I$ a compact interval containing supp ψ and $m = \int_\Omega \varrho \, \mathrm{d}x$ the total mass which is independent of t. Now, we have the following estimates:

(1)

$$|I^1_{\varepsilon,k}| \leq c(m) \int_J \int_\Omega \varrho^\gamma \, \mathrm{d}x \, \mathrm{d}t$$

provided $0 < \theta \leq 1$.

(2) Making use of the Hölder inequality, the Sobolev embedding theorem, and (37) we obtain

$$|I^2_{\varepsilon,k}| \leq c \sup_{t \in J} \|\varrho\|_{L^\gamma(\Omega)} \sup_{t \in J} \|\varrho^\theta\|_{L^p(\Omega)} \int_J \int_\Omega |\nabla \vec{u}|^2 \, \mathrm{d}x \, \mathrm{d}t$$

where

$$p > \frac{\gamma}{\gamma-1} \text{ for } N=2,\ p = \frac{3\gamma}{2\gamma-3} \text{ if } N=3.$$

Thus

$$|I^2_{\varepsilon,k}| \le c(m) \sup_{t\in J} \|\varrho\|_{L^\gamma(\Omega)} \int_J \int_\Omega |\nabla \vec{u}|^2 \, dx \, dt$$

provided

$$0 < \theta \le \frac{\gamma-1}{\gamma} \text{ for } N=2,\ 0<\theta \le \frac{2\gamma-3}{3\gamma} \text{ if } N=3.$$

(3) Similarly,

$$|I^3_{\varepsilon,k}| \le c(m) \int_J \|\nabla \vec{u}\|_{L^2(\Omega)} \, dt$$

if $0<\theta\le 1/2$, and

(4)

$$|I^4_{\varepsilon,k}| \le c(m) \int_J \|\nabla \vec{u}\|_{L^2(\Omega)},\ 0<\theta\le \frac{1}{2}.$$

(5) Now, seeing that $W^{1,4}(\Omega) \hookrightarrow C(\overline{\Omega})$ we can take $0<\theta\le \frac{1}{4}$ and make use of (37) to deduce

$$|I^5_{\varepsilon,k}| \le c(m)|J|F.$$

(6) Analogously,

$$|I^6_{\varepsilon,k}| \le c_1(m) \int_I |\partial_t \psi| \ \|\sqrt{\varrho}\|_{L^2(\Omega)} \|\sqrt{\varrho}\vec{u}\|_{L^2(\Omega)} \, dt \le c_2(m) \sup_{t\in J} \|\sqrt{\varrho}\vec{u}\|_{L^2(\Omega)}$$

for $0<\theta\le 1/4$.

(7) By virtue of the Hölder inequality, the Sobolev embedding theorem, and (38) we get

$$|I^7_{\varepsilon,k}| \le \sup_{t\in J} \|\varrho\|_{L^\gamma(\Omega)} \sup_{t\in J} \|\varrho^\theta\|_{L^p(\Omega)} \int_J \int_\Omega |\nabla \vec{u}|^2 \, dx \, dt$$

where, now,

$$p > \frac{\gamma}{\gamma-1} \text{ for } N=2,\ p=\frac{3\gamma}{2\gamma-3} \text{ if } N=3.$$

Consequently, choosing

$$0<\theta<\frac{\gamma-1}{\gamma} \text{ for } N=2,\ 0<\theta\le\frac{2\gamma-3}{3\gamma},\ N=3,$$

one obtains

$$|I^7_{\varepsilon,k}| \le c(m) \sup_{t\in J} \|\varrho\|_{L^\gamma(\Omega)} \int_J \int_\Omega |\nabla \vec{u}|^2 \, dx \, dt.$$

(8) In the same way, we obtain

$$|I^8_{\varepsilon,k}| \leq$$

$$c \sup_{t\in J} \|\varrho\|_{L^\gamma(\Omega)} \int_J \|\nabla \vec{u}\|_{L^2(\Omega)} \Big\| \mathcal{B}_i \Big[\vartheta_\varepsilon * \Big((\theta T_k'(\varrho^\theta)\varrho^\theta - T_k(\varrho^\theta)) \text{ div } \vec{u}\Big)\Big] \Big\|_{L^p(\Omega)} \text{ dt}$$

where, at this stage,

$$p > \frac{\gamma}{\gamma - 1} \text{ if } N = 2, \; p = \frac{6\gamma}{5\gamma - 6} \text{ if } N = 3.$$

By virtue of (37) and the Sobolev embedding theorem, we get

$$\Big\| \mathcal{B}_i \Big[\vartheta_\varepsilon * \Big((\theta T_k'(\varrho^\theta)\varrho^\theta - T_k(\varrho^\theta)) \text{ div } \vec{u}\Big)\Big] \Big\|_{L^p(\Omega)} \leq c\|\varrho^\theta \text{ div } \vec{u}\|_{L^q(\Omega)}$$

where

$$q = \max\Big\{1, \frac{pN}{p+N}\Big\}.$$

Consequently, we have

$$|I^8_{\varepsilon,k}| \leq c(m) \sup_{t\in J} \|\varrho\|_{L^\gamma(\Omega)} \int_J \int_\Omega |\nabla \vec{u}|^2 \text{ d}x \text{ dt}$$

provided

$$0 < \theta < \min\Big\{\frac{1}{2}, \frac{\gamma - 1}{\gamma}\Big\} \text{ if } N = 2, \; 0 < \theta < \min\{\frac{1}{2}, \frac{2\gamma - 3}{3\gamma}\Big\}.$$

(9) Finally, we have

$$|I^9_{\varepsilon,k}| \leq c \sup_{t\in J} \|\varrho\|_{L^\gamma(\Omega)} \Big(\int_J \int_\Omega |\nabla \vec{u}|^2 \text{ d}x \text{ dt}\Big)^{1/2} \Big(\int_J \|\mathcal{B}[r_\varepsilon]\|^2_{L^p(\Omega)} \text{ d}x \text{ dt}\Big)^{1/2}$$

where

$$p > \frac{\gamma}{\gamma - 1} \text{ if } N = 2, \; p = \frac{6\gamma}{5\gamma - 6} \text{ for } N = 3.$$

Note that all constants are independent of ε and k.

Obviously, the above estimates make sense provided $\gamma > 3/2$ for $N = 3$ (and $\gamma > 1$ for $N = 2$). Moreover, by virtue of (37), (40), we have

$$I^9_{\varepsilon,k} \to 0 \text{ for } \varepsilon \to 0 \text{ for any fixed } k.$$

Note that $6\gamma(5\gamma - 6)^{-1} < 6$ provided $\gamma > 3/2$. As the remaining integrals are bounded independently of ε, k, we can let first $\varepsilon \to 0$ and then $k \to \infty$ in (41) to obtain the following estimate:

$$\int_J \int_\Omega p(\varrho)\varrho^\theta \text{ d}x \text{ dt} \leq \tag{42}$$

$$c(m, \gamma, \theta, |J|, F)\Big(1 + \sup_{t\in J} \|\sqrt{\varrho}\vec{u}\|_{L^2(\Omega)} + [1 + \sup_{t\in J} \|\varrho\|_{L^\gamma(\Omega)}] \int_J \int_\Omega |\nabla \vec{u}|^2 \text{ d}x \text{ dt}\Big)$$

for any finite energy weak solution provided

$$\gamma > 1,\ 0 < \theta < \min\{\frac{1}{4}, \frac{\gamma - 1}{\gamma}\} \text{ if } N = 2,$$

$$\gamma > 3/2,\ 0 < \theta < \min\{\frac{1}{4}, \frac{2\gamma - 3}{3\gamma}\} \text{ for } N = 3.$$

Combining the estimate (42) with the energy inequality (35) we obtain the following result.

Proposition 3.1. *Assume $\Omega \subset R^N$, $N = 2, 3$ is a bounded Lipschitz domain. Let ϱ, $\vec{u}$ be a finite energy weak solution to the problem (1), (2), (12) on $I \times \Omega$ where the isentropic pressure p is given by (33) with*

$$\gamma > \frac{N}{2}.$$

Let

$$0 < \theta < \min\left\{\frac{1}{4}, \frac{1}{\gamma}\left(\frac{2\gamma}{N} - 1\right)\right\}$$

be given. Denote by $m = \int_\Omega \varrho$ the (conserved) total mass and let

$$F = \text{ess sup}_{I\times\Omega}|\vec{f}|.$$

Then for any bounded time interval $J \subset I$, we have

$$\int_J \int_\Omega \varrho^{\gamma+\theta} \,\mathrm{d}x \,\mathrm{d}t \le c(m, F, \theta, |J|)\left(1 + \text{ess sup}_{t\in J} \int_\Omega E(t) \,\mathrm{d}x\right)^{\frac{\gamma+1}{\gamma}}.$$

A local version of the above estimates was obtained by LIONS [10]. The validity of (42) up to the boundary was proved in [6] by means of the same technique as presented here. An alternative approach based on the properties of the Stokes operator was proposed by LIONS [11]. In fact, the bounds on θ in Proposition 3.1 are not optimal. Similarly as in the local case (see LIONS [10]), one could verify the best values for θ:

$$0 < \theta \le \frac{2}{N}\gamma - 1.$$

However, for the purposes of the next section, it is convenient to have the integrals containing ϱ^θ bounded in terms of the total mass m equivalent to the L^1–norm rather than the total energy $\int_\Omega E \,\mathrm{d}x$ proportional to the L^γ–norm.

3.4. Bounded absorbing sets

Before we start, let us say a few words about the total energy

$$\mathcal{E}(t) = \int_\Omega \frac{1}{2}\varrho|\vec{u}|^2 + \frac{a}{\gamma - 1}\varrho^\gamma \,\mathrm{d}x.$$

A priori $\mathcal{E}$ is defined only for almost any $t \in I$ and we would like it to be defined for any t. As far as the potential part $\int_\Omega a/(\gamma - 1)\varrho^\gamma \,\mathrm{d}x$, it can be considered as a lower continuous function of t since, by virtue of Proposition 2.2, ϱ is a continuous function of t with respect to the L^α–topology for any $1 \leq \alpha < \gamma$.

On the other hand, the time derivative of the momenta ϱu^i can be expressed by means of (2) to deduce

$$(\varrho u^i) \in C(J; L^{\frac{2\gamma}{\gamma+1}}_{weak}(\Omega)) \text{ for any compact } J \subset I,\ i = 1, ..., N.$$

Now, we show the seemingly obvious assertion, namely,

$$(\varrho u^i)(t) = 0 \text{ a.a. on the set } \{\varrho(t) = 0\}.$$

Indeed we have

$$\int_{\varrho(t)=0} |\varrho\vec{u}|(t)\ \mathrm{d}x \leq \liminf_{t_n \to t} \Big(\int_{\varrho(t)=0} \varrho(t_n)\ \mathrm{d}x \Big)^{\frac{1}{2}} \Big(\int_\Omega \varrho(t_n)|\vec{u}(t_n)|^2\ \mathrm{d}x \Big)^{\frac{1}{2}}$$

for *any* sequence $t_n \to t$. Now, the first term on the right-hand side tends to zero as ϱ is weakly continuous in t and the second one is bounded for a suitable choice of $t_n \to 0$.

Consequently, the total energy $\mathcal{E}$ may be defined for *any* $t \in I$ by the formula:

$$\mathcal{E}(t) = \frac{1}{2} \int_{\varrho(t)>0} \frac{|\varrho\vec{u}|^2}{\varrho}(t)\ \mathrm{d}x + \frac{a}{\gamma - 1} \int_\Omega \varrho^\gamma(t)\ \mathrm{d}x \text{ for all } t \in I.$$

It can be shown that $\mathcal{E}$ is a lower semi-continuous function on I (see [5]).

On the other hand, it follows from the energy inequality (35) that $\mathcal{E}$, redefined on a set of measure zero, is a function of bounded variation and

$$\mathcal{E}(t+) = \operatorname{ess} \lim_{s \to t+} \mathcal{E}(s) \leq \operatorname{ess} \lim_{s \to t-} \mathcal{E}(s) = \mathcal{E}(t-). \tag{43}$$

Moreover, a simple application of the Gronwall lemma yields

$$\mathcal{E}(t_2-) \leq (1 + \mathcal{E}(t_1+))\exp\Big(\sqrt{2m}F(t_2 - t_1)\Big) \text{ for all } t_1 < t_2. \tag{44}$$

Let us pause to prove an auxiliary result which will be crucial for showing the existence of a bounded absorbing set for the problem.

Proposition 3.2. *Assume that*

$$\gamma > 1 \textit{ for } N = 2,\ \gamma > \frac{5}{3} \textit{ if } N = 3. \tag{45}$$

Then there exists a constant B depending solely on γ, F and m enjoying the following property:

If

$$\mathcal{E}((T+1)-) > \mathcal{E}(T+) - 1 \textit{ for a certain } T \in I, \tag{46}$$

then

$$\sup_{t \in (T, T+1)} \mathcal{E}(t+) \leq B.$$

The proof of Proposition 3.2 will be carried over by a series of auxiliary lemmas:

Lemma 3.3. *Under the hypothesis (46), there exists a constant $c = c(F, m)$ such that*

$$\int_T^{T+1} \|\vec{u}(t)\|^2_{W_0^{1,2}(\Omega)} \,\mathrm{d}t \leq c(1 + \int_T^{T+1} \|\varrho(t)\|_{L^q(\Omega)} \,\mathrm{d}t)$$

where

$$q = \frac{3}{2} \textit{ for } N = 3 \textit{ and } q > 1 \textit{ arbitrary for } N = 2.$$

Proof. By virtue of (46) and the Poincaré inequality, we get from (35):

$$\int_T^{T+1} \|\vec{u}\|^2_{W_0^{1,2}(\Omega)} \,\mathrm{d}t \leq c(1 + \int_T^{T+1} \int_\Omega \varrho|\vec{u}| \,\mathrm{d}x \,\mathrm{d}t)$$

where the last term on the right-hand side may be estimated as

$$\int_\Omega \varrho|\vec{u}| \,\mathrm{d}x \leq \sqrt{m}\Big(\int_\Omega \varrho|\vec{u}|^2 \mathrm{d}x\Big)^{\frac{1}{2}} \leq \sqrt{m}\|\varrho\|^{\frac{1}{2}}_{L^q(\Omega)}\|\vec{u}\|_{L^p(\Omega)}$$

where

$$p = 6 \text{ for } N = 3 \text{ and } p = \frac{2q}{q-1} \text{ if } N = 2.$$

Using the Sobolev embedding theorem

$$W^{1,2}(\Omega) \hookrightarrow L^p(\Omega) \text{ for } p \text{ as above}$$

we get the desired conclusion.

□

Lemma 3.4. *Under the hypothesis (46), there exists a constant $c = c(F, m)$ such that*

$$\mathcal{E}(t+) \leq c(1 + \int_T^{T+1} \|\varrho(s)\|^\gamma_{L^\gamma(\Omega)} \,\mathrm{d}s) \textit{ for any } t \in [T, T+1].$$

Proof. Integrating (44) for the choice $t_2 = T + 1$ with respect to t_1 one gets

$$\mathcal{E}((T+1)-) \leq c(1 + \int_T^{T+1} \mathcal{E}(s) \,\mathrm{d}s).$$

On the other hand, (46) yields

$$\mathcal{E}(T+) < \mathcal{E}((T+1)+) + 1 \leq c(1 + \int_T^{T+1} \mathcal{E}(s) \,\mathrm{d}s). \tag{47}$$

Finally, taking $t_1 = T$ in (44) and using (47) we obtain

$$\mathcal{E}(t+) \leq c(1 + \int_T^{T+1} \mathcal{E}(s) \,\mathrm{d}s) \text{ for any } t \in [T, T+1].$$

Now, making use of Lemma 3.3 we get

$$\int_T^{T+1}\int_\Omega \varrho|\vec{u}|^2 \, dx \, dt \le c_1 \sup_{t\in[T,T+1]} \|\varrho(t)\|_{L^q(\Omega)} \int_T^{T+1} \|\vec{u}\|^2_{W_0^{1,2}(\Omega)}$$

$$\le c_2 \sup_{t\in[T,T+1]} \|\varrho(t)\|_{L^q(\Omega)} (1 + \int_T^{T+1} \|\varrho\|_{L^q(\Omega)} \, dt)$$

where q is the exponent from Lemma 3.3.

Interpolating between 1 and γ we get

$$\|\varrho\|_{L^q(\Omega)} \le \|\varrho\|^{1-\eta}_{L^1(\Omega)} \|\varrho\|^\eta_{L^\gamma(\Omega)}, \ \eta = \frac{\gamma(q-1)}{q(\gamma-1)}; \tag{48}$$

whence

$$\int_T^{T+1}\int_\Omega \varrho|\vec{u}|^2 \, dx \, dt \le c(m) \sup_{t\in[T,T+1]} \mathcal{E}(t+)^{\frac{q-1}{q(\gamma-1)}} (1 + \int_T^{T+1} \|\varrho(s)\|^{\frac{\gamma(q-1)}{q(\gamma-1)}}_{L^\gamma(\Omega)} \, ds).$$

Thus, finally,

$$\sup_{t\in[T,T+1]} \mathcal{E}(t+) \le$$

$$c(m)\Big(1 + \int_T^{T+1} \|\varrho(t)\|^\gamma_{L^\gamma(\Omega)} \, dt + \sup_{t\in[T,T+1]} \mathcal{E}(t+)^{\frac{q-1}{q(\gamma-1)}} (1 + \int_T^{T+1} \|\varrho(t)\|^{\frac{\gamma(q-1)}{q(\gamma-1)}}_{L^\gamma(\Omega)} \, dt\Big).$$

Under the hypothesis (45), we have always

$$\frac{q-1}{q(\gamma-1)} < \frac{1}{2}$$

and the desired conclusion follows.

□

At this stage, we are ready to complete the proof of Proposition 3.2. To this end, we are going to use the estimate (42). Interpolating between L^1 and $L^{\gamma+\theta}$ we get

$$\int_T^{T+1} \|\varrho(t)\|^\gamma_{L^\gamma(\Omega)} \, dt \le c(m)\Big(\int_T^{T+1}\int_\Omega \varrho^{\gamma+\theta} \, dx \, dt\Big)^{\frac{\gamma-1}{\gamma+\theta-1}}. \tag{49}$$

Moreover, assuming (46) we can use Lemma 3.3 and (48) to estimate

$$\Big| \sup_{t\in[T,T+1]} \|\varrho(t)\|_{L^\gamma(\Omega)} \int_T^{T+1} \|\vec{u}\|^2_{W_0^{1,2}(\Omega)} \, dt \Big|^{\frac{\gamma-1}{\gamma+\theta-1}}$$

$$\le c_1(F,m)\Big(\sup_{t\in[T,T+1]} \|\varrho(t)\|_{L^\gamma(\Omega)} (1 + \sup_{t\in[T,T+1]} \|\varrho(t)\|_{L^q(\Omega)})\Big)^{\frac{\gamma-1}{\gamma+\theta-1}} \tag{50}$$

$$\le c_2(F,m)\Big(1 + \sup_{t\in[T,T+1]} \|\varrho(t)\|_{L^\gamma(\Omega)}\Big)^{\frac{2\gamma q - q - \gamma}{q(\gamma+\theta-1)}}$$

where q is as in Lemma 3.3.

Combining Lemma 3.4 with the estimates (42), (49), and (50) we conclude

$$\sup_{t\in[T,T+1]} \mathcal{E}(t) \leq c(F,m)\Big(1 + \sup_{t\in[T,T+1]} \sqrt{\mathcal{E}(t+)} + \sup_{t\in[T,T+1]} \|\varrho(t)\|_{L^\gamma(\Omega)}^{\frac{2\gamma q - q - \gamma}{q(\gamma+\theta-1)}}\Big). \tag{51}$$

Now, if $N = 3$, γ as in (45), and $q = \frac{3}{2}$, it is a routine matter to find $\theta > 0$ compatible with (42) so that

$$\frac{2\gamma q - q - \gamma}{q(\gamma + \theta - 1)} < \gamma. \tag{52}$$

Similarly, (52) holds if $N = 2$, $\gamma > 1$, $\theta > 0$ are fixed and q taken sufficiently close to one. Consequently, in both cases, (51), (52) complete the proof of Proposition 3.2.

Now, we are ready to prove a result on the existence of *bounded absorbing sets* for isentropic flows.

Theorem 3.5. *Let $\Omega \subset R^N$, $N = 2,3$ be a bounded Lipschitz domain. Assume*

$$\gamma > 1 \textit{ if } N = 2, \ \gamma > \frac{5}{3} \textit{ for } N = 3$$

and let $\vec{f}$ be a bounded measurable function on $I \times \Omega$, $I = (a,b)$ such that

$$\operatorname{ess} \sup_{t\in I,\ x\in\Omega} |\vec{f}(t,x)| \leq F.$$

Then there exists a constant E_∞ depending solely on γ, F and the (given) total mass m having the following property:

Given E_0, there exists a time $T = T(E_0)$ such that

$$\mathcal{E}(t) = \mathcal{E}[\varrho,\vec{u}](t) \leq E_\infty \textit{ for all } t > a + T,\ t \in I$$

provided

$$\limsup_{t\to a+} \mathcal{E}[\varrho,\vec{u}](t) \leq E_0, \ \int_\Omega \varrho \,\mathrm{d}x = m,$$

and ϱ, $\vec{u}$ is a finite energy weak solution of the problem (1), (2), (33), and (12) on $I \times \Omega$.

Proof. To begin, observe there exists $T = T(E_0)$ such that

$$E(t_0+) \leq B \text{ for a certain } t_0 < a + T$$

where B is the quantity from Proposition 3.2. Indeed if it was not the case then, by virtue of Proposition 3.2, the total energy would become negative.

Next, we claim that

$$\mathcal{E}((t_0 + n)+) \leq B \tag{53}$$

for any integer $n \geq 0$. By induction, assuming

$$\mathcal{E}((t_0 + n)+) \leq B$$

we can use Proposition 3.2 to deduce that either

$$\sup_{t\in(t_0+n,t_0+n+1)} \mathcal{E}(t+) \leq L$$

and, consequently,

$$\mathcal{E}((t_0+n+1)-) \leq B,$$

or

$$\mathcal{E}((t_0+n+1)+) \leq \mathcal{E}((t_0+n+1)-) \leq \mathcal{E}((t_0+n) =) - 1 \leq B-1.$$

Finally, by virtue of (44) and (53), we can take

$$E_\infty = (1+B)\,\exp(\sqrt{2m}\;F) - 1.$$

Theorem 3.5 has been proved.

□

4. The effective viscous pressure and its properties

4.1. An application of compensated compactness

In this section, we shall assume that $\Omega \subset R^N$ is a bounded domain, $I \subset R$ a bounded interval, and ϱ_n, $\vec{u}_n$, p_n a sequence of distributional solutions of the system (1), (2) such that

$$\varrho_n \to \varrho \text{ weakly star in } L^\infty(I;L^\gamma),\ \vec{u}_n \to \vec{u} \text{ weakly in } L^2(I;W^{1,2}(\Omega)). \tag{54}$$

Moreover, we shall suppose the kinetic energy

$$\varrho_n|\vec{u}_n|^2 \text{ bounded in } L^\infty(I;L^1(\Omega)) \text{ uniformly in } n \tag{55}$$

and

$$\vec{f}_n \to \vec{f} \text{ weakly star in } L^\infty(I\times\Omega).$$

Finally, we assume the continuity equation to hold in the sense of renormalized solutions in $\mathcal{D}'(I\times\Omega)$ and the pressure p_n will satisfy

$$p_n \to p \text{ weakly in } L^r(I\times\Omega) \text{ for a certain } r>1. \tag{56}$$

The problem is not necessarily isentropic though (55) is strongly reminiscent of the estimates of Section 3.

Since $\partial_t\varrho$ can be expressed by means of (1), the convergence of the density in (54) can be strengthened to

$$\varrho_n \to \varrho \text{ in } C(\overline{I};L^\gamma_{weak}(\Omega))$$

which combined with (55) yields

$$\varrho_n\vec{u}_n \to \varrho\vec{u} \text{ weakly star in } L^\infty(I;L^{\frac{2\gamma}{\gamma+1}}(\Omega))$$

provided $\gamma>1$, $\gamma>\frac{6}{5}$ for $N=2$, $N=3$ respectively.

Since $\partial_t(\varrho_n\vec{u}_n)$ can be expressed from (2), we get

$$\varrho_n\vec{u}_n \to \varrho\vec{u} \text{ in } C(\overline{I}; L^{\frac{2\gamma}{\gamma+1}}_{weak}(\Omega)). \tag{57}$$

Now,, if $\gamma > N/2$, the relations (54), (57) together with Lions-Aubin Lemma yield

$$\varrho_n u_n^i u_n^j \to \varrho u^i u^j \text{ weakly in } L^2(I; L^p(\Omega)) \tag{58}$$

where

$$p < \frac{2\gamma}{\gamma+1} \text{ for } N = 2,\ p = \frac{6\gamma}{4\gamma+3} \text{ if } N = 3.$$

Finally, since ϱ_n is a renormalized solution of (1), we have

$$\mathcal{T}_k(\varrho_n) \to \overline{\mathcal{T}_k(\varrho)} \text{ in } C(\overline{I}; L^\alpha_{weak}(\Omega)) \text{ for any } \alpha > 1. \tag{59}$$

Consequently, the limit functions ϱ, $\vec{u}$, and p satisfy

$$\partial_t\overline{\mathcal{T}_k(\varrho)} + \operatorname{div}(\overline{\mathcal{T}_k(\varrho)}\vec{u}) + \overline{\Big(\mathcal{T}_k'(\varrho)\varrho - \mathcal{T}_k(\varrho)\Big)\operatorname{div}\vec{u}} = 0, \tag{60}$$

and

$$\partial_t(\varrho\vec{u}) + \operatorname{div}(\varrho\vec{u}\otimes\vec{u}) + \nabla p = \mu\nabla\vec{u} + (\lambda+\mu)\nabla \operatorname{div}\vec{u} + \overline{\varrho\vec{f}} \tag{61}$$

in $\mathcal{D}'(I\times\Omega)$ where

$$\varrho_n\vec{f}_n \to \overline{\varrho\vec{f}} \text{ weakly star in } L^\infty(I; L^\gamma(\Omega)),$$

$$(\mathcal{T}_k'(\varrho_n)\varrho_n - \mathcal{T}_k(\varrho_n))\operatorname{div}\vec{u}_n \to \overline{\Big(\mathcal{T}_k'(\varrho)\varrho - \mathcal{T}_k(\varrho)\Big)\operatorname{div}\vec{u}} \text{ weakly in } L^2(I\times\Omega).$$

Now, let us introduce the operators $\mathcal{A}_j$, $j = 1, ..., N$,

$$\mathcal{A}_j = \Delta^{-1}\partial_{x_j} \text{ or } \mathcal{A}_j[v] = \mathcal{F}^{-1}\Big[\frac{\mathrm{i}\xi_j}{|\xi|^2}\mathcal{F}[v](\xi)\Big]$$

where $\mathcal{F}$ is the Fourier transform.

Now, we take the quantities

$$\varphi_i(t,x) = \psi(t)\phi(x)\mathcal{A}_i[\phi(x)\mathcal{T}_k(\varrho)],\ i = 1, ..., N,\ \psi\in\mathcal{D}(I),\ \phi\in\mathcal{D}(\Omega)$$

as test functions for (2). Indeed $\mathcal{A}_i$ enjoy the regularizing properties:

$$\mathcal{A}_i : L^p\cap L^1(R^N) \mapsto W^{1,p}(\Omega) \text{ is are bounded linear operators} \tag{62}$$

and

$$\partial_t\Big(\mathcal{A}_i[\phi\mathcal{T}_k(\varrho_n)]\Big) = -\mathcal{A}_i[\operatorname{div}(\phi\mathcal{T}_k(\varrho_n)\vec{u}_n)] + \mathcal{A}_i\Big[\phi\Big(\mathcal{T}_k(\varrho_n) - \mathcal{T}_k'(\varrho_n)\varrho_n\Big)\operatorname{div}\vec{u}_n\Big] +$$

$$\mathcal{A}_i\Big[T_k(\varrho_n)\vec{u}_n^j\partial_{x_j}\phi\Big] \in L^2(I; L^q(\Omega))$$

with

$$q > 1 \text{ arbitrary for } N = 2 \text{ and } q = 6 \text{ if } N = 3.$$

The last assertion can be verified by the regularization procedure as in Lemma 2.1.

Consequently, using φ_i as test functions for (2) we obtain

$$\int_I \int_\Omega \psi\phi^2\Big(p_n - (\lambda + 2\mu) \operatorname{div} \vec{u}_n\Big)\mathcal{T}_k(\varrho_n) \, dx \, dt = \sum_{j=1}^{12} I_n^j \tag{63}$$

where

$$I_n^1 = -\int_I \int_\Omega \psi\partial_{x_i}\phi \; p_n \mathcal{A}_i[\phi\mathcal{T}_k(\varrho_n)] \, dx \, dt,$$

$$I_n^2 = (\lambda + \mu) \int_I \int_\Omega \psi\partial_{x_i}\phi \; (\operatorname{div} \vec{u}_n)\mathcal{A}_i[\phi\mathcal{T}_k(\varrho_n)] \, dx \, dt,$$

$$I_n^3 = \mu \int_I \int_\Omega \psi\partial_{x_j}\phi \; \partial_{x_j} u_n^i \; \mathcal{A}_i[\phi\mathcal{T}_k(\varrho_n)] \, dx \, dt,$$

$$I_n^4 = -\int_I \int_\Omega \psi\phi \; \varrho_n f_n^i \; \mathcal{A}_i[\phi\mathcal{T}_k(\varrho_n)] \, dx \, dt,$$

$$I_n^5 = -\int_I \int_\Omega \psi_t\phi \; \varrho_n u_n^i \; \mathcal{A}_i[\phi\mathcal{T}_k(\varrho_n)] \, dx \, dt,$$

$$I_n^6 = -\int_I \int_\Omega \psi\partial_{x_j}\phi \; \varrho_n u_n^i u_n^j \; \mathcal{A}_i[\phi\mathcal{T}_k(\varrho_n)] \, dx \, dt,$$

$$I_n^7 = \mu \int_I \int_\Omega \psi\phi\partial_{x_i}\phi \; u_n^i \mathcal{T}_k(\varrho_n) \, dx \, dt,$$

$$I_n^8 = -\int_I \int_\Omega \psi\phi \; \varrho_n u_n^i \; \mathcal{A}_i\Big[\mathcal{T}_k(\varrho_n) u_n^j \partial_{x_j}\phi\Big] \, dx \, dt,$$

$$I_n^9 = \int_I \int_\Omega \psi\phi \; \varrho_n u_n^i \; \mathcal{A}_i\Big[\phi\Big(\mathcal{T}_k'(\varrho_n)\varrho_n - \mathcal{T}_k(\varrho_n)\Big) \operatorname{div} \vec{u}_n\Big] \, dx \, dt,$$

$$I_n^{10} = -\mu \int_I \int_\Omega \psi\partial_{x_j}\phi \; u_n^i \partial_{x_j}\mathcal{A}_i[\phi\mathcal{T}_k(\varrho_n)] \, dx \, dt,$$

$$I_n^{11} = \int_I \int_\Omega \psi \; u_n^j \Big\{\Big(\phi\mathcal{T}_k(\varrho_n)\delta_{m,j} - \mathcal{R}_{m,l}[\phi\mathcal{T}_k(\varrho_n)\delta_{j,l}]\Big)\mathcal{R}_{m,i}[\phi\varrho_n u_n^i]\Big\} \, dx \, dt,$$

$$I_n^{12} = -\int_I \int_\Omega \psi \; u_n^j \Big\{\Big(\phi\varrho_n u_n^m - \mathcal{R}_{m,i}[\phi\varrho_n u_n^i]\Big)\mathcal{R}_{m,l}[\phi\mathcal{T}_k(\varrho_n)\delta_{j,l}]\Big\} \, dx \, dt$$

where

$$\mathcal{R}_{i,j} = \partial_{x_i}\Delta^{-1}\partial_{x_j} \text{ or } \mathcal{R}_{i,j}[v] = \mathcal{F}^{-1}\Big[\frac{\xi_i\xi_j}{|\xi|^2}\mathcal{F}[v](\xi)\Big]$$

and $\delta_{i,j}$ is the Kronecker symbol.

Now, similarly, we can use

$$\varphi_i = \psi\phi\mathcal{A}_i[\phi\overline{T_k(\varrho)}]$$

as test functions for (61) to deduce

$$\int_I\int_\Omega \psi\phi^2\Big(p - (\lambda+2\mu)\ \mathrm{div}\ \vec{u}\Big)\overline{T_k(\varrho)}\ \mathrm{d}x\ \mathrm{d}t = \sum_{j=1}^{12} I^j \tag{64}$$

where

$$I^1 = -\int_I\int_\Omega \psi\partial_{x_i}\phi\ p\mathcal{A}_i[\phi\overline{T_k(\varrho)}]\ \mathrm{d}x\ \mathrm{d}t,$$

$$I^2 = (\lambda+\mu)\int_I\int_\Omega \psi\partial_{x_i}\phi\ (\mathrm{div}\ \vec{u})\mathcal{A}_i[\phi\overline{T_k(\varrho)}]\ \mathrm{d}x\ \mathrm{d}t,$$

$$I^3 = \mu\int_I\int_\Omega \psi\partial_{x_j}\phi\ \partial_{x_j}u^i\ \mathcal{A}_i[\phi\overline{T_k(\varrho)}]\ \mathrm{d}x\ \mathrm{d}t,$$

$$I^4 = -\int_I\int_\Omega \psi\phi\ \overline{\varrho f^i}\ \mathcal{A}_i[\phi\overline{T_k(\varrho)}]\ \mathrm{d}x\ \mathrm{d}t,$$

$$I^5 = -\int_I\int_\Omega \psi_t\phi\ \varrho u^i\ \mathcal{A}_i[\phi\overline{T_k(\varrho)}]\ \mathrm{d}x\ \mathrm{d}t,$$

$$I^6 = -\int_I\int_\Omega \psi\partial_{x_j}\phi\ \varrho u^i u^j\ \mathcal{A}_i[\phi\overline{T_k(\varrho)}]\ \mathrm{d}x\ \mathrm{d}t,$$

$$I^7 = \mu\int_I\int_\Omega \psi\phi\partial_{x_i}\phi\ u^i\overline{T_k(\varrho)}\ \mathrm{d}x\ \mathrm{d}t,$$

$$I^8 = -\int_I\int_\Omega \psi\phi\ \varrho u^i\ \mathcal{A}_i\Big[\overline{T_k(\varrho)}u^j\partial_{x_j}\phi\Big]\ \mathrm{d}x\ \mathrm{d}t,$$

$$I^9 = \int_I\int_\Omega \psi\phi\ \varrho u^i\ \mathcal{A}_i\Big[\phi\overline{\Big(T_k'(\varrho)\varrho - T_k(\varrho)\Big)\ \mathrm{div}\ \vec{u}}\Big]\ \mathrm{d}x\ \mathrm{d}t,$$

$$I^{10} = -\mu\int_I\int_\Omega \psi\partial_{x_j}\phi\ u^i\partial_{x_j}\mathcal{A}_i[\phi\overline{T_k(\varrho)}]\ \mathrm{d}x\ \mathrm{d}t,$$

$$I^{11} = \int_I\int_\Omega \psi\ u^j\Big\{\Big(\phi\overline{T_k(\varrho)}\delta_{m,j} - \mathcal{R}_{m,l}[\phi\overline{T_k(\varrho)}\delta_{j,l}]\Big)\mathcal{R}_{m,i}[\phi\varrho u^i]\Big\}\ \mathrm{d}x\ \mathrm{d}t,$$

$$I^{12} = -\int_I\int_\Omega \psi\ u^j\Big\{\Big(\phi\varrho u^m - \mathcal{R}_{m,i}[\phi\varrho u^i]\Big)\mathcal{R}_{m,l}[\phi\overline{T_k(\varrho)}\delta_{j,l}]\Big\}\ \mathrm{d}x\ \mathrm{d}t$$

Our aim is to show that all the integrals on the right-hand side in (63) converge to their counterparts in (64).

(1–6) By virtue of (59) and the smoothing properties of $\mathcal{A}_i$ (62), it is easy to see that

$$\mathcal{A}_i[\phi\mathcal{T}_k(\varrho_n)] \to \mathcal{A}_i[\phi\overline{\mathcal{T}_k(\varrho)}] \text{ in } C(\overline{I \times \Omega});$$

whence $I_n^j \to I^j$ as $n \to \infty$ for $j = 1, ..., 6$.

(7) The relations (54), (59) yield $I_n^7 \to I^7$.

(8–9) It follows from (54), (59) that

$$\mathcal{T}_k(\varrho_n)u_n^j \to \overline{\mathcal{T}_k(\varrho)u^j} \text{ weakly in, say, } L^2(I \times \Omega)$$

and, consequently,

$$\partial_{x_j}\ \mathcal{A}_i[\phi\mathcal{T}_k(\varrho_n)u_n^j] \to \partial_{x_j}\ \mathcal{A}_i[\phi\overline{\mathcal{T}_k(\varrho)u^j}] \text{ weakly in } L^2(I; W_0^{1,2}(\Omega))$$

which, combined with (57) gives rise to $I_n^8 \to I^8$ provided $\gamma > \frac{N}{2}$. Analogously, one can show $I_n^9 \to I^9$.

(10) In accordance with (59), (62) we have

$$\partial_{x_j}\mathcal{A}_i[\phi\mathcal{T}_k(\varrho_n)] \to \partial_{x_j}\mathcal{A}_i[\phi\overline{\mathcal{T}_k(\varrho)}] \text{ in, say, } L^2(I; W^{-1,2}(\Omega))$$

which together with (54) implies $I_n^{10} \to I^{10}$.

(11–12) Making use of (59) we deduce

$$X_m^n(t) = \Big(\phi\mathcal{T}_k(\varrho_n)\delta_{m,j} - \mathcal{R}_{m,l}[\phi\mathcal{T}_k(\varrho_n)\delta_{j,l}]\Big) \to X_m(t) =$$

$$\Big(\phi\overline{\mathcal{T}_k(\varrho)}\delta_{m,j} - \mathcal{R}_{m,l}[\phi\overline{\mathcal{T}_k(\varrho)}\delta_{j,l}]\Big)$$

weakly in $L^\alpha(\Omega)$ for any fixed $t \in I$ and any fixed $\alpha \geq 1$. Similarly, (57) yields

$$Y_m^n(t) = \mathcal{R}_{m,i}[\phi\varrho_n u_n^i] \to Y_m(t) = \mathcal{R}_{m,i}[\phi\varrho u^i] \text{ weakly in } L^{\frac{2\gamma}{\gamma+1}}(\Omega)$$

for any $t \in I$.

Moreover, one has

$$\text{div } \vec{X}^n = \partial_{x_m}X_m^n = 0 \text{ and } \vec{Y}^n = \nabla\ \mathcal{A}_i[\phi\varrho_n u_n^i], \text{ i.e., curl } \vec{Y}^n = 0.$$

Thus we are allowed to apply the L^p- version of Div-Curl Lemma of the theory of compensated compactness (see e.g. YI [13]) to conclude $\vec{X}^n(t).\vec{Y}^n(t) \to \vec{X}(t)\vec{Y}(t)$ weakly in $L^p(\Omega)$ for any $p < 2\gamma/(\gamma+1)$. If $\gamma > \frac{n}{2}$ we have compact embedding $L^p \hookrightarrow [W^{1,2}(\Omega)]^*$ and, consequently,

$$\Big(\phi\mathcal{T}_k(\varrho_n)\delta_{m,j} - \mathcal{R}_{m,l}[\phi\mathcal{T}_k(\varrho_n)\delta_{j,l}]\Big)\mathcal{R}_{m,i}[\phi\varrho_n u_n^i] \to$$

$$\Big(\phi\overline{\mathcal{T}_k(\varrho)}\delta_{m,j} - \mathcal{R}_{m,l}[\phi\overline{\mathcal{T}_k(\varrho)}\delta_{j,l}]\Big)\mathcal{R}_{m,i}[\phi\varrho u^i] \text{ in, say, } L^2(I; [W^{1,2}(\Omega)]^*)$$

for any fixed j, which together with (54) implies $I_n^{11} \to I^{11}$. Similarly, one can prove $I_n^{12} \to I^{12}$.

Consequently, we have proved a remarkable property of the quantity $p - (\lambda + 2\mu)$ div $\vec{u}$ called usually the *effective viscous pressure*.

Proposition 4.1. *Let $\Omega \subset R^N$, $N = 2,3$ be a domain and $I \subset R$ an open time interval. Let ϱ_n, $\vec{u}_n$, p_n, and $\vec{f}_n$ satisfy (1), (2) in $\mathcal{D}'(I \times \Omega)$ and, moreover,*

$$\varrho_n \to \varrho \text{ weakly star in } L^\infty_{loc}(I; L^\gamma_{loc}(\Omega)),\ \gamma > \frac{N}{2},$$

$$\vec{u}_n \to \vec{u} \text{ weakly in } L^2_{loc}(I; W^{1,2}_{loc}(\Omega)),$$

$$p_n \to p \text{ weakly in } L^r_{loc}(I \times \Omega),$$

and

$$\vec{f}_n \to \vec{f} \text{ weakly star in } L^\infty_{loc}(I \times \Omega).$$

Finally, assume ϱ_n satisfy (1) in the sense of renormalized solutions and the kinetic energy $E_{kin} = 1/2\varrho_n|\vec{u}_n|^2$ is bounded in $L^\infty_{loc}(I; L^1_{loc}(\Omega))$ independently of n.

Then for any bounded open subset $Q \subset \overline{Q} \subset I \times \Omega$ we have

$$\lim_{n\to\infty} \int_Q \Big(p_n - (\lambda + 2\mu) \text{ div } \vec{u}_n\Big) T_k(\varrho_n) \, dx \, dt =$$

$$\int_Q \Big(p - (\lambda + 2\mu) \text{ div } \vec{u}\Big) \overline{T_k(\varrho)} \, dx \, dt$$

for any fixed $k \geq 1$ where $\overline{T_k(\varrho)}$ denotes the weak limit of $T_k(\varrho_n)$.

Proposition 4.1 was first proved by LIONS [10]. In fact, it is the heart of the existence theory as presented in [10]. In his proof, the last two integrals I_n^{11}, I_n^{12} are written as one expression

$$\int_I \int_\Omega \psi\phi T_k(\varrho_n)\Big(u_n^j \mathcal{R}_{i,j}[\phi\varrho_n u_n^i] - \mathcal{R}_{i,j}[\phi\varrho_n u_n^i u_n^j]\Big) \, dx \, dt$$

and convergence is obtained by means of the result of COIFMAN and MEYER [3] on certain regularity properties of commutators. A slight modification is needed to accommodate the case $\gamma \leq 3$. Interesting regularity properties of the effective viscous pressure were already observed by HOFF [8]. Assuming *a priori* validity of Proposition 4.1, SERRE [12] described possible oscillations of the density in terms of the corresponding Young measure.

4.2. On density oscillations for barotropic flows

We conclude this section by showing how Proposition 4.1 can be used to describe the amplitude of density oscillations for barotropic flows discussed in Section 2.3.

Accordingly, we shall assume $p = p(\varrho)$ where

$$p(0) = 0,\ P(\varrho) \geq 0 \text{ for } \varrho \geq 0,$$

$$p \text{ convex on } [0, \infty).$$

It is easy to verify that

$$p(y) - p(z) \geq p(y - z) \text{ for all } 0 \leq z \leq y$$

which yields immediately

$$\begin{aligned}&\Big(p(y)-p(z)\Big)\Big(\mathcal{T}_k(y)-\mathcal{T}_k(z)\Big)\\&\geq p\Big(|\mathcal{T}_k(y)-\mathcal{T}_k(z)|\Big)|\mathcal{T}_k(y)-\mathcal{T}_k(z)| \text{ for all } y,z\geq 0.\end{aligned} \tag{65}$$

Under the hypotheses of Proposition 4.1, we have $p=\overline{p(\varrho)}$ and we can write

$$\begin{aligned}&\lim_{n\to\infty}\int_Q p(\varrho_n)\mathcal{T}_k(\varrho_n)-\overline{p(\varrho)}\;\overline{\mathcal{T}_k(\varrho)}\;\mathrm{d}x\;\mathrm{d}t\\&=\lim_{n\to\infty}\int_Q\Big(p(\varrho_n)-p(\varrho)\Big)(\mathcal{T}_k(\varrho_n)-\mathcal{T}_k(\varrho))\;\mathrm{d}x\;\mathrm{d}t\\&\quad+\int_Q(\overline{p(\varrho)}-p(\varrho))(\mathcal{T}_k(\varrho)-\overline{\mathcal{T}_k(\varrho)})\;\mathrm{d}x\;\mathrm{d}t.\end{aligned} \tag{66}$$

As p is convex and $\mathcal{T}_k$ concave on $[0,\infty)$, the second integral on the right-hand side of (66) is nonnegative and, making use of (65) we infer

$$\begin{aligned}&\lim_{n\to\infty}\int_Q p(\varrho_n)\mathcal{T}_k(\varrho_n)-\overline{p(\varrho)}\;\overline{\mathcal{T}_k(\varrho)}\;\mathrm{d}x\;\mathrm{d}t\\&\geq\limsup_{n\to\infty}\int_Q p\Big(|\mathcal{T}_k(\varrho_n)-\mathcal{T}_k(\varrho)|\Big)|\mathcal{T}_k(\varrho_n)-\mathcal{T}_k(\varrho)|\;\mathrm{d}x\;\mathrm{d}t.\end{aligned} \tag{67}$$

On the other hand, we have

$$\begin{aligned}&\lim_{n\to\infty}\int_Q \operatorname{div}\vec{u}_n\mathcal{T}_k(\varrho_n)-\operatorname{div}\vec{u}\;\overline{\mathcal{T}_k(\varrho)}\;\mathrm{d}x\;\mathrm{d}t\\&=\lim_{n\to\infty}\int_Q\operatorname{div}\vec{u}_n(\mathcal{T}_k(\varrho_n)-\overline{\mathcal{T}_k(\varrho)})\;\mathrm{d}x\;\mathrm{d}t\\&\leq\sup_{n\geq 1}\|\operatorname{div}\vec{u}_n\|_{L^2(Q)}\limsup_{n\to\infty}\|\mathcal{T}_k(\varrho_n)-\overline{\mathcal{T}_k(\varrho)}\|_{L^2(Q)}.\end{aligned} \tag{68}$$

Thus if the pressure is superlinear at infinity, the relations (67), (68) together with Proposition 4.1 enable to estimate the amplitude of oscillations $\mathbf{osc}_p[\varrho_n-\varrho](Q)$ introduced in Section 2.3. In particular, the following result holds.

Corollary 4.2. *In addition to the hypotheses of Proposition 4.1, assume the flow is barotropic, i.e., $p=p(\varrho)$ is a function of the density only. Moreover, let $p(0)=0$, p convex and*

$$p(\varrho)\geq a\varrho^\gamma,\; a>0,\; \gamma>1.$$

Then

$$\mathbf{osc}_{\gamma+1}[\varrho_n-\varrho](Q)\leq c(Q)\Big(\sup_{n\geq 1}\|\operatorname{div}\vec{u}_n\|_{L^2(Q)}\Big)^{\frac{1}{\gamma}}$$

for any compact $Q\subset I\times\Omega$.

Finally, combining the conclusion of Corollary 4.2 with Proposition 2.4 we get:

Corollary 4.3. *Under the hypotheses of Corollary 4.2, the limit function ϱ is a renormalized solution of the equation (1) on $I \times \Omega$.*

5. Propagation of oscillations for isentropic flows

Let $\Omega \subset R^N$, $N = 2, 3$ be a bounded Lipschitz domain. In this section, we concentrate on isentropic flows where the pressure p is given by the constitutive relation

$$p(\varrho) = a\varrho^\gamma,\ a > 0,\ \gamma > \frac{N}{2}. \tag{69}$$

Moreover, we shall assume the velocity $\vec{u}$ to satisfy the no-slip boundary conditions (12). We restrict ourselves to the finite energy weak solutions introduced in Section 3.

Consider a sequence ϱ_n, $\vec{u}_n$ of finite energy weak solutions of the problem (1), (2) (with $\vec{f} = \vec{f}_n$), (69), (12) on $I \times \Omega$ such that

- $I = (a, b)$, $a \in R \cup \{-\infty\}$,

$$\text{esslim}\,\text{sup}_{t \to a+} \int_\Omega \frac{1}{2}\varrho_n|\vec{u}_n|^2 + \frac{a}{\gamma - 1}\varrho_n^\gamma \ \mathrm{d}x \leq E_0 \text{ independently of } n = 1, 2, ...;$$

- $\|\vec{f}_n\|_{L^\infty(I\times\Omega)} \leq F$ independently of $n = 1, 2, ...$

The issue we shall address here is the time propagation of possible oscillations in the density. To begin with, it seems worth-observing that any reasonable "solution operator" one might associate to finite energy weak solutions cannot be compact with respect to its density component. This is due to the hyperbolic character of the continuity equation (1). In accordance with the observation made by LIONS [9], oscillations of the density will propagate in time. This phenomenon was studied by SERRE [12] who showed that the amplitude of the Young measure associated to the oscillations is a non-increasing function of time. His proof is complete in the dimension $N = 1$ and formal for $N \geq 2$ anticipating validity of Proposition 4.1. Having shown Proposition 4.1, LIONS [10] completed the proof for $N \geq 2$. The fact that the oscillations cannot be created in the density plays a crucial role in the existence theory as presented in [10].

Here, we go a step further by showing that the amplitude of oscillations decays in time at uniform rate. Such a property brings about a lot of interesting corollaries. In particular, the time images of bounded sets of initial data are *asymptotically compact* with respect to the density component in the L^p–topology. This is precisely what is needed to develop a meaningful dynamical systems theory associated to the problem.

5.1. Defect measures

In accordance with the previous hypotheses and the pressure estimates obtained in Section 3, it can be shown that

$$\left.\begin{array}{c} \varrho_n \to \varrho \text{ in } C(J; L^\gamma_{weak}(\Omega)), \\ \\ \mathcal{T}_k(\varrho_n) \to \overline{\mathcal{T}_k(\varrho)} \text{ in } C(J; L^\alpha_{weak}(\Omega)) \text{ for any } \alpha \geq 1,\ k \geq 1, \\ \\ \vec{u}_n \to \vec{u} \text{ weakly in } L^2(J; W^{1,2}_0(\Omega)) \end{array}\right\} \tag{70}$$

for any compact time interval $J \subset \overline{I}$.

To measure possible oscillations of the sequence ϱ_n, we introduce a *defect measure* **dft**,

$$\mathbf{dft}[\varrho_n - \varrho](t) = \int_\Omega \nu(t,x)\ \mathrm{d}x \text{ where } \nu = \overline{\varrho\log(\varrho)} - \varrho\log(\varrho). \tag{71}$$

By virtue of Corollary 4.3, both ϱ_n and ϱ are renormalized solutions of (1) and we have

$$\left.\begin{array}{c} \varrho_n\log(\varrho_n) \to \overline{\varrho\log(\varrho)} \text{ in } C(J; L^\alpha_{weak}(\Omega)),\ 1 \leq \alpha < \gamma, \\ \\ \varrho\log(\varrho) \in C(J; L^\alpha_{weak}(\Omega)),\ 1 \leq \alpha < \gamma \end{array}\right\} \tag{72}$$

and, consequently, $\mathbf{dft}[\varrho_n - \varrho]$ is a continuous function on the time interval I.

5.2. Decay estimates

Let us introduce a family of functions:

$$L_k(z) = \begin{cases} z\log(z) \text{ for } 0 \leq z \leq k, \\ \\ z\log(k) + z\int_z^k \mathcal{T}_k(s)/s^2\ \mathrm{d}s \text{ for } z \geq k. \end{cases}$$

It is easy to observe that $L_k(z) = \beta_k z + b_k(z)$ where b_k satisfy (19) and

$$L'_k(z)z - L_k(z) = \mathcal{T}_k(z).$$

Since both ϱ_n, ϱ solve (1) and (18) in $\mathcal{D}'(I \times R^N)$ (cf. Lemma 2.5) we deduce

$$\int_\Omega \Big(L_k(\varrho_n) - L_k(\varrho)\Big)(t_2)\ \mathrm{d}x - \int_\Omega \Big(L_k(\varrho_n) - L_k(\varrho)\Big)(t_1)\ \mathrm{d}x =$$

$$\int_{t_1}^{t_2}\int_\Omega \mathcal{T}_k(\varrho)\ \mathrm{div}\ \vec{u} - \overline{\mathcal{T}_k(\varrho)}\ \mathrm{div}\ \vec{u}_n\ \mathrm{d}x\ \mathrm{d}t +$$

$$\int_{t_1}^{t_2}\int_\Omega \Big(\overline{\mathcal{T}_k(\varrho)} - \mathcal{T}_k(\varrho_n)\Big)\ \mathrm{div}\ \vec{u}_n\ \mathrm{d}x\ \mathrm{d}t.$$

Thus letting $n \to \infty$ and using Proposition 4.1 together with (67), we obtain

$$\int_\Omega \left(\overline{L_k(\varrho)} - L_k(\varrho)\right)(t_2)\ \mathrm{d}x - \int_\Omega \left(\overline{L_k(\varrho)} - L_k(\varrho)\right)(t_1)\ \mathrm{d}x + \tag{73}$$

$$\frac{a}{\lambda + 2\mu} \limsup_{n\to\infty} \int_{t_1}^{t_2} \int_\Omega |\mathcal{T}_k(\varrho_n) - \mathcal{T}_k(\varrho)|^{\gamma+1}\ \mathrm{d}x\ \mathrm{d}t \leq$$

$$\int_{t_1}^{t_2} \int_\Omega \left(\mathcal{T}_k(\varrho) - \overline{\mathcal{T}_k(\varrho)}\right) \mathrm{div}\ \vec{u}\ \mathrm{d}x\ \mathrm{d}t \text{ for any } t_1 \leq t_2,\ t_1, t_2 \in I.$$

Our aim is to pass to the limit for $k \to \infty$ in (73). Clearly,

$$\int_\Omega \left(\overline{L_k(\varrho)} - L_k(\varrho)\right)(t)\ \mathrm{d}x\ \mathrm{d}t \to \mathbf{dft}[\varrho_n - \varrho](t) \text{ for } k \to \infty$$

while

$$\|\mathcal{T}_k(\varrho) - \overline{\mathcal{T}_k(\varrho)}\|_{L^2(J\times\Omega)} \leq$$

$$\|\mathcal{T}_k(\varrho) - \overline{\mathcal{T}_k(\varrho)}\|^\beta_{L^1(J\times\Omega)} \Big(\mathbf{osc}_{\gamma+1}[\varrho_n - \varrho](J \times \Omega)\Big)^{1-\beta},\ \beta = \frac{\gamma - 1}{2\gamma}.$$

By virtue of Lemma 2.3 and Corollary 4.2, the right-hand side of the above inequality tends to zero for $k \to \infty$ and so does the right-hand side of (73).

Finally, we have

$$\limsup_{n\to\infty} \int_{t_1}^{t_2} \int_\Omega |\mathcal{T}_k(\varrho_n) - \mathcal{T}_k(\varrho)|^{\gamma+1}\ \mathrm{d}x\ \mathrm{d}t \geq$$

$$|\Omega|^{\frac{\alpha-\gamma+1}{\alpha}} \limsup_{n\to\infty} \int_{t_1}^{t_2} \|\mathcal{T}_k(\varrho_n) - \mathcal{T}_k(\varrho)\|^{\gamma+1}_{L^\alpha(\Omega)}\ \mathrm{d}x\ \mathrm{d}t,$$

and we deduce from (73):

$$\mathbf{dft}[\varrho_n - \varrho](t_2) - \mathbf{dft}[\varrho_n - \varrho](t_1) + \tag{74}$$

$$\left(\frac{a|\Omega|^{\frac{\alpha-\gamma+1}{\alpha}}}{\lambda + 2\mu}\right) \limsup_{n\to\infty} \int_{t_1}^{t_2} \|\varrho_n - \varrho\|^{\gamma+1}_{L^\alpha(\Omega)}\ \mathrm{d}x\ \mathrm{d}t \leq 0$$

for any $t_1 \leq t_2$, $t_1,\ t_2 \in I$ and $1 \leq \alpha < \gamma$.

To conclude, we shall need the following lemma (cf. [5, Lemma 2.1]).

Lemma 5.1. *Given $\alpha \in (1, \gamma)$ there exists $c = c(\alpha)$ such that*

$$z \log(z) - y \log(y) \leq (1 + \log^+(y))(z - y) + c(\alpha)\Big(|z - y|^{\frac{1}{2}} + |z - y|^\alpha\Big) \tag{75}$$

for any $y, z \geq 0$.

In accordance with Lemma 5.1, we can write

$$\int_\Omega \varrho_n \log(\varrho_n)\ \mathrm{d}x - \int_\Omega (1+\log^+(\varrho))(\varrho_n - \varrho)\ \mathrm{d}x \leq$$

$$c(\alpha)\Big(|\Omega|^{\frac{2\alpha-1}{2\alpha}} \|\varrho_n - \varrho\|^{2\alpha}_{L^\alpha(\Omega)} + \|\varrho_n - \varrho\|^{\alpha}_{L^\alpha(\Omega)} \Big)$$

which together with (74) yields

$$\mathbf{dft}[\varrho_n - \varrho](t_2) - \mathbf{dft}[\varrho_n - \varrho](t_1) + \int_{t_1}^{t_2} \Phi\Big(\mathbf{dft}[\varrho_n - \varrho](t)\Big)\ \mathrm{d}t \leq 0$$

where the nonlinear function Φ depends only on the structural properties of the logarithm and can be found to satisfy

$$\Phi : R \mapsto R \text{ is continuous and strictly increasing, } \Phi(0) = 0. \tag{76}$$

Our results can be formulated in the following way:

Theorem 5.2. *Let $\Omega \subset R^N$, $N = 2,3$ be a bounded domain. Let ϱ_n, $\vec{u}_n$ be a sequence of finite energy weak solutions of the problem (1), (2), (12) on $I \times \Omega$ where p is given by the relation*

$$p = a\varrho^\gamma,\ a > 0,\ \gamma > \frac{N}{2}$$

and $\vec{f} = \vec{f}_n$ are uniformly bounded on $I \times \Omega$. Finally, let

$$\mathrm{ess}\limsup_{t\to a+} \int_\Omega \frac{1}{2}\varrho_n|\vec{u}_n|^2 + \frac{a}{\gamma-1}\varrho_n^\gamma\ \mathrm{d}x \leq E_0 \text{ independently of } n = 1,2,...$$

where $a = \inf\{t \in I\}$.

Then

$$\mathbf{dft}[\varrho_n - \varrho](t_2) \leq \chi(t_2 - t_1) \text{ for any } t_1,\ t_2 \in I,\ t_1 \leq t_2,$$

where χ is the unique solution of the initial-value problem

$$\chi'(t) + \Phi(\chi(t)) = 0,\ \chi(0) = \mathbf{dft}[\varrho_n - \varrho](t_1)$$

where Φ is a fixed function having the property (76).

Remark: It can be shown that Φ has a polynomial growth for values close to zero and, consequently, the quantity $\mathbf{dft}[\varrho_n - \varrho](t)$ behaves like $t^{-\beta}$ for a certain $\beta > 0$ when $t \to \infty$.

References

[1] M. E. Bogovskii. Solution of some vector analysis problems connected with operators div and grad (in Russian). *Trudy Sem. S.L. Sobolev*, **80**(1):5–40, 1980.

[2] W. Borchers and H. Sohr. On the equation $rotv = g$ and $divu = f$ with zero boundary conditions. *Hokkaido Math. J.*, **19**:67–87, 1990.

[3] R. Coifman and Y. Meyer. On commutators of singular integrals and bilinear singular integrals. *Trans. Amer. Math. Soc.*, **212**:315–331, 1975.

[4] R.J. DiPerna and P.-L. Lions. Ordinary differential equations, transport theory and Sobolev spaces. *Invent. Math.*, **98**:511–547, 1989.

[5] E. Feireisl. Propagation of oscillations, complete trajectories and attractors for compressible flows. *NoDEA*, 2000. Submitted.

[6] E. Feireisl and H. Petzeltová. On integrability up to the boundary of the weak solutions of the Navier-Stokes equations of compressible flow. *Commun. Partial Differential Equations*, **25**(3-4):755–767, 2000.

[7] G. P. Galdi. *An introduction to the mathematical theory of the Navier - Stokes equations, I.* Springer-Verlag, New York, 1994.

[8] D. Hoff. Strong convergence to global solutions for multidimensional flows of compressible, viscous fluids with polytropic equations of state and discontinuous initial data. *Arch. Rational Mech. Anal.*, **132**:1–14, 1995.

[9] P.-L. Lions. Compacité des solutions des équations de Navier- Stokes compressible isentropiques. *C.R. Acad. Sci. Paris, Sér I.*, **317**:115–120, 1993.

[10] P.-L. Lions. *Mathematical topics in fluid dynamics, Vol.2, Compressible models.* Oxford Science Publication, Oxford, 1998.

[11] P.-L. Lions. Bornes sur la densité pour les équations de Navier- Stokes compressible isentropiques avec conditions aux limites de Dirichlet. *C.R. Acad. Sci. Paris, Sér I.*, **328**:659–662, 1999.

[12] D. Serre. Variation de grande amplitude pour la densité d'un fluid viscueux compressible. *Physica D*, **48**:113–128, 1991.

[13] Z. Yi. An L^p theorem for compensated compactness. *Proc. Royal Soc. Edinburgh*, **122 A**:177–189, 1992.

Institute of Mathematics,
Czech Academy of Sciences,
Žitná 25
115 67 Praha 1, Czech Republic
E-mail address: feireisl@math.cas.cz

Numerical Methods for Compressible Flow

Miloslav Feistauer

Abstract. In modern technologies one often meets the necessity to solve compressible flow (i. e. flow of gases) with a complicated structure. Let us mention, e. g., aviation design, turbomachinery, automobile, food and chemical industry, etc. There is a number of simplified models used for the simulation of gas flow, as inviscid irrotational subsonic flow, small disturbance transonic flow and full potential transonic flow. Modern computers allow us to use more and more complex models. Here we shall be concerned with the models of gas flow described by the Euler equations (inviscid flow) and the Navier–Stokes equations (viscous flow). We shall formulate initial-boundary value problems of gas dynamics and discuss numerical methods for their solution.

1. Basic Equations and Relations Describing Compressible Flow

Fluids, i. e., liquids and gases, have a discrete structure – they are formed by molecules and atoms. However, in mathematical description of fluid flow we usually ignore this microstructure and consider fluids as continua. This means that we assume that at every time instant some fluid particle passes through each point of the domain occupied by the fluid. Furthermore, we assume that functions describing fluid flow are arbitrarily continuously differentiable. This assumption is used for the derivation of basic equations of fluid dynamics. In the theoretical and numerical analysis this assumption is relaxed and we admit even discontinuities of these functions as we shall see in Section 2.

The gas flow will be described with the aid of the Eulerian approach. We use the following notation: t – time, $x = (x_1, x_2, x_3)$ – point of the space $I\!R^3$ expressed in a Cartesian coordinate system, $\mathbf{v} = (v_1, v_2, v_3)$ – velocity vector with components $v_i, i = 1, 2, 3$, ρ – density, p – pressure, θ – absolute temperature, S – entropy, e – total energy, $\mathbf{f} = (f_1, f_2, f_3)$ – density of outer volume force, τ_{ij} – components of the stress tensor, q – density of heat sources, $\mathbf{q}$ – the vector of heat flux, μ, λ – viscosity coefficients, k - heat conduction coefficient, c_p, c_v – specific heats at constant pressure and constant volume, R – gas constant, γ – Poisson adiabatic constant. Let us note that $c_p, c_v > 0, R = c_p - c_v > 0, \gamma = c_p/c_v > 1$, $\mu \geq 0$, $3\lambda + 2\mu \geq 0$ and $k \geq 0$.

We shall consider gas flow in a domain $\Omega \subset I\!R^3$ and time interval $(0, T)$. The basic equations describing the gas flow represent the fundamental physical conservation laws, namely,

- conservation of mass,
- conservation of momentum,
- conservation of the moment of momentum,
- conservation of energy,

in the form of partial differential equations:

$$\frac{\partial \rho}{\partial t} + \operatorname{div}(\rho \mathbf{v}) = 0 \tag{1}$$

(continuity equation),

$$\frac{\partial(\rho v_i)}{\partial t} + \operatorname{div}(\rho v_i \mathbf{v}) = \rho f_i + \sum_{j=1}^{3} \frac{\partial \tau_{ji}}{\partial x_j}, \quad i = 1,2,3 \tag{2}$$

(equations of motion),

$$\frac{\partial(\rho e)}{\partial t} + \operatorname{div}(\rho e \mathbf{v}) = \rho \mathbf{f} \cdot \mathbf{v} + \sum_{i,j=1}^{3} \frac{\partial}{\partial x_j}(v_i \tau_{ji}) + \rho q - \operatorname{div} \mathbf{q} \tag{3}$$

(energy equation). The continuity equation is equivalent to the conservation of mass, the equations of motion are equivalent to the conservation of momentum and the energy equation is equivalent to the conservation of energy. The conservation of the moment of momentum is equivalent to the symmetry of the stress tensor: $\tau_{ij} = \tau_{ji}$. This is hidden in the relation defining the stress tensor as a function of other quantities:

$$\begin{aligned} \tau_{ij} &= -p\delta_{ij} + \operatorname{div}(\lambda \mathbf{v})\,\delta_{ij} + 2\mu d_{ij}(\mathbf{v}) \\ d_{ij}(\mathbf{v}) &= \frac{1}{2}\left(\frac{\partial v_i}{\partial x_j} + \frac{\partial v_j}{\partial x_i}\right). \end{aligned} \tag{4}$$

(The dependence of the stress tensor τ_{ij} on the velocity deformation tensor $d_{ij}(\mathbf{v})$ is linear – gases behave as Newtonian fluids.) Here δ_{ij} is the Kronecker delta. Often we set

$$\lambda = -\frac{2}{3}\mu.$$

The heat flux is expressed according to the *Fourier law*:

$$q = -k \nabla \theta. \tag{5}$$

Furthermore, we close the system with the aid of thermodynamical relations. The density e of the total energy reads

$$e = u + \frac{1}{2}|\mathbf{v}|^2, \tag{6}$$

where the internal energy

$$u = c_v \theta. \tag{7}$$

Moreover, we consider the equation of state of perfect gas

$$p = R\rho\theta. \tag{8}$$

An important thermodynamical quantity is the entropy S defined by the relation

$$\theta dS = du + pdV, \tag{9}$$

where

$$V = \frac{1}{\rho} = \text{specific volume}. \tag{10}$$

For the perfect gas we have

$$S = c_v \ln \frac{p/p_0}{(\rho/\rho_0)^\gamma}, \tag{11}$$

where p_0 and ρ_0 are reference values of the pressure and density.

Not all possible solutions of the above system are acceptable from the physical point of view. It is necessary to use the selection principle called the *second law of thermodynamics* (or entropy condition) which can be formulated as the differential inequality:

$$\frac{\partial(\rho S)}{\partial t} + \operatorname{div}(\rho S\,\mathbf{v}) \geq \frac{\rho q}{\theta} - \operatorname{div}\left(\frac{\mathbf{q}}{\theta}\right). \tag{12}$$

It is possible to show that under a sufficient smoothness of quantities describing the flow and the conditions $\mu \geq 0$, $3\lambda + 2\mu \geq 0$, condition (12) is a consequence of the above relations.

Because gases are very light, the outer volume force is often neglected. Then we set

$$\mathbf{f} \equiv 0. \tag{13}$$

The viscosity coefficients μ, λ and the heat conduction coefficient are rather small and, therefore, they are sometimes neglected. In this case, usually heat sources and heat flux are also neglected. Then we obtain the model of *inviscid adiabatic flow*:

$$\mu = \lambda = 0, \quad k = 0, \quad q = 0, \quad \mathbf{q} = 0. \tag{14}$$

For perfect gas, in view of (11), we can write $p = p(\rho, S)$. We set

$$c = (\partial p/\partial \rho)^{1/2} = \text{local speed of sound} \tag{15}$$

(= the speed of propagation of pressure waves of small intensity). We define an important characteristics called the *Mach number*

$$M = |\,\mathbf{v}\,|/c \tag{16}$$

and distinguish the following regimes:

$M < 1$ – subsonic flow,
$M = 1$ – sonic flow,
$M > 1$ – supersonic flow.

These regimes have completely different character, because in supersonic flow the so-called shock waves or contact discontinuities can appear. They represent discontinuity of quantities describing flow (velocity, density, ...), in viscous flow smeared a little due to viscosity. In purely subsonic flow these discontinuities do not arise.

The complete system describing fluid flow is rather complex and we are not able to solve it in some general complicated situations. Therefore, we use various simplified models. Sometimes the viscosity effects are not important and the fluid can be considered inviscid. Le us mention, for example, inviscid irrotational subsonic flow or inviscid potential transonic flow. However, in highspeed flow with the so-called strong shock waves one cannot assume that the flow is irrotational, because the shock waves are source of vorticity and entropy. Then it is suitable to use the model formed by the system of the continuity equation, the Euler equations of motion and the energy equation. (This whole system is often also called the Euler equations.) The investigation and numerical solution of initial-boundary value problems for the Euler equations will be the subject of Section 2.

If the viscosity cannot be neglected, then it is necessary to use the complete system of viscous flow. Some computational methods for the solution of viscous compressible flow will be discussed in Section 3. Finally, in Section 4, we shall present results of the solution of some test problems and of a technically relevant flow problem.

2. The Compressible Euler Equations

This section will be concerned with some basic properties of the Euler equations and their numerical solution. This system is formed by equations (1), (2), (3) and relations (5) –(8) under assumptions (13) and (14). For simplicity we shall be concerned with 2D flow. This means that $x = (x_1, x_2), \mathbf{v} = (v_1, v_2)$ and $\Omega \subset \mathbb{R}^2$. The most of methods and techniques presented here can be extended to 3D flows.

2.1. Basic properties of the Euler equations

The 2D Euler equations can be written in the form

$$\frac{\partial w}{\partial t} + \sum_{j=1}^{2} \frac{\partial f_j(w)}{\partial x_j} = 0 \ \text{ in } Q_T = \Omega \times (0,T), \tag{17}$$

where

$$w = (w_1, \dots, w_4)^{\mathrm{T}} = (\rho, \rho v_1, \rho v_2, e)^{\mathrm{T}}, \tag{18}$$

is the so-called state vector and

$$\begin{aligned} f_j(w) &= (f_j^1(w), \dots, f_j^4(w))^{\mathrm{T}} \\ &= (\rho v_j, \rho v_1 v_j + \delta_{1j} p, \rho v_2 v_j + \delta_{2j} p, (e+p) v_j)^{\mathrm{T}}, j = 1, 2, \end{aligned} \tag{19}$$

are the so-called inviscid (Euler) fluxes defined in the set $D = \{w \in \mathbb{R}^4; w_1 > 0\}$. Obviously, $f_j \in [C^1(D)]^4$. The state equation implies that

$$p = (\gamma - 1)(e - \rho|\mathbf{v}|^2/2). \tag{20}$$

System (17) is *hyperbolic* in the following sense: If we set $\mathbf{A}_j(w) = Df_j(w)/Dw$ (= Jacobian matrix), then for any $w \in D$ and any $\nu \in \mathbb{R}^2$ the matrix

$$\mathbf{P} = \mathbf{P}(w, \nu) = \nu_1 \mathbf{A}_1(w) + \nu_2 \mathbf{A}_2(w)$$

has only real eigenvalues $\lambda_i = \lambda_i(w, \nu), i = 1, ..., 4$, and is diagonalizable: there exists a nonsingular matrix $\mathbf{T} = \mathbf{T}(w, \nu)$ such that

$$\mathbf{T}^{-1}\mathbf{P}\,\mathbf{T} = \mathbf{D} = \mathbf{D}(w, \nu) = \mathrm{diag}(\lambda_1, \dots, \lambda_4).$$

It is possible to show that for $\nu \in \mathbb{R}^2$ with $|\nu| = 1$, the eigenvalues λ_i have the form $\lambda_1 = \mathbf{v} \cdot \nu - c, \lambda_2 = \lambda_3 = \mathbf{v} \cdot \nu, \lambda_4 = \lambda_2 + c$. (The extensive treatment of nonlinear hyperbolic problems can be found, e.g., in [26], [36], [38] and [13].)

To system (17) we add the initial condition

$$w(x, 0) = g(x), \quad x \in \Omega, \tag{21}$$

and suitable boundary conditions

$$B(w) = 0 \text{ on } \partial\Omega \times (0, T), \tag{22}$$

chosen in such a way that problem (17) – (22) is linearly well–posed. See later.

Definition 2.1. *We say that a vector-valued function w is a classical solution of problem (17)–(22), if $w \in C^1(\overline{Q}_T)^4$ satisfies system (17) and conditions (21), (22) pointwise.*

It is well–known that nonlinear first order hyperbolic equations have a characteristic property: the possible rise of discontinuites in solutions even in the case when data are of the class C^∞. As an example we can consider a scalar equation

$$\frac{\partial w}{\partial t} + \frac{\partial f(w)}{\partial x} = 0, \quad x \in \mathbb{R}, \ t > 0,$$

which can be written as

$$\frac{\partial w}{\partial t} + a(w)\frac{\partial w}{\partial x} = 0, \quad a = f'.$$

We equip this equation by the initial condition (21) with $\Omega = \mathbb{R}$. If $a' > 0$ and $g' < 0$, then the characteristics intersect each other in finite time and the solution becomes discontinuous.

Therefore, the concept of a classical solution is very restrictive a it is suitable to introduce a weak solution. For simplicity we confine our considerations to the Cauchy problem with $\Omega = \mathbb{R}^2$ and $T = \infty$.

Let $C_0^\infty(\mathbb{R}^2 \times [0, \infty))$ be the space of all infinitely differentiable functions with compact support contained in $\mathbb{R}^2 \times [0, \infty)$. We proceed in a usual way. Assuming that $w \in [C^1(\mathbb{R}^2 \times [0, \infty))]^4$ is a classical solution, we multiply (17) by $\varphi \in$

$[C_0^\infty(\mathbb{R}^2 \times [0,\infty))]^4$, integrate over $\mathbb{R}^2 \times [0,\infty)$ and use Green's therorem. We obtain the identity

$$\int_0^\infty \int_{\mathbb{R}^2} \left(w\, \partial\varphi/\partial t + \sum_{j=1}^2 f_j(w)\, \partial\varphi/\partial x_j \right) \mathrm{d}x\mathrm{d}t + \tag{23}$$
$$+ \int_{\mathbb{R}^2} g(x)\, \varphi(x,0)\, \mathrm{d}x = 0$$
$$\forall \varphi \in [C_0^\infty(\mathbb{R}^2 \times [0,\infty))]^4.$$

Obviously, (23) has sense if $w \in L^\infty_{loc}(Q_T)^4$. Here $L^\infty_{loc}(Q_T)$ denotes the space of all measurable functions in Q_T which are bounded on every bounded subset of Q_T.

Definition 2.2. *We call w a weak solution of the Cauchy problem (17), (21), if $w \in L^\infty_{loc}(Q_T)^4$, $w(x,t) \in D$ for almost every $(x,t) \in Q_T$ and satisfies (23).*

Condition (23) implies that w is a solution of (17) in the sense of distributions.

Piecewise smooth solution is defined as a weak solution w such that $w \in C^1(Q_T - \Gamma)^4$, where Γ is formed by a finite number of smooth hypersurfaces in Q_T and w together with its first order derivatives has one–sided limits on Γ.

It is possible to show that a piecewise smooth function w is a weak solution of equation (17) if and only if it satisfies (17) in a classical sense in the region, where w is smooth and on any discontinuity the one-sided limits w^+ and w^- satisfy the so–called *Rankine–Hugoniot conditions*

$$(w^+ - w^-)n_t + \sum_{j=1}^2 (f_j(w^+) - f_j(w^-))n_j = 0, \tag{24}$$

where $n = (n_1, n_2, n_t)$ is the normal to Γ. If $(n_1, n_2) \neq 0$, then, under the notation $[w] = w^+ - w^-$, $[f(w)] = f(w^+) - f(w^-)$, it can be writen in the form

$$s[w] = \sum_{j=1}^2 \nu_j [f_j(w)], \tag{25}$$

where $\nu = (\nu_1, \nu_2)$ with $|\nu| = 1$ is the direction of the propagation of the discontinuity Γ and s is the velocity of the propagation of Γ.

2.1.1. ENTROPY CONDITION. Examples show that the weak solution to the nonlinear Cauchy problem (17), (21) need not be unique. In order to choose a physically admissible solution, it is necessary to formulate some *selection principle.* We proceed in the following way:

Definition 2.3. *We say that a convex function $U \in C^1(D)$ is the entropy of system (17), if there exist* entropy fluxes *$\Phi_j \in C^1(D)$, $j = 1,2$, so that*

$$\nabla_w U(w)\, \mathbf{A}_j(w) = \nabla_w \Phi_j(w), \quad j = 1,2. \tag{26}$$

If $w \in C^1(Q_T)^4$ is a classical solution of (17) and U is an entropy, then

$$\partial U(w)/\partial t + \sum_j \partial \Phi_j(w)/\partial x_j = 0 \text{ in } Q_T. \tag{27}$$

For a weak solution, (27) is not valid in general. We introduce the following *entropy condition:*

Definition 2.4. *We call a weak solution w of (17) an entropy solution, if*

$$\partial U(w)/\partial t + \sum_j \partial \Phi_j(w)/\partial x_j \leq 0 \quad \text{in } Q_T \tag{28}$$

in the sence of distributions, i.e.,

$$\begin{gathered}\int_0^\infty \int_{\mathbb{R}^2} \left[U(w)\varphi/\partial t + \sum_j \Phi_j(w)\partial\varphi/\partial x_j \right] dx dt \geq 0 \\ \forall \varphi \in C_0^\infty(Q_T)^+ = \{\psi \in C_0^\infty(Q_T);\ \psi \geq 0 \text{ in } Q_T\}.\end{gathered} \tag{29}$$

Remark 2.5. *For the system of the Euler equations we know only one entropy:*

$$U = -\rho S = -c_v \rho \ln \frac{p/p_0}{(\rho/\rho_0)^\kappa}. \tag{30}$$

Here S is the physical entropy defined by (11). This fact allows us to define physically admissible solutions of inviscid flow.

There is an important question: *How to obtain an entropy solution?* One possibility is to apply Lax's method of *artificial viscosity:* Consider

$$\frac{\partial w}{\partial t} + \sum_{j=1}^2 \frac{\partial f_j(w)}{\partial x_j} = \varepsilon \Delta w \tag{31}$$

with $\varepsilon > 0$. If $\{w_\varepsilon\}_{\varepsilon>0}$ is a system of "classical" solutions, uniformly locally bounded and $w^\epsilon \to w$ a.e. in Q_T as $\varepsilon \to 0$, then w is a weak entropy solution of equation (17).

Solvability of the Cauchy problem has been proved only in the case of a scalar equation by Kruzhkov in [37]. We consider the Cauchy problem

$$\partial w/\partial t + \sum_{j=1}^N \partial f_j(w)/\partial x_j = 0 \text{ in } \mathbb{R}^N \times (0,\infty), \tag{32}$$

$$w(x,0) = g(x), \quad x \in \mathbb{R}^N,$$

where $f_j \in C^1(\mathbb{R}),\ j = 1, \dots, N.$

Theorem 2.6. *If $g \in L^1(\mathbb{R}^N) \cap L^\infty(\mathbb{R}^N)$, then problem (32) has a unique entropy solution.*

For systems there are some solvability results only for problems with one space dimension:

$$\frac{\partial w}{\partial t} + \frac{\partial f(w)}{\partial x} = 0 \text{ in } \mathbb{R} \times (0,\infty), \tag{33}$$
$$w(x,0) = g(x), \quad x \in \mathbb{R}.$$

(Here $w, f(w) \in \mathbb{R}^m$.) For example, the existence of an entropy solution is proved for the so-called *Riemann problem* (33), where

$$g(x) = \begin{cases} w_L, & x > 0, \\ w_R, & x < 0, \end{cases} \tag{34}$$

and w_L, w_R are constant states. Under some *additional assumptions* on f, problem (33), (34) has a unique entropy solution provided the states w_L, w_R are close to each other. See [41]. E. g., for 1-D system of inviscid flow, which can be written as 33, the vector valued function f satisfies these assumptions.

The Riemann problem is important for the construction of numerical methods for the solution of hyperbolic systems.

2.2. Finite volume method

There exists a number of various approaches to the numerical solution of compressible flow. Basically, there are three fundamental techniques: finite difference method ([35], [34]), finite volume method ([13], [36], [12]) and finite element method ([1], [7], [28], [30], [39]). Here we will carry out the discretization of system (17) with the use of the finite volume method (FVM) which is now very popular because of its flexibility and applicability and because it reflects well some important characteristic features of compressible flow. We proceed in the following way:

Let Ω be a bounded domain. Every component of $\partial\Omega$ is approximated by a piecewise linear curve so that the domain Ω can be replaced by a polygonal domain Ω_h with boundary

$$\partial\Omega_h = \Gamma_{Ih} \cup \Gamma_{Oh} \cup \Gamma_{Wh}. \tag{35}$$

Here Γ_{Ih}, Γ_{Oh}, Γ_{Wh} are parts of $\partial\Omega_h$ approximating inlet, outlet and fixed impermeable wall. By $\mathcal{D}_h = \{D_i\}_{i\in J}$ we denote a partition of $\overline{\Omega}_h$ into a finite number of closed polygons such that their interiors are mutually disjoint and

$$\overline{\Omega}_h = \bigcup_{i\in J} D_i. \tag{36}$$

(J is a suitable index set of nonnegative integers.)

The so-called *finite volumes* D_i can be chosen as *triangles, quadrilaterals* or *dual finite volumes and barycentric finite volumes* over a triangular mesh $\mathcal{T}_h = \{T_i\}_{i\in I}$ = a triangulation of Ω_h with usual properties used in the finite element method ([5]). See Figure 1.

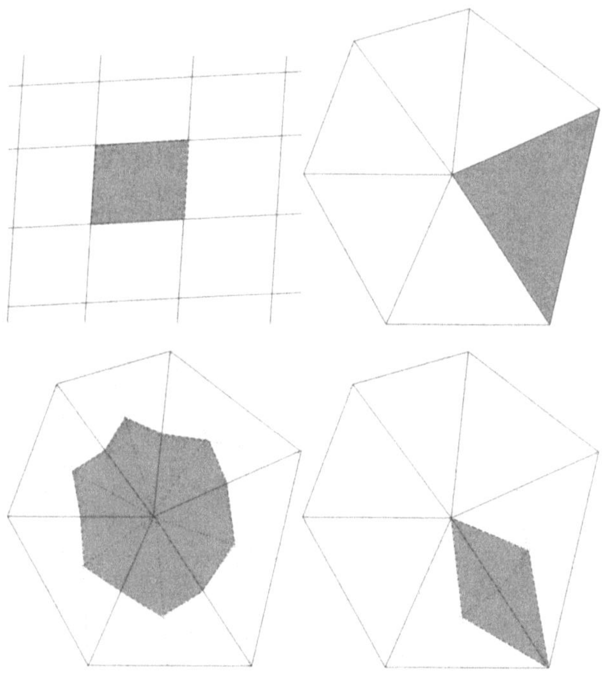

FIGURE 1. Quadrilateral, triangular, dual and barycentric finite volumes

If for two different finite volumes D_i and D_j their boundaries contain a common straight segment, we call them *neighbours*. Then we write

$$\Gamma_{ij} = \bigcup_{\alpha=1}^{\beta_{ij}} \Gamma_{ij}^{\alpha} = \partial D_i \cap \partial D_j = \Gamma_{ji}, \tag{37}$$

where $\Gamma_{ij}^{\alpha} (= \Gamma_{ji}^{\alpha})$ are straight segments. (If $\mathcal{D}_h$ is a triangular or rectangular grid with usual properties from the finite element method or if $\mathcal{D}_h$ is a grid of the barycentric finite volumes, then $\beta_{ij} = 1$. For a dual finite volume mesh over a triangular grid we have $\beta_{ij} = 2$ (for D_i or $D_j \subset \Omega_h$) or $\beta_{ij} = 1$ (for D_i and D_j adjacent to $\partial\Omega_h$.) For $i \in J$, let $s(i) = \{j \in J;\, D_j$ is a neighbour of $D_i\}$. If D_i is adjacent to $\Gamma_{Ih} \cup \Gamma_{Oh} \cup \Gamma_{Wh}$, then we denote the straight segments Γ_{ij}^{α}, $\alpha = 1, \dots, \beta_{ij}$, that form $\partial D_i \cap \partial\Omega_h$ with the aid of negative indices $j \in \gamma(i)$ (= a suitable index set). In this case we set $S(i) = s(i) \cup \gamma(i)$, otherwise we put $S(i) = s(i)$.

Obviously, for every $D_i \in \mathcal{D}_h$ we have

$$\partial D_i = \bigcup_{j\in S(i)} \Gamma_{ij} = \bigcup_{j\in S(i)} \bigcup_{\alpha=1}^{\beta_{ij}} \Gamma_{ij}^{\alpha}. \tag{38}$$

Furthermore, we introduce the following notation: $|D_i|$ = area of D_i, $\mathbf{n}_{ij}^{\alpha} = (n_{1ij}^{\alpha}, n_{2ij}^{\alpha})$ = unit outer normal to ∂D_i on Γ_{ij}^{α}, ℓ_{ij}^{α} = length of Γ_{ij}^{α}, and consider a partition $0 = t_0 < t_1 < \dots$ of the time interval $(0,T)$ and set $\tau_k = t_{t+1} - t_k$ for $k = 0, 1, \dots$.

The *finite volume (cell centred) discretization* is based on the integration of (17) over every set $D_i \times (t_k, t_{k+1})$, the use of Green's theorem, the approximation of the exact solution w averaged at time t_k over grid cells D_i by constant values w_i^k and the approximation of the flux

$$\int_{\Gamma_{ij}^{\alpha}} \sum_{r=1}^{2} f_r(w)\, n_r \, \mathrm{d}S$$

of the quantity w through the segment Γ_{ij}^{α} in the direction $\mathbf{n}_{ij}^{\alpha}$ on a time level t_k with the aid of the so-called *numerical flux* $H(w_i^k, w_j^k, \mathbf{n}_{ij}^{\alpha})$. In this way we obtain the following explicit numerical scheme:

$$\begin{aligned} w_i^{k+1} &= w_i^k - \frac{\tau_k}{|D_i|} \sum_{j\in S(i)} \sum_{\alpha=1}^{\beta_{ij}} H(w_i^k, w_j^k, \mathbf{n}_{ij}^{\alpha})\, \ell_{ij}^{\alpha}, \\ & \qquad D_i \in \mathcal{D}_h \ (\text{i.e., } i \in J), \ k = 0, 1, \dots . \end{aligned} \tag{39}$$

In case that $j \in S(i)$, $j < 0$ and, thus, $\Gamma_{ij}^{\alpha} \subset \partial\Omega_h$, it is necessary to determine the boundary state on Γ_{ij}^{α}. We will discuss the problem of boundary conditions later.

PROPERTIES OF H: We require that

1) H is locally Lipschitz continuous,
2) H is consistent: $H(w, w, \mathbf{n}) = f_1(w)n_1 + f_2(w)n_2$,
3) H is conservative: $H(u, v, \mathbf{n}) = -H(v, u, -\mathbf{n})$.

As a simple example we can write $H(u, v, \mathbf{n}) = \frac{1}{2}(f_1(u)+f_1(v))n_1+\frac{1}{2}(f_2(u)+f_2(v))n_2$. This numerical flux defines a central scheme, which is *unconditionaly unstable* and, therefore, for no use. Let us construct better methods. To this end, we shall use some special properties of the Euler equations. We set $\mathbf{n} = (n_1, n_2) = \mathbf{n}_{ij}^{\alpha}$ and introduce new Cartesian coordinates $\tilde{x}_1, \tilde{x}_2$ with the origin at the centre of Γ_{ij}^{α} such that

$$\begin{pmatrix} \tilde{x}_1 \\ \tilde{x}_2 \end{pmatrix} = \begin{pmatrix} n_1 & n_2 \\ -n_2 & n_1 \end{pmatrix} \begin{pmatrix} x_1 \\ x_2 \end{pmatrix} + \tilde{\sigma}, \tag{40}$$

i.e., $\tilde{x}_1$ is orthogonal to Γ_{ij}^{α} and $\tilde{x}_2$ is tangential to Γ_{ij}^{α}. Then the corresponding transformation of the state vector w is given by the matrix

$$\mathbf{Q} = \mathbf{Q}(\mathbf{n}) = \begin{pmatrix} 1 & 0 & 0 & 0 \\ 0 & n_1 & n_2 & 0 \\ 0 & -n_2 & n_1 & 0 \\ 0 & 0 & 0 & 1 \end{pmatrix} \tag{41}$$

and yields the transformed state vector

$$\begin{aligned} q &= \mathbf{Q}\, w = (\rho,\, \rho\tilde{u},\, \rho\tilde{v},\, e)^T \qquad (42) \\ \tilde{u} &= v_1\, n_1 + v_2\, n_2, \\ \tilde{v} &= v_1(-n_2) + v_2\, n_1. \end{aligned}$$

System (17) of the Euler equations is *rotationally symmetric*, i.e.,

$$f_1(w)n_1 + f_2(w)n_2 = \mathbf{Q}^{-1}\, f_1(\mathbf{Q}\, w). \tag{43}$$

Then the system for q has the form

$$\frac{\partial q}{\partial t} + \sum_{j=1}^{2} \frac{\partial f_j(q)}{\partial \tilde{x}_j} = 0. \tag{44}$$

Moreover, the Euler fluxes f_j are homogeneous of order 1, which implies that

$$f_j(w) = \mathbf{A}_j(w)w. \tag{45}$$

The transformed system (44), where we neglected the derivative $\partial/\partial\tilde{x}_2$, will be used for the construction of numerical schemes for the solution of system (17).

2.2.1. Upwind flux vector splitting methods of the Godunov type. In analogy to (43) we define

$$H(u, v, \mathbf{n}) = \mathbf{Q}^{-1}(\mathbf{n}) f_R(\mathbf{Q}(\mathbf{n})u, \mathbf{Q}(\mathbf{n})v) \tag{46}$$

where f_R is an *approximate Riemann solver* to the system

$$\frac{\partial q}{\partial t} + \frac{\partial f_1(q)}{\partial \tilde{x}_1} = 0 \text{ in } I\!R \times (0, \infty). \tag{47}$$

We define f_R in the following way. Let us equip (47) with the initial condition

$$q(\tilde{x}_1, 0) = \begin{cases} q_1 & \text{for } \tilde{x}_1 < 0, \\ q_2 & \text{for } \tilde{x}_1 > 0, \end{cases} \tag{48}$$

and obtain the Riemann problem. If the Riemann problem (47), (48) has a unique entropy weak solution, then it can be expressed as

$$q(\tilde{x}_1, t) = q_R\left(\frac{\tilde{x}_1}{t}; q_1, q_2\right). \tag{49}$$

Now we set

$$f_R(q_1, q_2) = f_G(q_1, q_2) = f_1(q_R(0; q_1, q_2)) \tag{50}$$

and obtain the well-known *Godunov's numerical flux*, which yields *Godunov's scheme.*

Let us demonstrate the construction of Godunov's scheme for a scalar linear equation

$$\frac{\partial q}{\partial t} + a\frac{\partial q}{\partial \tilde{x}_1} = 0 \text{ in } \mathbb{R} \times (0,\infty), \tag{51}$$

where $q : \mathbb{R} \times (0,\infty) \to \mathbb{R}$ and $a \in \mathbb{R},\ a \neq 0$. The Riemann problem (51), (48) can be solved with the aid of characteristics, which are curves $\tilde{x}_1 - at = \text{const}$ and along which the solution q is constant. This implies that

$$q(0,t) = q_1 \quad \text{and} \quad f_G(q_1,q_2) = a\,q_1, \quad \text{if} \quad a > 0, \tag{52}$$

and

$$q(0,t) = q_2 \quad \text{and} \quad f_G(q_1,q_2) = a\,q_2, \quad \text{if} \quad a < 0. \tag{53}$$

Thus, we can write Godunov's numerical flux in the form

$$f_G(q_1,q_2) = a^+ q_1 + a^- q_2, \tag{54}$$

which is also called the exact Riemann solver. (We set $a^+ = \max(0,a)$ and $a^- = \min(0,a)$).

As we see, Godunov's numerical flux $f_G(q_1,q_2)$ is defined with the use of the value q_i situated against the direction of the speed a of propagation of disturbances. Therefore, we speak about *upwinding.*

Now let us consider system (47), i.e.,

$$\frac{\partial q}{\partial t} + \frac{\partial f_1(q)}{\partial \tilde{x}_1} = 0.$$

The analytical solution of the Riemann problem is impossible in general. In this case we can determine 4 characteristic directions given by eigenvectors of the matrix $\mathbf{A}_1(q) = Df_1(q)/Dq$. The propagation speeds are represented by the eigenvalues $\lambda_1(q),\dots,\lambda_4(q)$ of $\mathbf{A}_1(q)$. This leads us to the idea to write

$$f_1(q) = f_1^+(q) + f_1^-(q) \tag{55}$$

where $f_1^{\pm}$ are such mappings that

$$\frac{Df_1^{\pm}(q)}{Dq} \text{ has only } \begin{matrix}\text{nonnegative}\\ \text{nonpositive}\end{matrix} \text{ eigenvalues.} \tag{56}$$

Then we set

$$f_R(u,v) = f_1^+(u) + f_1^-(v). \tag{57}$$

To derive $f_1^{\pm}$, we start from the relation

$$\frac{Df_1}{Dq}(q) = \mathbf{A}_1(q) = \mathbf{A}_1^+(q) + \mathbf{A}_1^-(q), \tag{58}$$

where $\mathbf{A}_1^{\pm}$ is defined in the following way: From the hyperbolicity it follows that there exists a nonsingular matrix $\mathbf{T}$ such that

$$\mathbf{A}_1 = \mathbf{T}\,\mathbf{D}\,\mathbf{T}^{-1}, \quad \mathbf{D} = \mathrm{diag}(\lambda_1 \dots , \lambda_4). \tag{59}$$

Then we set

$$\mathbf{D}^{\pm} = \mathrm{diag}(\lambda_1^{\pm}, \dots , \lambda_4^{\pm}) \quad \text{and} \quad \mathbf{A}_1^{\pm} = \mathbf{T}\,\mathbf{D}^{\pm}\,\mathbf{T}^{-1} \tag{60}$$

and symbolically write

$$f_1^{\pm}(w) = \frac{1}{2}\, f_1(q_0) + \int_{q_0}^{w} \mathbf{A}_1^{\pm}(w)\,\mathrm{d}w, \tag{61}$$

which together with (57) defines f_R.

Similarly we define the matrix $|\,\mathbf{A}_1\,| = \mathbf{T}\,|\,\mathbf{D}\,|\,\mathbf{T}^{-1}$, $|\,\mathbf{D}\,| = \mathrm{diag}(|\lambda_1|, \dots |\lambda_4|)$.

The integral in (61) has not sense in general, but (57) and (61) can be used to construct efficient numerical schemes.

One approach is to employ a suitable numerical quadrature for evaluating the integral in (61). Then we get, e.g.,

a) *Vijagasundaram scheme:*

$$f_V(q_1, q_2) = \mathbf{A}_1^{+}\left(\frac{q_1 + q_2}{2}\right) q_1 + \mathbf{A}_1^{-}\left(\frac{q_1 + q_2}{2}\right) q_2, \tag{62}$$

b) *Steger–Warming scheme:*

$$f_{SW}(q_1, q_2) = \mathbf{A}_1^{+}(q_1) q_1 + \mathbf{A}_1^{-}(q_2) q_2, \tag{63}$$

c) *VanLeer scheme:*

$$f_{VL}(q_1, q_2) = \left\{ f_1(q_1) + f_1(q_2) - \left|\mathbf{A}_1\left(\frac{q_1 + q_2}{2}\right)\right| (q_2 - q_1) \right\} \Big/ 2. \tag{64}$$

In the Osher-Solomon scheme ([40], [43]), the integral from (61) is evaluated along a a suitable integration path in the state space with the aid of Riemann invariants to the eigenvectors of the matrix $\mathbf{A}_1(q)$. The resulting approximate Riemann solver can be expressed as a linear combination of values of the vector function $f := f_1$ at uniquely determined and analytically expressed points as can be seen from the following table:

Table 1: Osher–Solomon Riemann solver $f_{\mathrm{OS}}(q_1, q_2)$

	$\tilde{u}_2 \geq -c_2$		$\tilde{u}_2 < -c_2$	
	$\tilde{u}_1 \leq c_1$	$\tilde{u}_1 > c_1$	$\tilde{u}_1 \leq c_1$	$\tilde{u}_1 > c_1$
$c_A \leq \tilde{u}_A$	$f(q_1^S)$	$f(q_1)$	$f(q_2) - f(q_2^S) + f(q_1^S)$	$f(q_1) - f(q_1^S) + f(q_2)$
$0 < \tilde{u}_A < c_A$	$f(q_A)$	$f(q_1) - f(q_1^S) + f(q_A)$	$f(q_2) - f(q_2^S) + f(q_A)$	$f(q_1) - f(q_2^S) + f(q_2) -$ $- f(q_1^S) + f(q_A)$
$-c_B \leq \tilde{u}_A \leq 0$	$f(q_B)$	$f(q_1) - f(q_1^S) + f(q_B)$	$f(q_2) - f(q_2^S) + f(q_B)$	$f(q_1) - f(q_2^S) + f(q_2) -$ $- f(q_1^S) + f(q_B)$
$\tilde{u}_A < -c_B$	$f(q_2^S)$	$f(q_1) - f(q_1^S) + f(q_2^S)$	$f(q_2)$	$f(q_1) + f(q_2) - f(q_1^S)$

The states q_A, q_B, q_1^S, q_2^S are defined in the following way. We set

$$\begin{aligned} &c_i = (\gamma p_i/\rho_i)^{1/2}, \quad s_i = p_i/\rho_i^\gamma, \quad i = 1,2, \quad \alpha = (s_2/s_1)^{1/2\gamma}, \\ &z_1 = (\gamma-1)\,\tilde{u}_1/2 + c_1, \quad z_2 = (\gamma-1)\,\tilde{u}_2/2 - c_2. \end{aligned} \tag{65}$$

Then

$$\begin{aligned} &c_A = (z_1-z_2)/(1+\alpha),\ c_B = \alpha c_A,\ \rho_A = (c_A/c_1)^{2/(\gamma-1)}\rho_1, \\ &\rho_B = \rho_A/\alpha^2,\ \tilde{u}_A = 2(z_1-c_A)/(\gamma-1),\ \tilde{u}_B = \tilde{u}_A, \\ &\tilde{v}_A = \tilde{v}_1,\ \tilde{v}_B = \tilde{v}_2, \\ &c_1^S = 2z_1/(\gamma+1),\ \rho_1^S = (c_1^S/c_1)^{2/(\gamma-1)}\rho_1,\ \tilde{u}_1^S = c_1^S,\ \tilde{v}_1^S = \tilde{v}_1, \\ &c_2^S = -2z_2/(\gamma+1),\ \rho_2^S = (c_2^S/c_2)^{2/(\gamma-1)}\rho_2,\ \tilde{u}_2^S = -c_2^S,\ \tilde{v}_2^S = \tilde{v}_2. \end{aligned} \tag{66}$$

A 3D version of the Osher-Solomon scheme was derived in [23]. Further Riemann solvers can be found in [46].

2.2.2. Boundary and Initial Conditions. If $\Gamma_{ij}^\alpha \subset \partial\Omega_h$, then a special attention must be paid to the determination of boundary conditions, in order to be able to compute the numerical flux through the segment Γ_{ij}^α, i.e., $H(w_i^k, w_j^k, \mathbf{n}_{ij}^\alpha)$. Namely, it is necessary to specify the *boundary state* $w_j^{k,\alpha} = w_j^k = w_B$. The question of the boundary conditions is rather delicate. We use here such boundary conditions which lead to a well-posed linearized Riemann initial-boundary value problem with one space variable that is orthogonal to Γ_{ij}^α (cf. [13], Par. 7.3.43]).

Let us assume that $\Gamma_{ij}^\alpha \subset \Gamma_{Ih} \cup \Gamma_{Oh} \cup \Gamma_{Wh}$. For the *Vijayasundaram, Steger–Warming and Van Leer* schemes we prescribe m_p components of the boundary state w_j^k and extrapolate m_e components of w_j^k from D_i onto Γ_{ij}^α, where m_p and m_e is the number of negative and nonnegative, respectively, eigenvalues of the matrix $\mathbf{P}(w_i^k, \mathbf{n}_{ij}^\alpha)$. It is suitable to distinguish several cases:

1. *Subsonic inlet* ($-c \le \mathbf{v}\cdot\mathbf{n} \le 0$ on D_i): We prescribe ρ, v_1, v_2 on Γ_{ij}^α and extrapolate p from D_i to Γ_{ij}^α.
2. *Supersonic inlet* ($-\mathbf{v}\cdot\mathbf{n} > c$ on D_i): We prescribe ρ, v_1, v_2, p on Γ_{ij}^α.
3. *Subsonic outlet* ($0 < \mathbf{v}\cdot\mathbf{n} < c$ on D_i): We prescribe p on Γ_{ij}^α and extrapolate ρ, v_1, v_2 from D_i to Γ_{ij}^α.
4. *Supersonic outlet* ($\mathbf{v}\cdot\mathbf{n} \ge c$ on D_i): We extrapolate ρ, v_1, v_2, p from D_i to Γ_{ij}^α.
5. *Solid impermeable boundary*: We extrapolate p from D_i to Γ_{ij}^α and (in virtue of the fact that $\mathbf{v}\cdot\mathbf{n} = 0$ on impermeable wall) we set

$$H(w_i^k, w_j^k, \mathbf{n}_{ij}^\alpha) = p_i^k(0, n_{1ij}^\alpha, n_{2ij}^\alpha, 0)^{\mathrm{T}}. \tag{67}$$

In the case of the *Osher–Solomon scheme* we use a modified approach worked out in [43]. We introduce the transformed boundary state $q_B = \mathbf{Q}(\mathbf{n}_{ij}^\alpha)\,w_B$ and then define the corresponding approximate Riemann solver $f_{\mathrm{OS}}(q_i, q_B) := f_1(q_B)$:

1. *Subsonic inlet*: We prescribe ρ_B, $\tilde{u}_B$, $\tilde{v}_B$ and compute p_B from the formulae

$$p_B = \rho_I c_I^2/\gamma, \; \rho_I = (c_I^2 \rho_i/\gamma p_i)^{1/(\gamma-1)}\rho_i, \; c_I = c_i + \frac{\gamma-1}{2}(\tilde{u}_i - \tilde{u}_B). \tag{68}$$

2. *Supersonic inlet*: We prescribe ρ_B, $\tilde{u}_B$, $\tilde{v}_B$, p_B.
3. *Subsonic outlet*: We prescribe p_B and use the formulae

$$\rho_B = \rho_i(p_B/p_i)^{1/\gamma}, \; \tilde{u}_B = \tilde{u}_i + \frac{2}{\gamma-1}\left(c_i - \sqrt{\gamma p_B/\rho_B}\right), \; \tilde{v}_B = \tilde{v}_i. \tag{69}$$

4. *Supersonic outlet*: $q_B = q_i$.
5. *Impermeable wall*:

$$\begin{aligned} &\tilde{u}_B = 0, \quad c_B = c_i + (\gamma-1)\,\tilde{u}_i/2, \\ &\rho_B = (c_B^2 \rho_i/\gamma\, p_i)^{1/(\gamma-1)}\rho_i, \quad p_B = \rho_B\, c_B^2/\gamma \end{aligned} \tag{70}$$

   and we have $f_{OS}(q_i, q_B) = (0,\, p_B,\, 0,\, 0)^{\mathrm{T}}$.

In the solution of a *cascade flow* (see Section 4), the domain Ω represents one period of the cascade. Then $\partial\Omega$ consists of the inlet, outlet, impermeable profile and artificial boundary formed by two arcs Γ^+ and Γ^- on which periodicity conditions are imposed

In the numerical computations we start from the *initial conditions*

$$w_i^0 = \frac{1}{|D_i|}\int_{D_i} w(x,0)\,\mathrm{d}x, \quad i \in J. \tag{71}$$

2.2.3. STABILITY. Since scheme (39) is explicit, it is necessary to apply a suitable stability condition. Using linearization and analogy with a scalar problem, we derived the stability condition ([13, Par. 7.3.116])

$$\frac{\tau_k}{|D_i|} \max_{\substack{j\in S(i)\\ \alpha=1,\dots,\beta_{ij}}} \left\{ |\partial D_i| \max_{m=1,\dots,4} \{|\lambda_m(w_i^k, \mathbf{n}_{ij}^\alpha)|\} \right\} \le \mathrm{CFL} \approx 0.85, \; i \in J, \tag{72}$$

applied to all above schemes.

Often our goal is to obtain a steady state solution of system (17) by *time marching process, i.e., time stabilization* for $t_k \to \infty$, using the described schemes and applying the stability condition (72).

2.3. Improvement of the accuracy via adaptive mesh refinement

In practical computations we try to achieve sufficiently accurate solutions with precise resolution of shock waves. There are two possibilities how to increase the accuracy of the method:

1. the use of MUSCL, ENO or WENO higher order approach;
2. the use of an adaptive refinement of the mesh.

Here we shall pay attention to several adaptive mesh refinement algorithms using the information of previously computed approximate solution. The most important tool in adaptive methods is the choice of a suitable *error indicator* and, for flows with shocks, also a *shock indicator*. We explain several mesh adaptation strategies developed for inviscid steady flow.

2.3.1. SHOCK INDICATOR. Here we are concerned with an adaptive shock capturing algorithm employing a divided differences approach and indicating admissible shock waves only. We have tested several shock indicators on *triangular meshes* (see [22]). The best results were obtained with the aid of the following shock indicator using the changes of the density::

$$g_{sh}(i) = \max_{j \in s(i)} [-(\rho_i - \rho_j)\, \mathbf{v}_i \cdot \mathbf{n}_{ij}]^+ / h_{ij}, \quad i \in J, \tag{73}$$

where h_{ij} is the distance of the barycentres of D_i and $D_j \in \mathcal{D}_h$ and $\mathbf{v}_i$ is the velocity vector on D_i. (We omit the superscript k.)

The proposed shock indicator can also be applied to viscous flow. Numerical experiments (see Section 4) show that the shock indicator safely indicates areas including shock waves. Unfortunately, it is not good for indicating contact discontinuities, boundary layers and wakes.

2.3.2. SUPERCONVERGENCE ERROR INDICATOR. The idea of the superconvergence indication goes from the indicator using superconvergence properties developed, e.g., in [48] for elliptic problems. We proceed in the following way:

1. Let us denote by $\mathbf{w}_h^k = \{w_i^k\}_{i \in \mathcal{D}_h}$ the piecewise constant finite volume approximate solution. From this function we compute its piecewise linear reconstruction $\mathbf{w}_h^{*k}$ which is determined uniquely by its values calculated at each vertex of the computational mesh. These values are weighted averages of the values w_i^k given on triangles which have the corresponding vertex in common. As weights the areas of these triangles are used.
2. We construct a function $\mathbf{g} = (g^1, g^2, g^3, g^4) := \mathbf{w}_h^{*k} - \mathbf{w}_h^k$, which is piecewise linear and discontinuous in the domain Ω. The array of the values $\| \mathbf{g} \|_{L^2(T_i)}$ is called *superconvergence error indicator.*

Numerical results obtained with the aid of this indicator show that the adaptive refinement based on it has a remarkable influence on the accuracy of the numerical solution and on the length of the computational time. The indicator is able to focus discontinuities quite sharply. Moreover, in contrast to the shock indicator, the superconvergence indicator indicates also areas, where the error is high, as in the vicinity of singular boundary points (corners, points with very large curvature etc.).

Another possibility is to use the *residual H^{-1}–norm error indicator*, which was developed for the solution of the Euler equations as well as the Navier–Stokes equations. Therefore, we shall mention it in Section 3 in the framework of the numerical simulation of viscous flow. In Section 3, also the anisotropic mesh refinement will be briefly explained.

The above indicators (let us denote them by $g(i), i \in J$, are applied in the following way:

1. We calculate the *relative indicator*
$$\eta(i) = g(i) / \max_{j \in J} g(j), \quad i \in J. \tag{74}$$
2. If for some $D_i \in \mathcal{D}_h$ we have $\eta(i) \geq \delta =$ a given tolerance, $\delta \in (0,1)$, then the "mother" triangle D_i is divided into four equal "daughter" subtriangles (red refinement).
3. The resulting partition of Ω_h is modified so that the conforming triangulation of Ω_h is obtained (green refinement).
4. Initial approximation on the new mesh is defined by taking the same values on the daughter triangles as on the mother triangle.

The above process is repeated in several steps until the desired accuracy of the solution and sharpness of discontinuities is obtained.

3. Numerical Solution of Viscous Compressible Flow

In many cases viscosity effects resulting, e. g., in boundary layer separation and backward flows cannot be neglected and it is necessary to work with the complete viscous compressible system consisting of the continuity equation, the Navier–Stokes equations and the energy equation. We shall present here the numerical method for the solution of viscous gas flow based on the combination of the finite volume method with the finite element method. We confine our considerations again to 2D flow.

The complete system describing viscous compressible flow in a plane bounded domain Ω and time interval $(0,T)$ is written in the form
$$\frac{\partial w}{\partial t} + \sum_{s=1}^{2} \frac{\partial f_s(w)}{\partial x_s} = \sum_{s=1}^{2} \frac{\partial R_s(w, \nabla w)}{\partial x_s} \quad \text{in } Q_T = \Omega \times (0,T), \tag{75}$$
where
$$\begin{aligned} & w = (\rho, \rho v_1, \rho v_2, e)^{\mathrm{T}}, \quad w = w(x,t), \quad x \in \Omega, \ t \in (0,T), \\ & f_s(w) = (f_s^1, \dots, f_s^4)^{\mathrm{T}} = (\rho v_s, \rho v_s v_1 + \delta_{s1} p, \rho v_s v_2 + \delta_{s2} p, (e+p)\, v_s)^{\mathrm{T}}, \\ & R_s(w, \nabla w) = (R_s^1, \dots, R_s^4)^{\mathrm{T}} = (0, \tau_{s1}, \tau_{s2}, \tau_{s1} v_1 + \tau_{s2} v_2 + k \partial\theta/\partial x_s)^{\mathrm{T}}, \\ & \tau_{ij} = \lambda \operatorname{div} \mathbf{v}\, \delta_{ij} + \mu(\partial v_i/\partial x_j + \partial v_j / \partial x_i), \\ & p = (\gamma - 1)\,(e - \rho |\mathbf{v}|^2/2), \quad e = \rho(c_v \theta + |\mathbf{v}|^2/2). \end{aligned} \tag{76}$$
We use the same notation as in Section 1 and assume that $\lambda = -2\mu/3$.

The above system is equipped with *initial conditions* prescribing the state w on Ω at time $t = 0$ and *boundary conditions*: On inlet we prescribe ρ, v_1, v_2 and set $\partial\theta/\partial n = 0$, on fixed walls we assume that $v_1 = v_2 = 0$, $\partial\theta/\partial n = 0$ and on outlet we set $-pn_j + \sum_{i=1}^{2} \tau_{ij} n_i = 0$, $j = 1, 2$, and $\partial\theta/\partial n = 0$. Here $\partial/\partial n$ is the derivative in the direction of the unit outer normal $\mathbf{n} = (n_1, n_2)$ to the boundary $\partial\Omega$.

3.1. Combined finite volume - finite element method for viscous flow

One possible way of the discretization of the above problem can be based on *inviscid-viscous operator splitting*. We consider viscous terms R_s as a perturbation of the inviscid system of Euler equations and split system (75) (and the boundary conditions) into the inviscid system and the purely viscous system

$$\frac{\partial w}{\partial t} + \sum_{s=1}^{2} \frac{\partial f_s(w)}{\partial x_s} = 0, \tag{77}$$

$$\frac{\partial w}{\partial t} = \sum_{s=1}^{2} \frac{\partial R_s(w, \nabla w)}{\partial x_s} \tag{78}$$

and discretize them separately.

The time discretization of (77) and (78) is carried out on a partition $0 = t_0 < t_1 < t_2 < \dots$ of the time interval $(0, T)$. We set $\tau_k = t_{k+1} - t_k$.

The inviscid system (77) is discretized by the cell-centred *finite volume (FV) method* on a mesh $\mathcal{D}_h = \{D_i\}_{i \in J}$ described in Section 2. Now let us deal with the discretization of the purely viscous system.

The standard approach discretizes the viscous terms also by the finite volume method as it is carried out, e.g., in [50], [31]. However, the structure of the viscous system (78) offers us the application of the *finite element (FE) method*.

Similarly as above, let Ω_h be a polygonal approximation of the domain Ω. By $\mathcal{T}_h$ we denote a triangulation of the domain Ω_h with standard properties (see [5]) and by P_i, $i \in J$, the vertices of all triangles $T \in \mathcal{T}_h$. We use conforming piecewise linear finite elements. This means that the components of the state vector w are approximated by functions from the finite dimensional space

$$X_h = \left\{\varphi_h \in C(\overline{\Omega}_h);\ \varphi_h|_T \text{ is linear for each } T \in \mathcal{T}_h\right\}. \tag{79}$$

Further, we set $\mathbf{X}_h = [X_h]^4$ and

$$\begin{aligned} \mathbf{V}_h &= \{\varphi_h = (\varphi_1, \varphi_2, \varphi_3, \varphi_4) \in \mathbf{X}_h,\ \varphi_i = 0 \\ &\quad \text{on the part of } \partial\Omega_h \text{ approximating the part} \\ &\quad \text{of } \partial\Omega \text{ where } w_i \text{ satisfies the Dirichlet condition}\}, \end{aligned} \tag{80}$$

$$\begin{aligned} \mathbf{W}_h &= \{w_h \in \mathbf{X}_h;\ \text{its components satisfy} \\ &\quad \text{the prescribed Dirichlet boundary conditions}\}. \end{aligned} \tag{81}$$

Multiplying (78) considered on time level t_k by any $\varphi_h \in \mathbf{V}_h$, integrating over Ω_h, using Green's theorem, taking into account the boundary conditions and approximating the time derivative by a forward finite difference, we obtain the following *explicit scheme* for the calculation of an approximate solution w_h^{k+1} on the $(k+1)$-st time level:

$$\text{a)} \quad w_h^{k+1} \in \mathbf{W}_h, \tag{82}$$

$$\text{b)} \quad \int_{\Omega_h} w_h^{k+1} \varphi_h \,\mathrm{d}x = \int_{\Omega_h} w_h^k \varphi_h \,\mathrm{d}x - $$
$$-\tau_k \int_{\Omega_h} \sum_{s=1}^{2} R_s(w_h^k, \nabla w_h^k) \frac{\partial \varphi_h}{\partial x_s} \,\mathrm{d}x \quad \forall \varphi_h \in \mathbf{V}_h .$$

Replacing $R_s(w_h^k, \nabla w_h^k)$ by $R_s(w_h^{k+1}, \nabla w_h^{k+1})$, we obtain an *implicit scheme*. However, its application is much more complicated from the algorithmization point of view.

The integrals are approximated by a numerical quadrature, called *mass lumping*, using the vertices of triangles as integration points:

$$\int_T F \,\mathrm{d}x \approx \frac{1}{3}|T| \sum_{i=1}^{3} F(P_T^i) \tag{83}$$

for $F \in C(T)$ and a triangle $T \in \mathcal{T}_h$ with vertices P_T^i, $i = 1, 2, 3$. Numerical integration in (82, b) yields the identity

$$(w_h^{k+1}, \varphi_h)_h = (w_h^k, \varphi_h)_h - \tau_k \, a_h(w_h^k, \varphi_h) \quad \forall \varphi_h \in \mathbf{V}_h, \tag{84}$$

where

$$(w_h, \varphi_h)_h = \frac{1}{3} \sum_{T \in \mathcal{T}_h} |T| \sum_{i=1}^{3} w_h(P_T^i)\, \varphi_h(P_T^i), \qquad w_h, \varphi_h \in \mathbf{X}_h \tag{85}$$

and

$$a_h(w_h, \varphi_h) = a_h^1(w_h, \varphi_h) + \ldots + a_h^4(w_h, \varphi_h),$$

$$a_h^1 \equiv 0,$$

$$a_h^2(w_h, \varphi_h) = \sum_{T \in \mathcal{T}_h} |T| \left\{ 2\mu \left.\frac{\partial v_{h,1}}{\partial x_1}\right|_T \left.\frac{\partial \varphi_{h,2}}{\partial x_1}\right|_T + \right.$$
$$\left. \lambda(\operatorname{div} \mathbf{v}_h)|_T \left.\frac{\partial \varphi_{h,2}}{\partial x_1}\right|_T + \mu \left(\left.\frac{\partial v_{h,2}}{\partial x_1}\right|_T + \left.\frac{\partial v_{h,1}}{\partial x_2}\right|_T \right) \left.\frac{\partial \varphi_{h,2}}{\partial x_2}\right|_T \right\},$$

$$a_h^3(w_h,\varphi_h) = \sum_{T\in\mathcal{T}_h} |T| \left\{ \mu \left(\frac{\partial v_{h,2}}{\partial x_1}\bigg|_T + \frac{\partial v_{h,1}}{\partial x_2}\bigg|_T \right) \frac{\partial \varphi_{h,3}}{\partial x_1}\bigg|_T + \right.$$

$$\left. 2\mu \frac{\partial v_{h,2}}{\partial x_2}\bigg|_T \frac{\partial \varphi_{h,3}}{\partial x_2}\bigg|_T + \lambda(\operatorname{div} \mathbf{v}_h)|_T \frac{\partial \varphi_{h,3}}{\partial x_2}\bigg|_T \right\},$$

$$a_h^4(w_h,\varphi_h) = \sum_{T\in\mathcal{T}_h} \left\{ \frac{1}{3}|T| \left(\tau_{h,11}|_T \sum_{i=1}^{3} v_{h,1}(P_T^i) + \right. \right. \tag{86}$$

$$\left. \tau_{h,12}|_T \sum_{i=1}^{3} v_{h,2}(P_T^i) \right) \frac{\partial \varphi_{h,4}}{\partial x_1}\bigg|_T +$$

$$+ \frac{1}{3}|T| \left(\tau_{h,21}|_T \sum_{i=1}^{3} v_{h,1}(P_T^i) + \tau_{h,22}|_T \sum_{i=1}^{3} v_{h,2}(P_T^i) \right) \frac{\partial \varphi_{h,4}}{\partial x_2}\bigg|_T +$$

$$\left. k|T| \sum_{j=1}^{2} \frac{\partial \theta_h}{\partial x_j}\bigg|_T \frac{\partial \varphi_{h,4}}{\partial x_j}\bigg|_T \right\},$$

$$\tau_{h,rs}|_T = \frac{1}{2} \left(\frac{\partial v_{h,r}}{\partial x_s} + \frac{\partial v_{h,s}}{\partial x_r} \right)\bigg|_T = \text{const.}$$

Here $v_{h,s}$ and θ_h represent functions from the space X_h approximating the velocity components and temperature.

Another possibility of the finite element discretization of the purely viscous system is to use the well-known *Crouzeix-Raviart nonconforming piecewise linear finite elements* (commonly used for the incompressible Navier–Stokes equations, see, [6]). In this case the spaces $X_h, \mathbf{X}_h, \mathbf{V}_h, \mathbf{W}_h$ are defined as follows:

$$\begin{aligned} X_h &= \Big\{ v_h \in L^2(\Omega_h);\, v_h \text{ is linear on each } T \in \mathcal{T}_h \\ &\qquad \text{and continuous at all midpoints of sides of } T \in \mathcal{T}_h \Big\}, \\ \mathbf{X}_h &= [X_h]^4, \\ \mathbf{V}_h &= \Big\{ \varphi_h = (\varphi_1,\dots,\varphi_4) \in \mathbf{X}_h;\, \varphi_i(Q_j) = 0 \text{ for midpoints } Q_j \\ &\qquad \text{of sides of } T \in \mathcal{T}_h,\ Q_j \text{ lying on } \partial\Omega_h, \text{ where } w_i \\ &\qquad \text{satisfies the Dirichlet condition} \Big\}, \\ \mathbf{W}_h &= \Big\{ w_h \in \mathbf{X}_h;\ \text{its components satisfy the prescribed} \\ &\qquad \text{Dirichlet boundary condition at midpoints of sides on } \partial\Omega_h \Big\}. \end{aligned} \tag{87}$$

The numerical integration uses midpoints of sides of triangles $T \in \mathcal{T}_h$ as integration points.

Finally, we come to the *discretization of the complete viscous system.* We combine the finite volume discretization of the inviscid system (77) on a finite volume mesh $\mathcal{D}_h$ with the finite element discretization of the purely viscous system (78) on a triangulation $\mathcal{T}_h$. Of course, the grids $\mathcal{D}_h$ and $\mathcal{T}_h$ must be *compatible* so that there exists a mapping L_h (the so-called lumping operator) that associates piecewise linear finite elements with piecewise constant finite volume functions: If we denote by $\tilde{w}_h \in \mathbf{X}_h$ a finite element function, then $w_h = L_h\tilde{w}_h$ is a finite volume function. We distinguish the following cases.

a) Combination of conforming piecewise linear finite elements on a FE mesh $\mathcal{T}_h$ with dual finite volumes $D_i \in \mathcal{D}_h$ associated with vertices P_i of triangles $T \in \mathcal{T}_h$, see Section 2. In this case the lumping operator L_h is defined by $L_h\tilde{w}_h|_{D_i} = \tilde{w}_h(P_i)$ for $\tilde{w}_h \in X_h$.

b) Combination of nonconforming piecewise linear Crouzeix–Raviart finite elements with barycentric finite volumes $D_i \in \mathcal{D}_h$ associated with midpoints Q_i of sides of triangles $T \in \mathcal{T}_h$, see again Section 2. Now we set $L_h\tilde{w}_h|_{D_i} = \tilde{w}_h(Q_i)$ for $\tilde{w}_h \in X_h$.

c) Combination of conforming piecewise linear triangular finite elements with triangular finite volumes. In this case we start from a basic triangular FV mesh $\mathcal{D}_h$ and construct an adjoint triangular FE mesh $\mathcal{T}_h$. The vertices of $T \in \mathcal{T}_h$ are barycentres of $D_i \in \mathcal{D}_h$ and vertices of $D_i \in \mathcal{D}_h$ lying on $\partial\Omega_h$. See Figure 2. In this case $L_h\tilde{w}_h|_{D_i} = \tilde{w}_h(P_i)$, $\tilde{w}_h \in X_h$. Here P_i is the barycenter of D_i and simultaneously a vertex of some $T \in \mathcal{T}_h$.

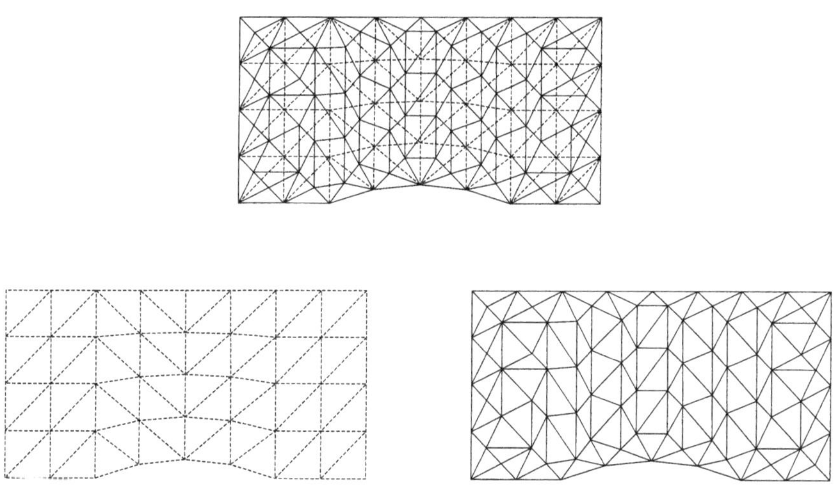

FIGURE 2. Triangular and adjoint mesh, separately and together.

The algorithm for the combined finite volume - finite element schemes for the complete Navier–Stokes system can be realized in two ways:

1. *Inviscid-viscous operator splitting.* We start from the approximation w_i^0, $D_i \in \mathcal{D}_h$, of the initial condition given by (71). Let us assume that we have already obtained the values w_i^k, $D_i \in \mathcal{D}_h$, of the approximate solution on the mesh $\mathcal{D}_h$ at time t_k. The transition to the next time level is carried out in two *fractional steps*:

(i) Compute

$$w_i^{k+1/2} = w_j^k - \frac{\tau_k}{|D_i|} \sum_{j\in S(i)} \sum_{\alpha=1}^{\beta_{ij}} H(w_i^k, w_j^k, \mathbf{n}_{ij}^\alpha)\, \ell_{ij}^\alpha,$$

$D_i \in \mathcal{D}_h$, & inviscid boundary conditions;

(88)

(ii) α) Define $\tilde{w}_h^{k+1/2} \in \mathbf{X}_h$ such that $\tilde{w}_h^{k+1/2}(P_i) = w_i^{k+1/2}$, $i \in J$.

β) Compute $\tilde{w}_h^{k+1} \in \mathbf{W}_h$ such that

$$(\tilde{w}_h^{k+1}, \varphi_h)_h = (\tilde{w}_h^{k+1/2}, \varphi_h)_h - \tau_k\, a_h(\tilde{w}_h^{k+1/2}, \varphi_h)\ \forall \varphi_h \in \mathbf{V}_h,$$

γ) Set $w_i^{k+1} = \tilde{w}_h^{k+1}(P_i)$, $i \in J$, $k := k+1$, go to (i).

(88, i) is obtained from scheme (39). In (88, ii), P_i has the following meaning: If $D_i \in \mathcal{D}_h$ are dual finite volumes, then P_i are vertices of triangles $T \in \mathcal{T}_h$. In the case of the combination of barycentric finite volumes with nonconforming finite elements, P_i denote midpoints of sides of triangles $T \in \mathcal{T}_h$. (Hence, $\tilde{w}_h^{k+1/2} = L_h^{-1} w_h^{k+1/2}$, with $w_h^{k+1/2}$ denoting the piecewise constant finite volume function with values $w_i^{k+1/2}$ on D_i.) It is also possible to combine triangular finite volumes with conforming piecewise linear triangular finite elements. Then P_i are vertices of $T \in \mathcal{T}_h$. The values $\tilde{w}_h^{k+1/2}(P_i)$ are defined for $P_i \in \partial\Omega_h$ from the Dirichlet boundary conditions and by suitable extrapolation.

2. *Direct discretization of the viscous problem.* Let us define the form

$$b_h(w_h, \varphi_h) = \sum_{D_i\in\mathcal{D}_h} \varphi_h(P_i) \sum_{j\in S(i)} \sum_{\alpha=1}^{\beta_{ij}} H(w_h(P_i), w_h(P_j), \mathbf{n}_{ij}^\alpha)\, \ell_{ij}^\alpha \qquad (89)$$

$$w_h \in \mathbf{X}_h,\ \varphi \in \mathbf{V}_h .$$

(The finite volume D_i is assocoated with the node P_i.) Then we can write down the direct discretization of the complete system (75) in the fully explicit form:

a) $w_h^{k+1} \in \mathbf{W}_h$, (90)

b) $(w_k^{k+1}, \varphi_h)_h =$

$$= (w_h^k, \varphi_h)_h - \tau_k[b_h(w_h^k, \varphi_h) + a_h(w_h^k, \varphi_h)],\ \varphi_h \in \mathbf{V}_h .$$

A combination of the finite volume and finite element methods was also used in [3].

The above schemes can be applied only under some stability conditions. In virtue of the explicit discretization of inviscid terms, we apply the stability condition (72). In the case of explicit discretization of the viscous terms (scheme (88, ii) or (90)) we consider, moreover, the additional stability condition in the form

$$\frac{3}{4}\frac{h(T)}{\sigma(T)}\frac{\tau_k}{|T|}\max(\mu, k) \leq \mathrm{CFL}, \quad T \in \mathcal{T}_h, \tag{91}$$

where $h(T)$ is the length of the maximal side in T and $\sigma(T)$ is the radius of the largest circle inscribed into T. This condition was derived as an $L^\infty-$ stability condition for the explicit scheme used for the solution of the heat conduction equation and is applied heuristically with success to our complicated flow problem (see [32]).

Remark 3.1. *The attempts to prove the convergence of numerical methods for the solution of the complete system describing viscous flow have not yet been successful. Therefore, in order to support our computational results obtained by the combined finite volume – finite element method, the convergence analysis was carried out for a model scalar nonlinear convention – diffusion conservation law equation*

$$\frac{\partial u}{\partial t} + \sum_{s=1}^{2} \frac{\partial f_s(u)}{\partial x_s} = \nu \, \Delta \, u + g \quad \textit{in } Q_T = \Omega \times (0, T), \; T < \infty, \tag{92}$$

equipped with initial and boundary conditions

$$\begin{aligned} u(x, 0) &= u^0(x), \quad x \in \Omega, \\ u|_{\partial\Omega \times (0,T)} &= 0. \end{aligned} \tag{93}$$

In [18] and [20], the convergence of the combined finite volume – finite element method was proved for piecewise linear conforming finite elements and dual finite volumes, for semiimplicit (explicit convection, implicit diffusion) as well as explicit scheme. In [19], error estimates have been established for a semiimplicit scheme. In [2] and [11], semiimplicit combination of barycentric finite volumes with nonconforming finite elements is analyzed.

3.2. Mesh refinement

Numerical experiments proved that it is possible to use the shock indicator (73) for the detection of shock waves in inviscid as well as viscous compressible flow. Here we shall explain two more strategies for the identification of areas for the mesh refinement.

3.2.1. RESIDUAL H^{-1}-NORM ERROR INDICATOR. Let us explain this indicator for the case of stationary viscous compressible flow. The complete system is formed by an inviscid part and a viscous part. The weak form of the inviscid part can be written componentwise as

$$b^\ell(w_h, \varphi) = \int_{\Omega_h} \sum_{s=1}^{2} f_s^\ell(w_h) \frac{\partial \varphi}{\partial x_s} \, \mathrm{d}x, \quad \varphi \in H_0^1(\Omega_h), \quad \ell = 1, \ldots, 4. \tag{94}$$

Here w_h is the piecewise constant finite volume approximation over $\mathcal{D}_h$ of a steady solution. By $H_0^1(\Omega_h)$ we denote the Sobolev space of functions square integrable over Ω_h together with their first order distribution derivatives and having zero traces on $\partial\Omega_h$. The viscous part reads

$$\tilde{b}^\ell(\tilde{w}_h,\varphi) = -\int_{\Omega_h} \sum_{s=1}^{2} R_s^\ell(\tilde{w}_h,\nabla\tilde{w}_h)\,\frac{\partial\varphi}{\partial x_s}\,\mathrm{d}x, \quad \varphi\in H_0^1(\Omega_h), \quad \ell=1,\dots,4, \tag{95}$$

where $\tilde{w}_h$ is the piecewise linear finite element approximation of the solution over the mesh $\mathcal{T}_h$. The sum $b^\ell(w_h,\varphi)+\tilde{b}^\ell(\tilde{w}_h,\varphi)$ represents the residual in the ℓ-th equation tested by $\varphi\in H_0^1(\Omega_h)$. Hence, the mapping "$\varphi\in H_0^1(\Omega)\to b^\ell(w_h,\varphi)+\tilde{b}(\tilde{w}_h,\varphi)\in \mathbb{R}$" defines a continuous linear functional $r_h^\ell\in H^{-1}(\Omega_h)$ = dual of $H_0^1(\Omega)$. Its norm is defined by

$$\|r_h^\ell\|_{H^{-1}(\Omega_h)} = \sup_{0\neq\varphi\in H_0^1(\Omega_h)} \frac{|b^\ell(w_h,\varphi)+\tilde{b}^\ell(\tilde{w}_h,\varphi)|}{|\varphi|_{H_0^1(\Omega_h)}} \tag{96}$$

where

$$|\varphi|_{H_0^1(\Omega_h)} = \left(\int_{\Omega_h} |\nabla\varphi|^2\,\mathrm{d}x\right)^{1/2}. \tag{97}$$

The approximation and localization of (96) yields the local error indicator

$$\begin{aligned} g_{er}^\ell(i) &= \max_{j\in N(i)} \frac{|b^\ell(w_h,\varphi_{ij})+\tilde{b}^\ell(\tilde{w}_h,\varphi_{ij})|}{|\varphi_{ij}|_{H_0^1(\Omega_h)}} \\ g_{er}(i) &= \left\{\sum_{\ell=1}^{4} g_{er}^\ell(i)\right\}, \quad T_i\in\mathcal{T}_h. \end{aligned} \tag{98}$$

In (98) we use the following notation: $N(i)$ represents the index set of neighbouring elements $T_j\in\mathcal{T}_h$ to $T_i\in\mathcal{T}_h$. This means that T_i and T_j have a common side Γ_{ij}. Further, φ_{ij} is the test function defined in the following way: We construct the refinement $\mathcal{T}_{h/2}$ of $\mathcal{T}_h$ dividing each $T_i\in\mathcal{T}_h$ into four equal triangles. The function φ_{ij} is continuous in $\overline{\Omega}_h$, linear on each triangle $K\in\mathcal{T}_{h/2}$, $\varphi_{ij}=1$ at the midpoint of Γ_{ij} and vanishes at all other vertices of $K\in\mathcal{T}_{h/2}$.

The residual H^{-1}-norm error indicator is applied as described in Section 2. If the flow is inviscid, then the forms $\tilde{b}^\ell, \ell=1,\dots,4$, vanish and we obtain a residual H^{-1}-norm error indicator applicable to the Euler equations.

3.2.2. ANISOTROPIC MESH REFINEMENT. This approach allows grid alignment along shock waves and boundary layers. Let ξ be a suitable quantity (e. g. the density). Then we proceed in several steps.

1. For each node P_i we construct an approximate Hessian

$$\widetilde{H}_i = \begin{pmatrix} \partial^2\xi/\partial x_1^2, & \partial^2\xi/\partial x_1\partial x_2 \\ \partial^2\xi/\partial x_1\partial x_2, & \partial^2\xi/\partial x_2^2 \end{pmatrix} = R_i \begin{pmatrix} \lambda_{i1} & 0 \\ 0 & \lambda_{i2} \end{pmatrix} R_i^{-1}, \tag{99}$$

where R_i is a nonsingular matrix.

2. We set
$$H_i = R_i \begin{pmatrix} |\lambda_{i1}| & 0 \\ 0 & |\lambda_{i2}| \end{pmatrix} R_i^{-1}. \tag{100}$$
3. We set
$$M_i = c\,(I + \alpha(\|H_i\|)\,H_i)\,, \tag{101}$$
$$\|H\| = \max_{\alpha,\beta=1,2} |h_{\alpha\beta}|, \quad c = \frac{3\sqrt{3}}{4}\,\frac{\text{nelem}}{|\Omega|},$$
$$\alpha\,(\|H\|) = \frac{\varepsilon_1}{\varepsilon_2 + \|H\|}, \quad \varepsilon_1, \varepsilon_2 > 0,$$
($h_{\alpha\beta}$ = elements of the matrix H, nelem = number of elements of $\mathcal{T}_h$, $\varepsilon_1, \varepsilon_2$ = suitable parameters).
4. For each side S_ℓ with nodes P_{i_1}, P_{i_2} we denote
$$\widetilde{M}_\ell = \frac{1}{2}\,(M_{i_1} + M_{i_2})\,, \quad \mathbf{v}_\ell = P_{i_2} - P_{i_1}, \tag{102}$$
and define the norm
$$\|\,\mathbf{v}_\ell\,\|_{\widetilde{M}_\ell} = \left(\mathbf{v}_\ell^{\mathrm{T}}\,\widetilde{M}_\ell\,\mathbf{v}_\ell\right)^{1/2}. \tag{103}$$
5. Finally, for each element $K \in \mathcal{T}_h$ with side vectors $\mathbf{v}_j^K$ and corresponding matrices $\widetilde{M}_j^K$, $j = 1, 2, 3$, we set
$$q_K = \sum_{j=1}^{3} \left(\|\,\mathbf{v}_j^K\,\|_{\widetilde{M}_j^K} - \sqrt{3}\right)^2 \tag{104}$$
and define the quality parameter of $\mathcal{T}_h$:
$$Q_{\mathcal{T}_h} = \frac{1}{\text{nelem}} \sum_{K \in \mathcal{T}_h} q_K. \tag{105}$$

The triangulation $\mathcal{T}_h$ is called *optimal*, if $Q_{\mathcal{T}_h} = 0$. If a triangulation $\mathcal{T}_h^{\text{old}}$ is given, then our goal is to find $\mathcal{T}_h^{\text{new}}$ such that
$$Q_{\mathcal{T}_h^{\text{new}}} < Q_{\mathcal{T}_h^{\text{old}}}. \tag{106}$$
This is realized with the aid of a sequence of operations: adding a node, removing an edge, swapping a diagonal and moving a node. This procedure is applied iteratively. For details see [9].

4. Applications

In this section we present some computational results demonstrating the applicability of the methods explained above. In all cases the Osher–Solomon numerical flux is used. The computational results were obtained by A. Kliková [32] and V. Dolejší [8].

4.1. Flow through the GAMM channel

As a simple test problem we use the flow through the GAMM channel, i.e., past a 10 % circular bump. The inlet Mach number is 0.671. First we consider inviscid steady-state flow obtained with the aid of scheme (39) via time stabilization for $t_k \to \infty$. In Figure 3, the computational mesh refined four times on the basis of the shock indicator and Mach number isolines are shown. In Figure 4, there are similar results obtained with the aid of the superconvergence error indicator. We see that the shock indicator leads to the mesh refinement in the vicinity of the shock wave, whereas the superconvergence error indicator also indicates the areas close to corners, i.e., boundary singular points.

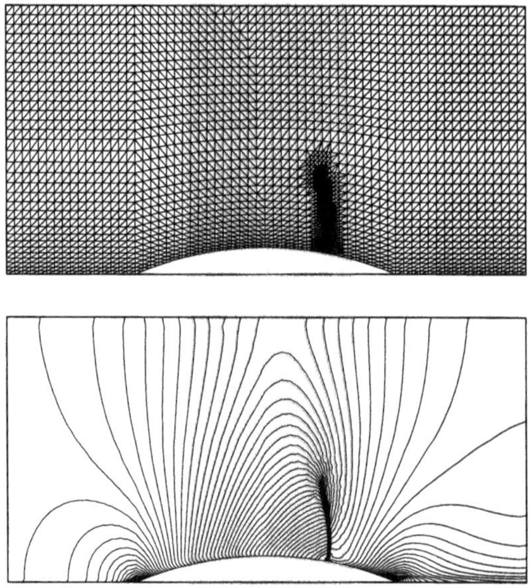

FIGURE 3. Inviscid flow in GAMM chanel: Computational mesh refined on the basis of the shock indicator. Mach number isolines computed on the refined mesh.

Finally, Figure 5 shows the computational results for viscous flow with Reynolds number $= 10^5$ and Prandtl number $= 0.72$. The mesh was refined on the basis of the residual H^{-1}-norm error indicator. At the inlet, $M = 0.671$, the density and velocity component v_1 are assumed constant and $v_2 = 0$. At the outlet the normal component of the stress tensor vanishes. On the walls the no-slip condition is used. The temperature satisfies the condition $\partial\theta/\partial n = 0$ on the whole boundary.

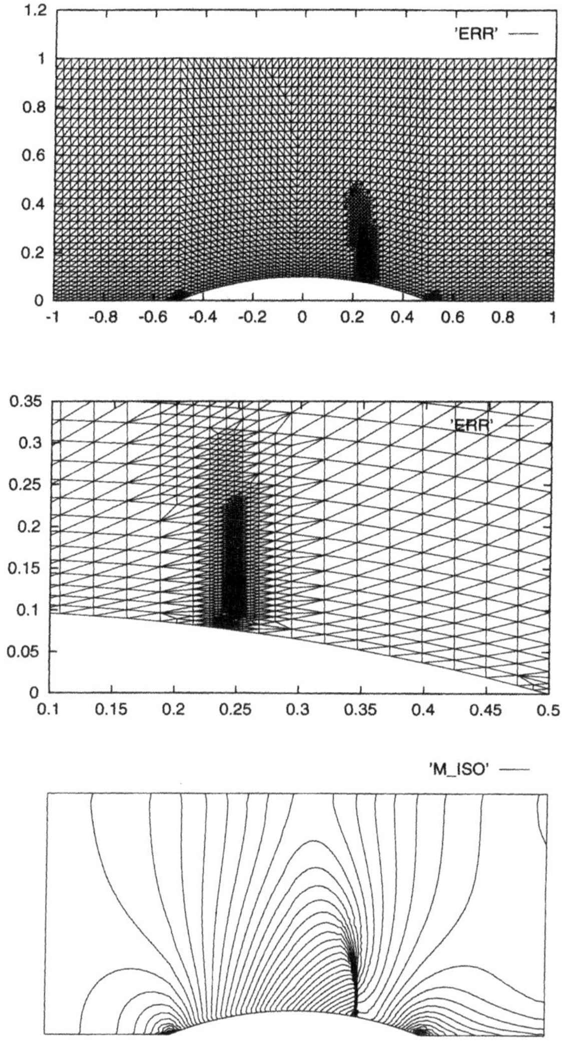

FIGURE 4. Inviscid flow in GAMM chanel: Computational mesh refined on the basis of the supercovergence indicator. Detail of the mesh in the vicinity of a shock wave. Mach number isolines computed on the refined mesh.

4.2. Flow past a cascade of profiles

Now we will present the computational results obtained for a technically relevant problem of flow past a cascade of profiles. (This model is used for the simulation of flow in blade rows of turbines or compressors.)

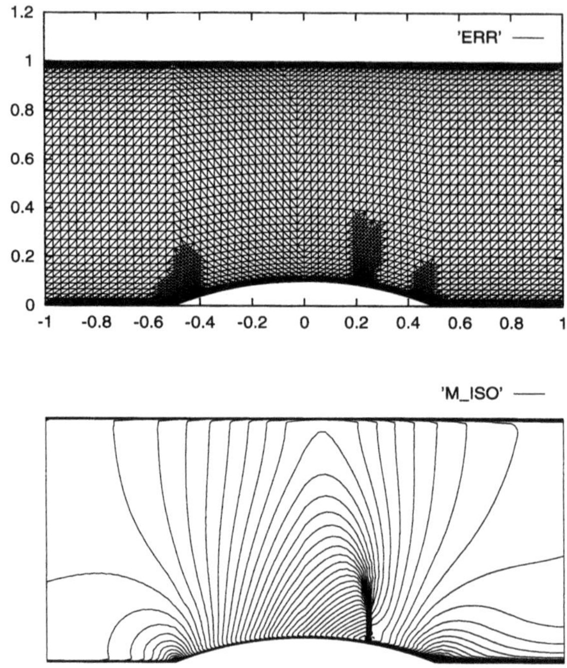

FIGURE 5. Viscous flow in GAMM chanel: Computational mesh refined on the basis of the H^{-1}-norm error indicator. Mach number isolines computed on the refined mesh.

In this case the region occupied by the fluid is represented by a plane infinitely connected domain $\tilde{\Omega}$ bounded in the direction x_1 and unbounded but periodic in the direction x_2. Assuming also the periodicity of the flow field, the computational domain Ω can be chosen in the form of one period of the domain $\tilde{\Omega}$. The boundary $\partial\Omega$ is formed by Γ_I, Γ_O, Γ_W, Γ^- and Γ^+, representing the inlet, outlet, impermeable profile and artificial cuts, respectively. The arcs Γ^- and Γ^+ satisfy the condition

$$\Gamma^+ = \left\{(x_1, x_2 + \vartheta); (x_1, x_2) \in \Gamma^-\right\},$$

where $\vartheta > 0$ is the width of one period of the cascade in the direction x_2. See Figure 7. This cascade flow was solved as inviscid ([22]) as well as viscous ([16], [11]) for the following physical data: angle of atack $= 19^\circ\, 18'$, inlet Mach number $= 0.32$, outlet Mach number $= 1.18$, $\gamma = 1.4$, Reynolds number $= 1.5 \cdot 10^6$, Prandtl number $= 0.72$. The computational results are compared with a wind tunnel experiment (by courtesy of the Institute of Thermomechanics of the Czech Academy of Sciences in Prague, see [44]). Figure 6 represents the wind tunnel interferogram showing density isolines. In Figure 7, a triangular mesh refined with the aid of superconvergence error indicator are plotted. Figure 8 shows computed density isolines plotted for inviscid flow. (See also [32]. In [22] the shock indicator was used.)

The solution of the viscous cascade flow presented here was computed with the use of the combination of the nonconforming piecewise linear triangular finite elements with the barycentric finite volumes. The computational mesh was obtained with the aid of the anisotropic mesh refinement. In Figures 9 and 10 the final triangulation and the corresponding barycentric mesh are plotted. Figure 11 shows the computed density isolines. In the computation, the unsplit scheme (90) was used and steady flow was achieved via the time stabilitation process. Comparing this figure with the interferogram of the density, we see that there is a good agreement of computational and experimental results.

Remark 4.1. *As we see from the above examples, the combined finite volume – finite element method applied together with adaptive mesh refinement can be quite efficient, accurate and robust. There is, of course, a number of challenging open problems connected with the numerical solution of compressible flow. Beside theoretical analysis of schemes for the solution of inviscid as well as viscous systems, it is the development of efficient, robust, high-order accurate methods for the solution of complex multidimensional problems. A very perspective technique is probably the high-order discontinuous Galerkin method that has been already successfully applied to viscous compressible flow, e.g., in* [4]. *For this method a number of questions and problems must be solved, as, e.g., stabilization in the case of problems with discontinuous solutions, the effects of the approximation of the boundary and numerical integration, algorithmization of the solution of the discrete problem, etc. and the theoretical analysis of the convergence and error estimates.*

Acknowledgements. A number of results presented here were obtained under the grant No. 201/99/0267 of the Czech Grant Agency and the grant No. MSM 113200007. The author gratefully acknowledges this support.

References

[1] S. K. Aliabadi and A. C. Galeao, *An adaptive Petrov-Galerkin formulation for the compressible Euler and Navier-Stokes equations*, Comput. Methods Appl. Mech. Engrg., **129** (1996) 157–176.

[2] Ph. Angot, V. Dolejší, M. Feistauer and J. Felcman, *Analysis of a combined barycentric finite volume–nonconforming finite element method for nonlinear convection-diffusion problems*, Appl. Math., **43** (1998), 263–310.

[3] P. Arminjon and A. Madrane, *A mixed finite volume/finite element method for 2-dimensional compressible Navier–Stokes equations on unstructured grids*, Hyperbolic problems: Theory, Numerics, Applications, Edited by M. Fey and R.Jeltsch, Birkhäuser (Basel) **1999**, 11–20.

FIGURE 6. Interferogram of density isolines for a cascade flow (courtesy of the Institute of Thermomechanics in Prague)

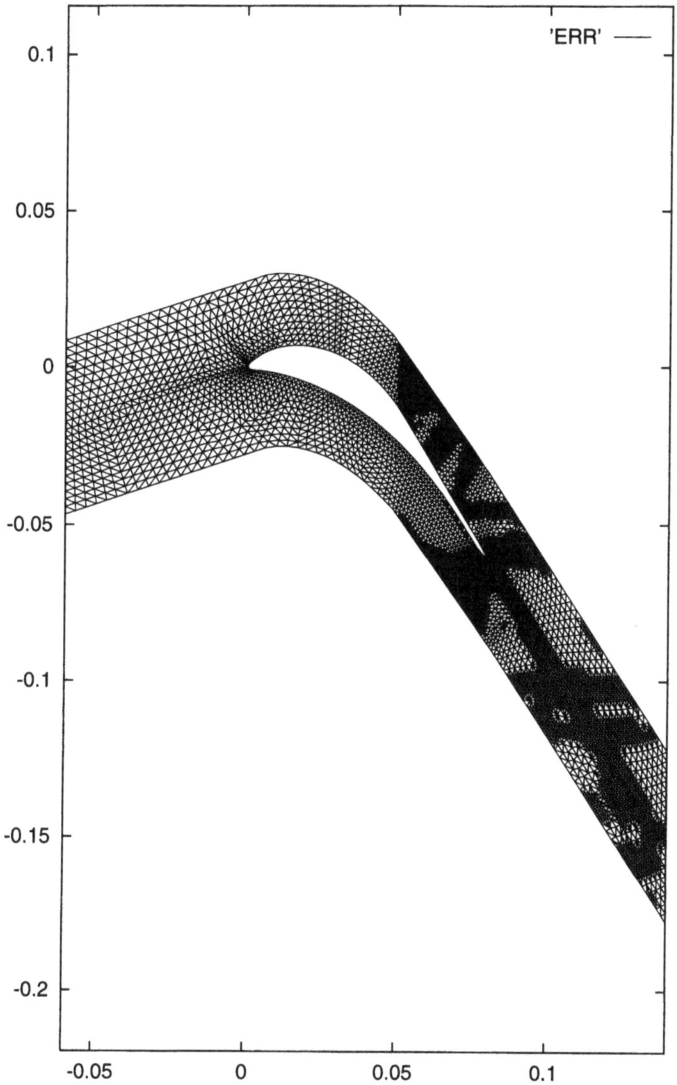

FIGURE 7. Mesh for inviscid cascade flow refined with the aid of the superconvergence error indicator

[4] F. Bassi and S. Rebay, *A high-order accurate discontinuous finite element method for the numerical solution of the compressible Navier-Stokes equations*, J. Comput. Phys., **131** (1997), 267–279.

[5] P. G. Ciarlet, *The Finite Element Method for Elliptic Problems*, North-Holland (Amsterdam) **1979**.

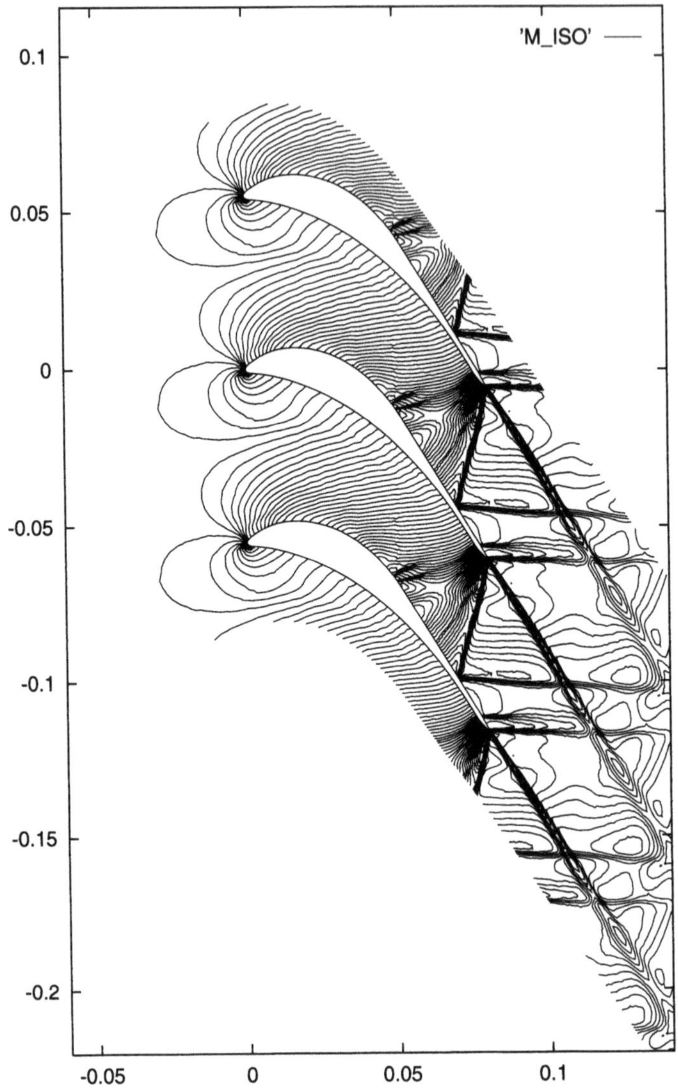

FIGURE 8. Mach number isolines for inviscid cascade flow

[6] M. Crouzeix and P.-A. Raviart, *Conforming and nonconforming finite element methods for solving the stationary Stokes equations*, RAIRO, Anal. Numér., **7** (1973), 33–76.

[7] L. Demkowicz, J. T. Oden and W. Rachowicz, *A new finite element method for solving compressible Navier-Stokes equations based on an operator splitting method and $h-p$ adaptivity*, Comput. Methods Appl. Mech. Engrg., **88** (1991), 363–396.

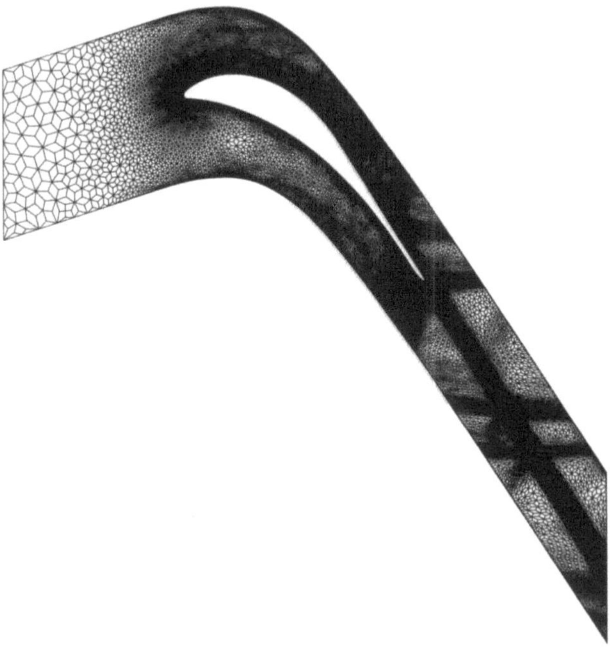

FIGURE 9. Triangulation for viscous cascade flow obtained with the aid of anisotropic mesh refinement

[8] V. Dolejší. *Sur des méthodes combinant des volumes finis et des éléments finis pour le calcul d'écoulements compressibles sur des maillages non structurés*, PhD Dissertation, Charles University (Prague) – L'Université Méditerranée (Marseille) **1998**.

[9] V. Dolejší, *Anisotropic mesh adaptation for finite volume and finite element methods on triangular meshes*, Comput Visual Sci, **1** (1998), 113–121.

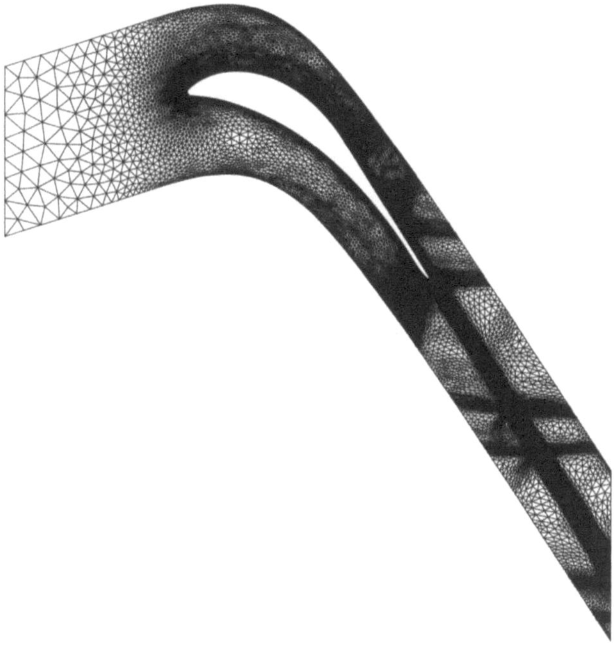

FIGURE 10. Barycentric finite volume grid for viscous cascade flow

[10] V. Dolejší and P. Angot, *Finite volume methods on unstructured meshes for compressible flows*, Finite Volumes for Complex Applications (Problems and Perspectives), Edited by F. Benkhaldoun and R. Vilsmeier, Hermes (Rouen) **1996**, 667–674.

[11] V. Dolejší, M. Feistauer, J. Felcman and A. Kliková, *Error estimates for barycentric finite volumes combined with nonconforming finite elements applied to nonlinear convection-diffusion problems*, Appl. Math. (to appear).

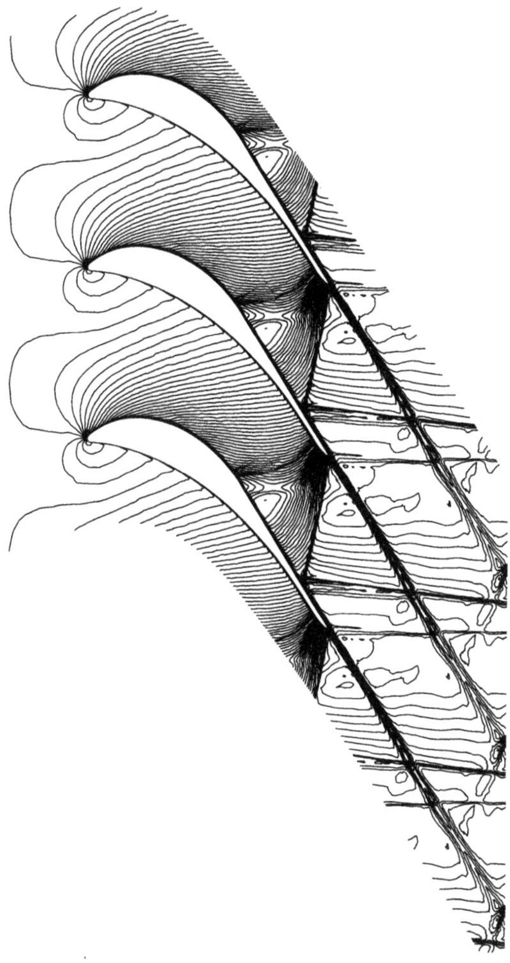

FIGURE 11. Density isolines for viscous cascade flow

[12] R. Eymand, T. Gallouet and R. Herbin, *Finite Volume Methods*, Technical Report 97-19, Centre de Mathématiques et d'Informatique, Université de Provence (Marseille) **1997**.

[13] M. Feistauer, *Mathematical Methods in Fluid Dynamics*, Pitman Monographs and Surveys in Pure and Applied Mathematics 67, Longman Scientific & Technical (Harlow) **1993**.

[14] M. Feistauer, V. Dolejší, J. Felcman and A. Kliková, *Adaptive mesh refinement for problems of fluid dynamics*, Proc. of Colloquium Fluid Dynamics'99, Edited by P. Jonáš, V. Uruba, Institute of Thermomechanics, Academy of Sciences (Prague) **1999**, 53–60.

[15] M. Feistauer and J. Felcman, *Convection-diffusion problems and compressible Navier–Stokes equations*, The Mathematics of Finite Elements and Applications, Edited by J. R. Whiteman, Wiley (Chichester) **1996**, 175–194.

[16] M. Feistauer, J. Felcman and V. Dolejší, *Numerical simulation of compresssible viscous flow through cascades of profiles*, ZAMM, **76** (1996), 297–300.

[17] M. Feistauer, J. Felcman and M. Lukáčová, *Combined finite element–finite volume solution of compressible flow*, Journal of Comput. and Applied Math., **63** (1995), 179–199.

[18] M. Feistauer, J. Felcman and M. Lukáčová, *On the convergence of a combined finite volume–finite element method for nonlinear convection–diffusion problems*, Numer. Methods Partial Differential Eq, **13** (1997), 163–190.

[19] M. Feistauer, J. Felcman, M. Lukáčová and G. Warnecke, *Error estimates of a combined finite volume–finite element method for nonlinear convection–diffusion problems*, SIAM J. Numer. Anal., **36** (1999), 1528–1548.

[20] M. Feistauer, J. Slavík and P. Stupka, *On the convergence of the combined finite volume–finite element method for nonlinear convection–diffusion problems. ii Explicit schemes*, Numer. Methods Partial Differential Eq, **15** (1999), 215–235.

[21] J. Felcman and V. Dolejší, *Adaptive methods for the solution of the Euler equations in elements of the blade machines*, ZAMM, **76** (1996), 301–304.

[22] J. Felcman, V. Dolejší and M. Feistauer, *Adaptive finite volume method for the numerical solution of the compressible Euler equations*, Computational Fluid Dynamics '94, Edited by J. Périaux, S. Wagner, E. H. Hirschel and R. Piva, Wiley (Stuttgart) **1994**, 894–901.

[23] J. Felcman and P. Šolín, *On the construction of the Osher-Solomon scheme for the 3D Euler equations*, East-West J. Numer. Math., **6** (1998), 43–64.

[24] J. Felcman and G. Warnecke. *Adaptive computational methods for gas flow*, Proceedings of the Prague Mathematical Conference'96, Edited by K. Segeth, ICARIS (Prague) **1996**, 99–104.

[25] J. Fořt, M. Huněk, K. Kozel and M. Vavřincová, *Numerical simulation of steady and unsteady flows through plane cascades*, Numerical Modeling in Continuum Mechanics II, Edited by R. Ranacher, M. Feistauer and K. Kozel, Faculty of Mathematics and Physics, Charles Univ. (Prague) **1995**, 95–102.

[26] E. Godlewski and P. A. Raviart, *Hyperbolic Systems of Conservation Laws*, Springer (New York) **1996**.

[27] S. K. Godunov, *Finite difference method for numerical computation of discontinuous solutions of the equations of fluid dynamics*, Matem. Sbornik, **47** (1959), 271–306 (Russian; also English translation: Cornell Aeronautical Lab.).

[28] O. Hassan, K. Morgan and J. Peraire, *An adaptive implicit/explicit finite element scheme for compressible viscous high speed flows*, AIAA 27th Aerospace Science Meeting, Reno, Nevada, 9–12 January 1989.

[29] Ch. Hirsch, *Numerical Computation of Internal and External Flows*, Wiley (Chichester) **1989**.

[30] T. J. R. Hughes, L. P. Franca and M. Mallet, *A new finite element formulation for computational fluid dynamics: I. Symmetric forms of the compressible Euler and Navier-Stokes equations and the second law of thermodynamics*, Comput. Methods Appl. Mech. Engrg. **54** (1986), 223–234.

[31] T. Hůlek, M. Huněk and K. Kozel, *Numerical solution of Euler and Navier–Stokes equations for 2D transonic flow*, Computational Fluid Dynamics'92, I, Edited by Ch. Hirsch, J. Périaux and W. Kordula, Elsevier Science Publishers (Amsterdam) **1992**, 61–68.

[32] A. Kliková, *Finite Volume – Finite Element Solution of Compressible Flow*, Doctoral Thesis, Faculty of Mathematics and Physics, Charles University (Prague) **2000**.

[33] A. Kliková, M. Feistauer and J. Felcman, *Adaptive methods for problems of fluid dynamics*, Software and Algorithms of Numerical Mathematics'99, Edited by J. Holenda and I. Marek, West - Bohemian University (Pilsen) **1999**.

[34] V. M. Kovenya, G. A. Tarnavskii and S. G. Chornyi, *Application of Splitting–Up Method in Problems of Aerodynamics.* Nauka (Novosibirsk) **1990** (Russian).

[35] V. M. Kovenya and N. N. Yanenko, *Splitting–Up Method in Problems of Gas Dynamics*, Nauka (Novosibirsk) **1981** (Russian).

[36] D. Kröner, *Numerical Schemes for Conservation Laws*, Wiley – Teubner (Chichester) **1997**.

[37] S. N. Kruzhkov, *First order quasilinear equations in several independent variables*, Math. USSR Sbornik, **10** (1970), 217–243.

[38] R. J. LeVeque, *Numerical Methods for Conservation Laws*, Birkhäuser (Basel – Boston – Berlin) **1990**.

[39] J. T. Oden, T. Strouboulis and Ph. Devloo, *Adaptive finite element methods for high-speed compressible flows*, Internat. J. Numer. Methods in Engrg., **7** (1987), 1211–1228.

[40] S. Osher and F. Solomon, *Upwind difference schemes for hyperbolic systems of conservation laws*, Math. Comp., **38** (1982), 339–374.

[41] J. Smoller, *Shock Waves and Reaction-Diffusion Equations*, Springer (New York – Heidelberg – Berlin) **1983**.

[42] Th. Sonar, *Mehrdimensionale ENO-Verfahren*, Teubner (Stuttgart) **1993**.

[43] S. P. Spekreijse, *Multigrid Solution of the Steady Euler Equations*, Centrum voor Wiskunde en Informatica (Amsterdam) **1987**.

[44] M. Šťastný and P. Šafařík, *Experimental analysis data on the transonic flow past a plane turbine cascade*, ASME Paper 90-GT-313 (New York) **1990**.

[45] J. L. Steger and R. F. Warming, *Flux vector splitting of the inviscid gas dynamics equations with applications to finite difference methods*, J. Comp. Phys., **40** (1981), 263–293.

[46] E. F. Toro, *Riemann Solvers and Numerical Methods for Fluid Dynamics*, Springer (Berlin) **1997**.

[47] B. Van Leer, *Towards the ultimate conservative difference scheme III. Upstream centered finite–difference schemes for ideal compressible flow*, J. Comp. Phys., **23** (1975), 263–275.

[48] R. Verfürth, *A Review of A Posteriori Error Estimation and Adaptive Mesh-Refinement Techniques*, Wiley – Teubner (Chichester) **1996**.

[49] G. Vijayasundaram, *Transonic flow simulation using an upstream centered scheme of Godunov in finite elements*, J. Comp. Phys., **63** (1986), 416–433.

[50] R. Vilsmeier and D. Hänel, *Adaptive solutions of the conservation equations on unstructured grids*, Notes on Numerical Fluid Mechanics Vol. 35, Proceedings of the Ninth GAMM-Conference on Numerical Methods in Fluid Mechanics, Edited by J. B. Vos, A. Rizzi and I. L. Ryhming, Vieveg (Braunschweig – Wiesbaden) **1992**, 321–330.

Charles University Prague
Faculty of Mathematics and Physics
Sokolovská 83, 186 75 Praha 8
Czech Republic
E-mail address: `feist@karlin.mff.cuni.cz`

Instability of Steady Flows of an Ideal Incompressible Fluid

Susan Friedlander and Alexander Shnirelman

Abstract. This article is devoted to recent developments and open questions concerning instabilities in ideal fluid flows. It is argued that in some appropriate sense almost all steady flows of an ideal incompressible fluid are unstable. However there are different kinds of instability. Many of the instabilities that are described could be termed "slow" and technically they are associated with the Jordan cell structure of the governing operator as opposed to the "fast" instabilities associated with isolated unstable eigenvalues. Numerous examples are given to stress the importance for the existence of instabilities of the norm in which the growth of disturbance is measured.

1. Introduction

The topic of this article is linear and nonlinear instability of steady (i.e. time independent) flows of an ideal incompressible fluid. Our goal is to discuss our understanding of this subject, rather than to try to survey all works in this domain. We admit that our list of references is quite incomplete and personal. The ideal incompressible fluid is a basic model in fluid dynamics. It is believed to describe correctly the motion of real fluids in conditions such that the effects of viscosity and compressibility are negligible. The stability and instability of steady flows is an old subject with many impressive achievements and many well-developed methods. However there remain significant unsolved problems, and we may say that the main problems of instability are open.

Let us start from the notion of stability and instability of steady flows. The motion of an inviscid incompressible fluid in a domain $M \subset \mathbf{R}^n$ is described by the Euler equations:

$$\frac{\partial u}{\partial t} + (u, \nabla)u + \nabla p = 0; \tag{1.1}$$

$$\nabla \cdot u = 0. \tag{1.2}$$

Here $u(x,t)$ is the velocity field of the fluid, $p(x,t)$ is the pressure. The boundary conditions are $u_n\Big|_{\partial M} = 0$, i.e. the flow is tangent to the boundary; the initial condition is $u(x,0) = u_0(x)$.

Let $U_0(x)$ be the velocity field of a steady flow of an ideal incompressible fluid. U_0 satisfies the steady Euler equation

$$(U_0, \nabla)U_0 + \nabla p_0 = 0; \tag{1.3}$$

$$\nabla \cdot U_0 = 0. \tag{1.4}$$

Loosely speaking, this flow is unstable if a small change of initial velocity field results in a considerable change of the actual flow field $u(x,t)$ at some time T. If, on the other hand, small perturbations result in small changes of the flow during arbitrarily long time, the flow is stable.

Both stability and instability are important features of fluid flows and deserve much attention. It appears that only stable flows may exist for a long time, while the unstable ones break down, or require some external stabilization or some feedback to persist. Note that both breaking down of unstable flows, and their stabilization and control are of great interest for numerous applications in natural sciences and engineering. There exists an inverse relation between the stability and the controllability. The system which is excessively stable is difficult to control, and conversely, unstable system is usually easier to control if one manages to stabilize it. For example, today's jet fighters are intentionally built unstable, so that they cannot fly without the stabilization done by an on-board computer. But this instability makes these planes very maneuverable.

The breakdown of unstable flows itself may be very important; for example, the action of the most of musical instruments is based on some sort of instability. The impression that only stable flows persist for a long time without external stabilization may be wrong. There exist different kinds of instability and some of them, which may be called "slow instabilities", are delicate enough and, in their turn, very unstable themselves; they are described in more detail below.

The question of instability falls naturally into two parts, linear and nonlinear instability. The majority of the classical work is connected with the linear problem and studies of properties of the eigenvalues of the operator obtained by linearizing (1.1), (1.2) about a specific steady flow [DR], [Cha]. Much of the work has concentrated on a relatively small number of special fluid configurations, and even in these open questions remain. In section 2 we describe some recent results obtained through a study of the evolution operator for the linearized Euler equations which shed light on the linear instability of rather general Euler flows. We also discuss the existence of unstable discrete eigenvalues in a few specific cases. Linear instability where growth is measured in the energy norm appears to be ubiquitous for 3-dimensional Euler flows.

The problem of nonlinear stability/instability is even more subtle and perplexing than the questions assocated with the linear theory. We address certain aspects of this problem in section 3. We do something that is rather unusual: we work with two definitions of nonlinear stability/instability and discuss ramifications that follow from these two related but nonequivalent definitions. That we proceed in this fashion is at least partly dictated by our lack of full understanding of the nature of fluid instability and partly due to the subtleties of a system which exhibits fascinating degrees of instability. The first definition allows us to discuss a general result of instability in function spaces that are "correct spaces" for the Euler equations, i.e. where local existence and uniqueness of solutions is known. The second definition applies to spaces that are not "correct" in this sense but are natural spaces from a physical point of view. Although we discuss results concerning fairly general flows, much of the section 3 concentrates on one of the most basic classes of flows, namely plane parallel shear flows. The simplest nontrivial example, i.e. a shear flow with a linear profile, illustrates the complexity of fluid stability:

(i) it is linearly (spectrally) stable;
(ii) it is nonlinearly stable in the sense of Arnold in the vorticity norm;
(iii) it is nonlinearly unstable in any norm that includes derivatives of vorticity;
(iv) it is nonlinearly unstable in the L^2-norm (with no conditions on the derivatives of velocity);
(v) it is linearly and nonlinearly unstable as a 3-dimensional flow in a norm which includes the magnitude of vorticity.

The energy and strength of these instabilities will be different from the "fast" exponential instabilities that exist for shear flows with inflection points in the profile.

Throughout this article we hope to communicate several key observations, namely the crucial importance of the norm in which growth of disturbances is measured and the existence of different types of instability that all influence the evolution of a fluid configuration.

2. Linear instability

We consider the linearized Euler equations for a small perturbation $v(x,t)$ about a steady flow U_0, satisfying (1.3), (1.4):

$$\frac{\partial v}{\partial t} = -(U_0, \nabla)v - (v, \nabla)U_0 - \nabla P; \tag{2.1}$$

$$\nabla \cdot v = 0, \tag{2.2}$$

with initial condition

$$v(x,0) = v_0(x). \tag{2.3}$$

These equations may be recast in the following more convenient form:

$$\frac{\partial v}{\partial t} = Lv, \tag{2.4}$$

where

$$Lv = -P\Big((U_0 \cdot \nabla)v + (v \cdot \nabla)U_0\Big); \tag{2.5}$$

here P is the orthogonal projector in $L^2(M, \mathbf{R}^n)$ onto the space J of incompressible vector fields tangent to ∂M.

The stability problem for the linearized Euler equations may be posed as follows. Let us choose a function space X of vector fields where the problem (1.1), (1.2) is well posed. We ask if there exists an initial condition $v_0 \in X$ such that $||v(t)||_X$ is *unbounded* on the whole t-axis; in this case the zero solution of linearized equations is *unstable*. Otherwise it is stable. The classical approach to linear stability of fluid motion is based on an investigation of the spectrum of the operator L in a given space X. Much of the discussion in such texts as [Cha] or [DR] concerns properties of eigenvalues of L in the case of of specific, relatively simple flows U_0. There is a more powerful approach in which the spectrum of the evolution operator is studied (see [VF], [V], [FV]).

The spectrum σ of the operator e^{tL} is naturally decomposed into a discrete part consisting of isolated eigenvalues of finite multiplicity and an essential spectrum:

$$\sigma = \sigma_{\text{disc}} \cup \sigma_{\text{ess}}. \tag{2.6}$$

The operator L is a degenerate non-selfadjoint non-elliptic operator. For an arbitrary steady flow $U_0(x)$ the structure of the spectrum σ is remarkably little understood. Although, as we will describe, we have some general results concerning the essential spectrum, the problem of the existence of discrete eigenvalues is at present too difficult for any general results and must be treated on a case by case basis.

2.1. The unstable essential spectrum

Recently Friedlander and Vishik [VF],[FV] developed a useful tool for investigating the unstable essential spectrum of the linearized Euler equation (2.1), (2.2). One of the main ideas in [VF] is to replace the study of the spectrum of L by the study of the spectrum of the evolution operator e^{tL} for $t > 0$. This permits the development of an explicit formula for the growth rate of a small perturbation due to the essential spectrum. The following theorem proved by Vishik [V] gives an expression for the essential spectral radius $r_{\text{ess}}(e^{tL})$ in terms of a geometric quantity that can be considered as a Lyapunov exponent for fluid flow. The results are proved for free space or periodic boundary conditions and are valid in any spatial dimension, although spatial dimensions 2 and 3 are, of course, physically the most interesting. We consider perturbations with $v_0 \in L^2$, $\nabla \cdot v_0 = 0$.

Theorem 2.1

$$r_{\text{ess}}(e^{tL}) = e^{\mu t}, \tag{2.7}$$

where

$$\mu = \lim_{t \to \infty} \frac{1}{t} \log \sup |b(x_0, \xi_0, b_0; t)|. \tag{2.8}$$

Here supremum is taken over all triplets (x_0, ξ_0, b_0), such that $|b_0| = 1, |\xi_0| = 1, \xi_0 \cdot b_0 = 1$, and the vector $b = b(x_0, \xi_0, b_0; t)$ is determined by the following system of ODE's which we call the bicharacteristic amplitude equations:

$$
\left.
\begin{aligned}
&(a) \quad \dot{x} = U_0(x) \\
&(b) \quad \dot{\xi} = -\left(\frac{\partial U_0}{\partial x}\right)^T \xi \\
&(c) \quad \dot{b} = -\left(\frac{\partial U_0}{\partial x}\right) b + 2\left[\left(\frac{\partial U_0}{\partial x}\right) b \cdot \xi\right] \xi / |\xi|^2
\end{aligned}
\right\} \tag{2.9}
$$

with $x(0) = x_0$, $\xi(0) = \xi_0$, $b(0) = b_0$.

This theorem is proved by writing the evolution operator e^{tL} as a product of a pseudo-differential operator and a shift operator along the trajectories of the equilibrium flow U_0. This allows the growth of the evolution operator to be studied to precise exponential asymptotics. A heuristic derivation of this result is obtained by applying a "geometric optics" treatment based on high frequency solutions of equations (2.1), (2.2). In the language of geometric optics, equation (2.9b) is the Eikonal equation and equation (2.9c) is the transport equation. Equation (2.9c) is the evolution equation for the amplitude of a high frequency wavelet initially localized at x_0, with initial wave number vector ξ_0. The quantity on the RHS of (2.8) is the "fluid Lyapunov exponent" corresponding to the maximal exponential growth rate of such an amplitude vector $b(t)$.

The result of Theorem 2.1 gives one piece of information concerning the stability spectrum for inviscid flows, namely the maximum growth rate of instability in the essential spectrum. Moreover it implies that any point z in the spectrum $\sigma(e^{tL})$ such that $|z| > e^{\mu t}$ is necessarily an eigenvalue of finite multiplicity. A positive lower bound for the value of the Lyapunov exponent μ can be explicitly computed in many examples [FV],[FV1]. Furthermore theorem 1 provides an effective sufficient condition for instability of large classes of inviscid fluid flows. Since expression (2.8) involves the supremum over initial conditions (x_0, ξ_0, b_0), it is only necessary to show there exists at least one initial condition for which the solution to the system (2.9) of ODE gives

$$\lim_{t \to \infty} \frac{1}{t} \log |b| > 0 \tag{2.10}$$

to conclude that $\mu > 0$, and hence the unstable essential spectrum is nonempty.

We comment briefly on some other results obtained in the context of fluid instability using a "geometric optics" approach. This method has been employed successfully by a number of authors with the earliest work being on the hyperbolic systems for *compressible* fluids by G. Friedlander in 1958 [Fr] and Ludwig in 1960 [L]. Later Eckhoff [E] and Eckhoff and Storesletten [ES] studied the stability of azimuthal shear flows of a compressible fluid and more generally symmetric hyperbolic systems using an approach based on a generalized progressive wave expansion. Eckhoff shows that local instability problems for hyperbolic systems can be essentially reduced to a local analysis involving ODEs. We note that the *incompressible* Euler equations (1.1), (1.2) do not form a strictly hyperbolic system, and the proof of Theorem 1 does not follow directly from Eckhoff. Results using a system of ODEs equivalent to (2.8) to detect instabilities in incompressible flows include those of Bayly et al [BHL], who obtained the growth rate of instabilities for columnar and elliptic vortices, and Lifschitz and Hameiri [LH] who obtained instability conditions for vortex rings.

We will now discuss in a little more detail some of the explicit instability results of Friedlander and Vishik that follow from Theorem 2.1. The idea that exponential stretching of fluid particles could imply instability for the Euler equations is originally due to Arnold [A1]. Friedlander and Vishik [FV] use Theorem 2.1 to prove that every flow with exponential stretching, even at one point, is linearly unstable.

Theorem 2.2 *Consider a steady solution U_0 of 3-dimensional Euler equations. Suppose the flow U_0 has a positive classical Lyapunov exponent at some point x_0. Then $\mu > 0$, and hence the flow U_0 is unstable.*

The proof of this theorem follows from a result obtained from the system (2.9) of ODEs, namely:

$$\frac{d}{dt}(b_1 \times b_2 \cdot \xi) = 0, \tag{2.11}$$

where b_1 and b_2 are two linearly independent solutions of (2.9) corresponding to a cotangent vector ξ that satisfies (2.9). Since the flow is volume preserving the existence of a positive Lyapunov exponent (i.e. an exponentially growing tangent vector) implies the existence of an exponentially decaying cotangent vector ξ. From (2.10) we then conclude the existence of at least one exponentially growing amplitude vector b, which implies $\mu > 0$. Therefore we note that in 3 dimensions the classical Lyapunov exponent provides a lower bound on μ.

In 2 dimensions, the system (2.9) of ODEs provides an even stronger constraint on the relation between $b(t)$ and $\xi(t)$, namely

$$\frac{d}{dt}(|b(t)||\xi(t)|) = 0. \tag{2.12}$$

Hence in 2 dimensions the fluid Lyapunov exponent μ and the maximal classical Lyapunov exponent for the dynamical system $\dot{x} = U_0(x)$ are the same.

It follows from Theorem 2 that any flow U_0 (in 2 or 3 dimensions) with a hyperbolic stagnation point x_0 is unstable (i.e. $U(x_0) = 0$ and there exists an eigenvalue of $(\frac{\partial U_0}{\partial x})_{x_0}$ with positive real part). There are large classes of fluid flows U_0 with such stagnation points. For all such flows $r_{\mathrm{ess}} > 1$.

A class of flows with presumably chaotic stream lines was identified by Arnold [A2]. An example is the so-called ABC flow $U_0 = (\dot{x}, \dot{y}, \dot{z})$ where

$$\begin{aligned} \dot{x} &= A \sin z + C \cos y \\ \dot{y} &= B \sin x + A \cos z \\ \dot{z} &= C \sin y + B \cos x. \end{aligned} \tag{2.13}$$

For general values of the constants A, B and C numerical investigations [OFGHMS] indicate that ABC flows exhibit the phenomenon of Lagrangian chaos which suggests strong exponential stretching. Analytic treatment of ABC flows [FGV], [Chi] proves that for certain ranges of A, B and C there is exponential stretching either at hyperbolic points or associated with hyperbolic closed trajectories. The result of Theorem 2 then proves that these ABC flows are hydrodynamically unstable.

In 3 dimensions the mechanism of vortex tube stretching, which is absent in strictly 2-dimensional flows, can give rise to values of μ which are greater than the classical Lyapunov exponent. For example, in [F] analysis of a model equation for 3-dimensional Euler known as the surface quasi-geostrophic equation leads to a quantity analogous to μ that tends to infinity for flows with hyperbolic structures: i.e. there exist perturbations that grow like e^{t^2}. This result suggests that a 3-dimensional flow with hyperbolic structure is strongly unstable. Furthermore in 3 dimensions it is possible to have flows U_0 for which the classical Lyapunov exponents are all zero yet the fluid exponent μ is positive. Such an example is constructed in [FV]. It is proved that the integrable flow $U_0 \times \text{curl} U_0 = -\nabla H$ with $\nabla H \neq 0$ has μ positive provided that a certain geometric condition is satisfied by the stream lines. The following (non sharp) condition ensures $\mu > 0$:

$$\int_0^T \{\kappa \hat{n} \cdot \nabla H - \tau_g U_0 \cdot \text{curl} U_0 / |\nabla H|^2\} dt \geq 0 \tag{2.14}$$

where, for any stream line of the flow as it wraps around the toroidal surface $H = H_0$, T is the period, κ the curvature, τ_g the geodesic torsion and $\hat{n}$ the principal unit normal to the stream line.

We have described many fluid flows where it can be shown that $\mu > 0$. In a few cases μ can be computed explicitly. For example, the 2-dimensional cellular flow

$$U_0 = (-\sin x \cos y, \cos x \sin y).$$

In this case μ is given by the positive real eigenvalue of the matrix $(\frac{\partial U_0}{\partial x})$ at a hyperbolic stagnation point. Thus $\mu = 1$ for this simple cellular flow. There are certain classes of 2-dimensional flows for which it follows from (2.8) that $\mu = 0$ [FSV]. In particular $\mu = 0$ (i.e. there is no unstable essential spectrum) for 2-dimensional flows with no stagnation points or any 2-dimensional plane parallel shear flow.

2.2. Examples of Instability in the Discrete Spectrum

We now turn to the question of existence and distribution of unstable eigenvalues in the discrete spectrum of equation (2.1), (2.2). The linearized Euler operator is degenerate, non-elliptic, and there are no general theorems that may be applied to prove the existence of unstable discrete eigenvalues. However in certain rather special examples it is possible to construct unstable eigenvalues.

The spectral problem for the linearized Euler operator is considerably simpler in 2 dimensions rather than in 3 dimensions. In particular, in 2 dimensions we can define a scalar stream function to replace the divergence free vector field. We write

$$U_0 = \hat{k} \times \nabla\Psi(x,y), \;\; v = \hat{k} \times \nabla\phi(x,y,t). \tag{2.15}$$

Hence,

$$\nabla \times U_0 = \hat{k}\nabla^2\Psi(x,y), \;\; \nabla \times v = \hat{k}\nabla^2\phi(x,y,t). \tag{2.16}$$

Here $\hat{k}$ is the unit vector perpendicular to the 2-dimensional plane with Cartesian coordinates (x, y). The 2-dimensional steady equations (1.3), (1.4) will be satisfied when Ψ satisfies an elliptic equation of the form

$$\nabla^2\Psi = -F(\Psi). \tag{2.17}$$

Taking the curl of equation (2.1) gives the equation for the evolution of the perturbation vorticity $\omega \equiv \nabla \times v$:

$$\frac{\partial\omega}{\partial t} = \{U_0, \omega\} + \{v, \nabla \times U_0\}, \tag{2.18}$$

where $\{,\}$ denotes the Poisson bracket of two vector fields, i.e.

$$\{A, B\} = (B \cdot \nabla)A - (A \cdot \nabla)B. \tag{2.19}$$

In general the second Poisson bracket on the RHS of (2.19) is very difficult to analyse. However in 2 dimensions the problem greatly simplifies because $\hat{k} \cdot \nabla(\cdot) \equiv 0$. The vorticity equation (2.19) reduces to

$$\lambda\omega + (U_0 \cdot \nabla)\omega + (v \cdot \nabla)(\nabla \times U_0) = 0. \tag{2.20}$$

We consider the eigenfunction ϕ and the eigenvalue λ for equation (2.19); after substituting (2.18) and (2.21) into (2.19) we obtain equation

$$\lambda\nabla^2\phi = (\Psi_y\frac{\partial}{\partial x} - \Psi_x\frac{\partial}{\partial y})(\nabla^2\phi + F'(\Psi)\phi). \tag{2.21}$$

We take the boundary conditions to be 2π-periodicity in (x, y).

A simple and very classical example that has received much attention in the literature of the past 100 years is plane parallel shear flow (see, for example, [DR],

[Cha]). In this case $U_0 = (U(y), 0)$ and (2.22) becomes the so-called Rayleigh equation:

$$(\frac{\lambda}{ik} + U(y))(\frac{d^2}{dy^2} - k^2)\Phi(y) - U''(y)\Phi(y) = 0, \tag{2.22}$$

where we have written

$$\phi(x, y, t) = \Phi(y)e^{ikx}e^{\lambda t}. \tag{2.23}$$

The celebrated Rayleigh stability criterion [DR] says that a necessary condition for instability is the presence of an inflection point in the profile $U(y)$. As we remarked in section 2.1, the concept of the "fluid Lyapunov exponent" μ given by expression (2.8) can be used to prove that equation (2.22) with periodic boundary conditions has no unstable essential spectrum for any profile $U(y)$. It remains to discuss the possibility of discrete unstable eigenvalues (i.e. λ such that $\mathrm{Re}\lambda > 0$) associated with equation (2.23) for profiles $U(y)$ that contain at least one inflection point.

Meshalkin and Sinai [MS], followed by Yudovich [Y1] investigated the instability of a *viscous* shear flow $U(y) = \sin my$ using techniques of continued fractions. More recently Friedlander et al [FSV], [BFY], [FH] showed that these techniques could be used for the inviscid equation (2.23) with $U(y) = \sin my$. Eigenfunctions are constructed in terms of Fourier series that converge to C^∞-smooth functions for eigenvalues λ that satisfy the characteristic equation. We write

$$\Phi(y) = \sum_{n=-\infty}^{\infty} a_n e^{iny}. \tag{2.24}$$

The recurrence relation equivalent to (2.23) yields a tridiagonal infinite algebraic system which is analyzed using continued fractions to yield the characteristic equation relating the eigenvalues to the wavenumbers k and m. The Fourier coefficients a_n decay exponentially with n for each root λ of the characteristic equation. In the example $U(y) = \sin my$ this procedure exhibits the complete unstable spectrum in L^2 of the linearized Euler equation.

The existence of unstable eigenvalues for shear flows with a general rapidly oscillating profile $U(my), m >> 1$, was demonstrated in [BFY] using homogenization techniques to compute the spectral asymptotics. Gordin [G] has solved numerically an interesting problem of finding a "maximally unstable" profile $U(y)$, provided its enstrophy $\int |U'(y)|^2 dy$ is fixed.

On the other hand, if the profile $U(y)$ is close enough to the linear one, i.e. $U(y) = y + \varepsilon f(y)$ for arbitrary smooth function $f(y)$ and sufficiently small $\varepsilon > 0$, then there are no unstable eigenvalues (for a fixed wave number k). This may be deduced from the paper of L. Faddeev [Fa]. We can regard this result as relating to flows on the side surface of a cylinder, in which case the wave number of perturbation cannot be smaller than some constant. We shall discuss these flows in section 3.5.

There are a few results concerning the unstable eigenvalues of (2.22) for somewhat more general flows U_0. A specific "cats-eye" type flow was studied by Friedlander et al [FVY] and the method of averaging was used to construct the formal asymptotic expansion for eigenfunctions of a class of unstable eigenvalues for equation (2.22) with $F(\Psi)$ corresponding to the "cats-eye" flow. We remark that the existence of hyperbolic stagnation points in this flow means that, in contrast with parallel shear flow, μ is positive. Hence both the discrete and the essential unstable spectrum are nonempty.

3. Nonlinear Instability

Problems connected with stability and instability of the full nonlinear Euler equations (2.1)-(2.2) are even more complex than those related to the spectrum of the linearized equation discussed in Section 2. Hence many questions remain open. However some results have been obtained recently. We will describe these "small steps of progress" and indicate some promising paths for future development.

A steady state is called nonlinearly stable if every disturbance that is "small" initially generates a solution to the nonlinear Euler equation which stays "close" to the steady state for all time. There are several natural precise definitions of nonlinear stability and its converse instability. To a certain extent these definitions incorporate a concept of "degrees" of instability. The definitions reflect the crucial dependence of a stable or unstable state on the norm in which growth with time of disturbances is measured. The first definition we give allows us to consider nonlinear stability/instability in function spaces for which it is known that there is local existence and uniqueness. Later in this section we prove a theorem under this definition relating linear instability in L^2 with nonlinear instability in $H^s, s > n/2 + 1$. In the second definition we consider nonlinear stability in L^2 and H^1 which are natural spaces to measure growth of a disturbance but are not "correct" spaces for the Euler equation in terms of proven properties of the solutions of the nonlinear equation. The elegant nonlinear stability results of Arnold [A3] fall under this second definition, as does the concept of minimal flows introduced by Shnirelman [S1].

3.1. Definitions of Nonlinear Stability/Instability

First definition of nonlinear instability. We define nonlinear stability for a general evolution equation of the form

$$u_t = Lu + N(u), \quad u(0) = u_0, \tag{3.1}$$

where L and N are respectively the linear and nonlinear terms. Let X and Z be a fixed pair of Banach spaces with $X \subset Z$ being a dense embedding. We assume that for any $u_0 \in X$ there exists a $T > 0$ and a unique solution $u(t)$ to (3.1) with

$$u(t) \in L^\infty((0,t);X)\bigcap C([0,T],Z) \tag{3.2}$$

in the sense that for any $\phi \in D(0,T)$

$$\int_0^T \{u(\tau)\phi'(\tau) + (Lu(\tau) + N(w(\tau))\Phi(\tau)\}d\tau = 0. \tag{3.3}$$

The initial condition is assumed in the sense of strong convergence in Z:

$$\lim_{\tau \to 0+} ||u(\tau) - u_0||_Z = 0. \tag{3.4}$$

Definition 3.1 *The trivial solution $u_0 = 0$ of (3.1) is called* nonlinearly stable *in X (i.e. Lyapunov stable) if for all $\varepsilon > 0$ there exists $\delta > 0$ so that $||u_0||_X < \delta$ implies*

(a) *we can choose T in (3.2) to be $T = \infty$, and*
(b) $||u)t)||_X < \varepsilon$ *for a.e. $t \in [0, \infty)$.*

The trivial solution is called nonlinearly unstable *in X if it does not satisfy the conditions stated above.*

We remark that by this definition finite time "blow up" (i.e. a maximal *finite* $T > 0$ in (3.1)) is a special case of nonlinear instability. This is valuable in the context of the Euler equations (1.1), (1.2) since in 3 dimensions the possibility of finite time blow up has not yet been ruled out. In the context of the Euler equations the "natural" choice for the spaces X and Z are H^s with $s > n/2 + 1$ where n is the space dimension and L^2 respectively.

Second definition of nonlinear instability. The second notion of (in)stability is what we call Z-(in)stability. In this definition we do not split the operator into linear and nonlinear parts, because such splitting makes no sense for strong perturbations we are dealing with. Let $X \subset Z$ be a pair of Banach spaces with dense and compact embedding. Consider an operator equation having the form

$$\frac{du}{dt} = A(u). \tag{3.5}$$

Suppose that for every $u_0 \in X$ and every $T > 0$ there exists unique solution $u(t) \in L^\infty((0,T); X) \bigcap((0,T); Z)$. Let $U_0 \in X$ be a fixed point, i.e. $A(U_0) = 0$.

Definition 3.2 *The constant solution $u(t) \equiv U_0$ is called stable in Z, or Z-stable, if for every $\varepsilon > 0$ there exists $\delta > 0$ such that if $v_0 \in X$ and $||v_0 - U_0||_Z < \delta$, and if $v(t)$ is a solution of (3.5), satisfying $v(0) = v_0$, then $||v(t) - U_0|| < \varepsilon$ for all $t \in$* **R**. *Otherwise steady solution U_0 is called Z-unstable.*

3.2. Instability may depend on the functional space

In infinite-dimensional systems like a fluid the choice of functional space may be crucial for the stability/instability of the system, as was emphasized by Yudovich [Y2]. This is both a linear and a nonlinear phenomenon.

To illustrate the dependence of stability on the choice of norm consider the following simple example given in [Y2], namely the Cauchy problem

$$\frac{\partial v}{\partial t} = x\frac{\partial v}{\partial x}, \quad v(0) = \phi(x). \tag{3.6}$$

The unique solution for an arbitrary smooth function $\phi(x)$ is

$$v(x,t) = \phi(xe^t). \tag{3.7}$$

A simple calculation shows that

$$||\frac{\partial^k v(\cdot,t)}{\partial x^k}||_{L^p(\mathbf{R}} = e^{(k-p^{-1})t}||\phi^{(k)}||_{L^p(\mathbf{R})}. \tag{3.8}$$

Hence this linear equation is asymptotically stable in L^p for any $p, 1 \leq p < \infty$ and disturbances decay exponentially. The solution is stable but not asymptotically in $L^\infty(\mathbf{R})$, $C(\mathbf{R})$ and $W^{1,1}(\mathbf{R})$. In any space $W^{k,p}(\mathbf{R})$ with $k > 1$ or $k = 1, p > 1$, the solution is exponentially unstable.

The following simple example of Shnol [Sh] shows that an Euler flow may be linearly stable in one natural function space and unstable in another one. Let $U_0(x)$ be a plane-parallel flow in a strip with a linear profile: $U_0(x,y) = (ay, 0)$, $a \neq 0$. Then its vorticity is constant, and the linearized equation for the perturbation vorticity is

$$\frac{\partial \omega}{\partial t} + ay\frac{\partial \omega}{\partial x} = 0, \tag{3.9}$$

i.e. the vorticity perturbation is transported by the flow. The zero solution of this equation is stable in the space J_1 where $||u||_{J_1} = ||u||_{L^2} + ||\nabla \times u||_{L^2}$. On the other hand, it is unstable in Sobolev spaces H^s for $s > 1$, because the derivatives of vorticity grow in time. Interesting enough, the zero solution is unstable also in $J = L^2$, but the reason for this instability is quite different. Namely, every smooth solution of (3.9) tends weakly to some function depending only on y, as $t \to \infty$; this means that the velocity field tends in L^2 to some field, parallel to the x-axis and depending only on y, and this field is zero, if the mean value of the initial perturbation along the x-axis is identically zero. Thus the subspace of parallel flows in the space J appears to be an attracting set, consisting of fixed points. But (and this is the original reasoning of Shnol) the equation (3.9) is time reversible; this means that there exist arbitrarily small in L^2 perturbations of velocity, which grow arbitrarily big after some time; this means that this simplest nontrivial parallel flow is linearly unstable in L^2. The same is true for arbitrary nontrivial parallel flow. A similar situation may be observed for other classes of steady flows, for example for a potential flow in a multiconnected domain with one or more hyperbolic stagnation points.

As Yudovich [Y3] observes, there exists a class of exact solutions to the full nonlinear Euler equations which imply that 2-dimensional steady flows are, with very few exceptions, unstable with respect to 3-dimensional perturbations in any norm which includes the maximum of vorticity modulus. In particular, consider the plane parallel shear flow $U_0 = (f(y), 0, 0)$ with $y \in [0,1]$. This flow is well known to be linearly stable in L^2 if there are no inflection points in the profile of $f(y)$ in $[0,1]$ (see, for example, [DR]). In an appropriate sense that we discuss in the next section it is also nonlinearly stable to 2-dimensional perturbations. It is easy to check that the following is an exact solution to the full nonlinear Euler equations (1.1), (1.2) for any smooth functions f and w:

$$u = (f(y), 0, w(x - tf(y))). \tag{3.10}$$

The corresponding vorticity

$$\nabla \times u = -(tf'(y)w'(x - tf(y)), w'(x - tf(y)), f'(y)). \tag{3.11}$$

Hence the vorticity of a perturbation $(0, 0, w)$ to the steady shear flow grows linearly with time provided only that f and w are nonconstant functions. Thus a shear flow, even with no inflection points, is nonlinearly unstable to 3-dimensional perturbations in any norm that incorporates the magnitude of vorticity.

This set of exact solutions to the 3-dimensional Euler equation can be generalized to a suitable z-independent perturbation of any 2-dimensional steady flow (see [Y3], [F]), namely

$$u = (-\Psi_x, \Psi_y, w(x, y, t)), \tag{3.12}$$

where $\Psi(x, y)$ is a stream function for a steady 2-dimensional flow (see Section 2.2) and w satisfies

$$(\frac{\partial}{\partial t} + ((\hat{k} \times \nabla)\Psi) \cdot \nabla)w = 0. \tag{3.13}$$

The evolution equation for the vorticity gives

$$(\frac{\partial}{\partial t} + ((\hat{k} \times \nabla)\Psi) \cdot \nabla)(\hat{k} \times \nabla w) = (\hat{k} \times \nabla)w \cdot \nabla((\hat{k} \times \nabla)\Psi), \tag{3.14}$$

i.e. the vorticity component $(\hat{k} \times \nabla)w$ evolves as a tangent vector to the 2-dimensional flow $u_H = (\hat{k} \times \nabla)\Psi$. Hence for almost all choices of Ψ there exists a perturbation $w(x, y, t)$ such that the vorticity of the 3-dimensional Euler flow (3.12) grows with time. This growth can be exponential on a set of measure zero if the flow u_H has a hyperbolic fixed point.

Note that these instabilities are associated with the essential spectrum and thus have different nature than instabilities connected with discrete eigenvalues. If the perturbation is an eigenfunction of the linearized equations (2.1), (2.2) with an eigenvalue having a positive real part, then, of course, it grows exponentially in any norm.

3.3. A nonlinear instability theorem

In this section we describe a result which applies to nonlinear instability in the sense of our first definition (see section 3.1). In the context of the Euler equation the result relates spectral instability of the evolution operator in L^2 with nonlinear instability in the Sobolev space $H^s, s > n/2+1$.

We formulate the relevant theorem in a general setting. We consider the stability of the zero solution of an evolution equation

$$\frac{dv}{dt} = Lv + N(v), \tag{3.15}$$

where L and N are respectively the linearized and nonlinear parts of the governing equation. Once the spectrum of the linear part L is analyzed and shown to have an unstable component (i.e. the zero solution is linearly unstable) then the question arises whether the zero solution is nonlinearly unstable. It is well known (see, for example, Lichtenberg and Lieberman [LL] that the linear instability implies nonlinear instability in the finite-dimensional case (i.e. if (3.15) is an ODE). In the infinite-dimensional case (PDE) such general result is not known, although for many particular types of evolution PDE's it has been shown that linear instability implies nonlinear instability (e.g. such a result for the incompressible Navier-Stokes equations in a bounded domain has been proved by Yudovich [Y2]). Difficulties with deriving the nonlinear instability from the linear one usually appear whenever the essential spectrum of L is non-empty as it generally is for the Euler equation.

In [FSV] Friedlander, Strauss, and Vishik proved the following abstract nonlinear instability theorem under the spectral gap condition.

Theorem 3.1 *Fix a pair of Banach spaces $X \hookrightarrow Z$ with a dense embedding. Let (3.15) admit a local existence theorem in X. Let N and L satisfy the following conditions.*

(1) $\|N(v)\|_Z \leq C_0\|v\|_X\|v\|_Z$ *for $v \in X$ with $\|v\|_X < \rho$ for some $\rho > 0$.* (3.16)

(2) *A spectral "gap" condition, i.e.$\sigma(e^{tL}) = \sigma_+ \cup \sigma_-$ with $\sigma_+ \neq \phi$* (3.17)

where

$$\begin{aligned} \sigma_+ &\subset \{z \in \boldsymbol{C} | e^{Mt} < |z| < e^{\Lambda t}\} \\ \sigma_- &\subset \{z \in \boldsymbol{C} | e^{\lambda t} < |z| e^{\alpha t}\} \end{aligned} \tag{3.18}$$

with

$$-\infty < \lambda < \alpha < M < \Lambda < \infty \text{ and } M > 0. \tag{3.19}$$

Then the trivial solution $v = 0$ to equation (3.15) is nonlinearly unstable.

The main idea of the proof of this theorem is as follows. We assume the contrary, namely that the trivial solution $v = 0$ is nonlinearly stable. Let $\varepsilon > 0$ sufficiently small be given: it will be specified later. From the definition 1 of nonlinear stability it follows that there exists a global solution $v(t), t \in [0, \infty)$ such that $\|v(t)\|_X < \varepsilon$ provided $\|V(0)\|_X < \delta(\varepsilon)$.

We project $v(t)$ onto two subspaces using the spectral gap condition (3.17), (3.18). We denote by $P_\pm$ the Riesz projection corresponding to the partition of the spectrum created by the gap and introduce a new norm $|||\cdot|||$ on Z. For any $x \in Z$ let

$$|||x||| = |||P_+x||| + |||P_-x||| \\ = \int_0^\infty ||e^{-\tau L}P_+x||_Z e^{\tau M} d\tau + \int_0^\infty ||e^{\tau L}P_-x||_Z e^{-\tau\alpha} d\tau. \qquad (3.20)$$

The norm $|||\cdot|||$ is equivalent to $||\cdot||_Z$, i.e. there exists $C > 0$ such that

$$C^{-1}||x||_Z \leq |||X||| \leq C||x||_Z. \qquad (3.21)$$

Since $v(t)$ is a solution to (3.15) it can be shown that

$$(|||P_+v(t)||| - |||P_-v(t)|||)|_{t_1}^{t_2} \\ \geq \int_{t_1}^{t_2} \{M|||P_+v(\tau)||| - \alpha|||P_-v(\tau)||| \\ +C^{-1}|||v(\tau)||| - |||N(v(\tau))|||\}d\tau \qquad (3.22)$$

for any interval $0 \leq t_1 \leq t_2$.

We choose the initial condition $v_0 = \delta\bar{v}_0$, where $\bar{v}_0 \in X$ is an arbitrary vector satisfying

$$|||P_+\bar{v}_0||| > |||P_-\bar{v}_0|||, \ ||\bar{v}_0||_X < 1. \qquad (3.23)$$

Since $||v_0||_X < \delta$ our assumption of nonlinear stability implies

$$||v(t)||_X < \varepsilon \quad \text{for a.e. } t \in [0,\infty), \qquad (3.24)$$

and from condition (3.23)

$$|||N(v(t))||| \leq Cc_0||v(t)||_Z \leq C^2c_0\varepsilon|||v(t)||| \quad \text{for a.e. } t \in [0,\infty). \qquad (3.25)$$

Now the inequalities (3.24)-(3.25) plus Gronwall's inequality give

$$|||P_+v(t)||| - |||P_-v(t)||| \\ \geq \delta(|||P_+\bar{v}_0||| - P_-\bar{v}_0|||)e^{MT}, \ t \in [0,\infty), \qquad (3.26)$$

provided ε is chose so that $\varepsilon < \min(C^{-3}c_0^{-1}, \rho)$. Since $M > 0$, for sufficiently large t the inequality (3.26) contradicts our assumption that $||v(t)||_X < \varepsilon$. Hence the trivial solution to (3.15) is nonlinearly unstable in X.

We now consider Theorem 3.1 in the context of the Euler equations (1.1), (1.2). We write

$$u = U_0 + v, \\ Lv = -((U_0\cdot\nabla)v - (v\cdot\nabla)U_0 - \nabla p, \\ N(v) = -(v\cdot\nabla)v - \nabla q; \qquad (3.27)$$

thus the notation of the general theorem applies to instability of the steady flow U_0. The local existence requirement and condition (1) of Theorem 3.1 are easy to satisfy by making the natural choice for the spaces X and Z, namely

$$X = H_s,\ s > \frac{n}{2} + 1 \text{ and } Z = L^2 \tag{3.28}$$

with the restriction to vector fields that are divergence free and satisfy appropriate boundary conditions. However the spectral gap condition is much more difficult to verify for a given steady solution U_0 because, as we have discussed, the essential spectrum of e^{tL} is non-empty and at least in some examples fills the whole annulus.

One piece of information we have about the structure of the spectrum is the essential spectral radius theorem discussed in section 2.1. In some examples the "fluid Lyapunov exponent" μ can be explicitly calculated. Also the theorem implies, in particular, that any $z \in \sigma(e^{tL})$ with $|z| > e^{\mu t}$ is a point of the discrete spectrum (i.e. an isolated point with finite multiplicity where the range of $(z - e^{tL})$ is closed). Any accumulation point of $\sigma_{\text{disc}}(e^{tL})$ necessarily belongs to $\sigma_{\text{ess}}(e^{tL})$. Thus if

$$\sigma(e^{tL}) \bigcap \{|z| > e^{\mu t}\} \neq \emptyset, \tag{3.29}$$

then there exists a partition

$$\sigma(e^{tL}) = \sigma_+ \bigcup \sigma_- \tag{3.30}$$

satisfying the gap condition (3.17), (3.18).

There are several examples of 2-dimensional flows where μ can be computed and discrete unstable eigenvalues calculated to show that (3.18) holds. These are the examples of discrete unstable eigenvalues discussed in section 2.2. As we remarked in section 2.1, in 2 dimensions the fluid Lyapunov exponent and the classical Lyapunov exponent are equal. Hence $\mu = 0$ for any plane-parallel shear flow. It therefore follows from Theorem 3.1, plus the result of [BFV] that there exist unstable discrete eigenvalues for any shear flow with a rapidly oscillating profile, that all such shear flows are nonlinearly unstable in H^2.

Other recent results concerning nonlinear instability of 2-dimensional shear flows include the work of Grenier [G] who proves nonlinear instability in L^∞ for piecewise linear profiles. Koch [K] proves in 2 dimensions that nonlinear stability in $C^{1,\alpha}$ requires uniform boundedness of the derivatives of the flow map which implies that all steady shear flows are nonlinearly unstable in $C^{1,\alpha}$.

A more general 2-dimensional flow than parallel shear flow that can be shown to be nonlinearly unstable is the "cats-eye" flow studied in [FVY]. In this case the existence of hyperbolic stagnation points implies that $\mu > 0$. The exact value of μ can be calculated as the positive eigenvalue of the gradient matrix of U_0 at the hyperbolic point. The results of [FV] show that there exist discrete unstable eigenvalues with real part $> \mu$, hence again we can invoke theorem 3.1 to prove that such "cats-eye" flows are nonlinearly unstable.

The problem of verifying the gap condition for the spectrum corresponding to 3-dimensional flows is more difficult and the structure of the spectrum remains an open question. As we discussed in Section 2.1, there is some evidence that in 3 dimensions the combination of vortex tube stretching and hyperbolic stagnation points may provide a situation in which $\mu \to \infty$ and hence it would be very difficult to verify the gap condition. Proving instability is then beyond the tools we have presently available.

3.4. Arnold stable and minimal flows

In this section we discuss some results and open questions concerning nonlinear stability in the sense of our second definition.

The most frequently employed method to prove the nonlinear stability of particular flows and classes of flows was developed by Arnold [AK]. He used the *Energy-Casimir method* based on the existence of two different integrals of motion. The simplest example of an application of this method is to the stability of rigid rotation of a fluid in a disk. Here we have two integrals, the energy E and the angular momentum Ω. Consider the space J of all square integrable incompressible vector fields in the disk tangent to its boundary. The fields with given momentum Ω_0 form a hyperplane in this space, and the functional E achieves an absolute minimum E_0 on this hyperplane at the field $U(x)$, which is the velocity field of the rigid rotation. This critical point is nondegenerate (its second variation is positive definite in L^2). Hence, every vector field $u(x)$ of the space J with the energy E and the angular momentum Ω, close resp. to E_0 and Ω_0, is close in the space J to the field $U(x)$ and remains close forever, because the functionals E and Ω do not depend on time.

For more general flows the Energy-Casimir method assumes more sophisticated forms. We use the fact that the vorticity ω is transported by the fluid. In other words, the fluid moves in such a manner that its vorticity field at every moment is obtained from the vorticity at the initial moment by some volume-preserving diffeomorphism depending on t: $\omega(x,t) = \omega(g_t^{-1}(x), 0)$ (in the terminology of Arnold, the velocity fields of the flow at any two moments are *isovortical*). The relation of equivorticity defines partition of the space of velocity fields into equivalence classes which may be regarded as a sort of generalized Casimir. For a given field $U_0(x)$ the class of isovortical fields is an infinite-dimensional manifold V, which is the orbit of the group $\mathcal{D}$ of volume-preserving diffeomorphisms in the space X of incompressible vector fields. Arnold has proved [AK] that the steady solutions of the Euler equations are exactly the fields of V which are critical points of the energy functional E, restricted on V. If the critical point is a point of a strict local maximum or minimum of E, then the flow is nonlinearly stable in the space J_1, whose elements are incompressible vector fields $u(x)$ in the flow domain, tangent to the boundary and having a finite norm $\|u\|_{J_1} = \|u\|_{L^2} + \|\nabla \times u\|_{L^2}$. The development of this idea gives rise to the celebrated results of Arnold concerning the nonlinear stability of certain classes of steady flows. In particular, Arnold's methods show that for plane-parallel shear flow in 2 dimensions the Rayleigh criterion (i.e. no inflection points in the profile) guarantees not only spectral stability but also nonlinear stability in J_1 (see [AK] for more details).

This theory has several weak points. Firstly, on the manifold V of isovortical fields there may be no critical points at all. At least, the functional E usually does not assume its maximum and minimum on the surface V. Here is the typical example of situation, where the minimum and/or maximum of E on V cannot be achieved. Consider some velocity field u_0 in a strip $0 < y < 1$ with the period L along the x-axis. Let $\omega_0(x, y)$ be its vorticity, and suppose that $\int\int \omega dxdy = 0$. We are looking for a flow $u(x, y)$ with vorticity $\omega(x, y)$ such that ω is obtained from ω_0 by a volume preserving permutation of points, i.e. element of the group $\mathcal{D}$; in other words, u and u_0 are on the same manifold V and the energy $E(u)$ should be minimal (maximal). But the minimum cannot be achieved, because it is zero. In fact, we can construct a sequence $g_1, g_2, \cdots$ of diffeomorphisms, becoming more and more "mixing", which transform ω_0 into $\omega_1, \omega_2, \cdots$, and this sequence of vorticities tends weakly to 0. Corresponding velocity fields $u_1, u_2, \cdots$ tend to 0 strongly in L^2; thus, $E(u_i) \to 0$. The maximum value of E is also not always achievable. In fact, the supremum and infimum of energy on V depend only on the value distribution of ω_0, and not on the topology of its level lines. Thus, if for some u_0 the maximum is achieved (say, at the same u_0), then it cannot be achieved for any other function u_0' with the same value distribution of vorticity, but with different topology. For example, if the flow domain is a disk, and ω_0 is positive, depends only on the radius, decreases when the radius is growing, and is concentrated in a small neighborhood of the center of the disk, and ω_0' is another vorticity, having the form of *two* such spots whose size is $\sqrt{2}$ times smaller, then for the corresponding flow u_0' the maximum of E on its surface V' cannot be achieved. In all cases we see no evidence of the existence of other local maxima and minima, different from the global ones.

Secondly, as Sadun and Vishik [SV] observe, there is a serious drawback in applying the Arnold method in 3 or more dimensions. A natural way to prove that a critical point is a strict local maximum or minimum of E is to show that the second variation of the energy (the Hessian) defined on the tangent space to V is negative or positive definite at a critical point u_0. However Sadun and Vishik show in [SV] that in 3 or more dimensions the spectrum of the Hessian is not only *never* definite, but is generally unbounded from below as well as from above. The only exception are harmonic flows (i.e. both divergence and curl are zero) in which case the Hessian is identically zero. This result is suggestive, but does not prove that most flows in 3 dimensions are likely to be nonlinearly *unstable* in the Arnold's sense.

Another approach to nonlinear stability in 2 dimensions introduced by A. Shnirelman [S1] is based on the following ideas. Consider the space J_1 of incompressible vector fields in the flow domain M, which are tangent to the boundary and have a finite norm $||u||^2_{J_1} = ||u||^2_{L^2} + ||\nabla \times u||^2_{L^2}$. Consider the group $\mathcal{D}$ of volume preserving diffeomorphisms of M; it acts in $L^2(M)$ by the formula $g \cdot f(x) = f(g^{-1}(x))$. This is a unitary operator in $L^2(M)$, and we shall identify the group $\mathcal{D}$ with the group of these unitary transformations. Now, for every

$g \in \mathcal{D}$ and every $u \in J_1$ we define $g \cdot u$ as a unique field $v \in J_1$, such that $\nabla \times v(x) = \nabla \times u(g^{-1}(x))$ (assume for simplicity that the domain M is simply connected). Thus, we have defined an action of the group $\mathcal{D}$ in the space J_1. Hence we see that the manifolds of isovortical fields are just the orbits of this action. Consider now an extension of the group $\mathcal{D}$. It consists of linear operators in L^2, having the form $Kf(x) = \int K(x,y)f(y)dy$, where the kernel $K(x,y)$ satisfies the following conditions:

$$\begin{aligned} &(1) \qquad K(x,y) \geq 0 \quad \text{(i.e. } K(x,y)dxdy \text{ is a positive measure);} \\ &(2) \qquad \int_M K(x,y)dx \equiv 1; \qquad\qquad (3.31) \\ &(3) \qquad \int_M K(x,y)dy \equiv 1. \end{aligned}$$

Such operators, usually called bistochastic, or, in the terminology of A. Vershik, polimorphisms, form a semigroup $\mathcal{P}$; it consists of contracting operators in L^2, and the group $\mathcal{D}$ is dense in $\mathcal{P}$ in a weak operator topology.

Let us define the action of the semigroup $\mathcal{P}$ in the space J_1. For every $u \in J_1$ and $K \in \mathcal{P}$ we define $K \cdot u$ as a unique field $v \in J_1$, such that $\nabla \times v = K(\nabla \times u)$. We can define a partial order relation in the space J_1: suppose $u, v \in J_1$; we say that $u \prec v$, if $u = K \cdot v$ for some $K \in \mathcal{P}$.

For any $u_0 \in J_1$ let us consider the set S_{u_0} of vector fields $u \in J_1$, such that $u \prec u_0$, and $||u||_{L^2} = ||u_0||_{L^2}$. This set is a lattice with respect to the binary relation $\prec$; using Zorn's lemma, we prove that there exists a minimal element $v \in S_{u_0}$ (not necessarily unique). It turns out that this minimal element, which is a vector field, is a velocity field of a steady flow in M. We call such flows *minimal flows.*

Minimal flows have a clear physical meaning. The flow of an ideal incompressible fluid transports its own vorticity. It is natural to assume that the vorticity is permanently distorted by the flow, and effectively mixed. But the mixing operators are just polimorphisms. So, at any remote time moment $t >> 1$ the vorticity is (presumably) close in a weak sense to the result of the action of some operator $K_t \in \mathcal{P}$ on the initial vorticity $\omega_0 = \nabla \times u_0$, which means that the velocity field $u(x,t)$ is L^2-close to $K_t \cdot u_0(x)$.

The mixing of the vorticity field by the flow is practically irreversible, and we may assume that it proceeds until some constraint makes further mixing impossible. These constraints may be any integral of the motion, the primary one being the energy $E(u)$; further mixing is impossible, if every operator $K \in \mathcal{P}$ changes the kinetic energy of the flow. Thus minimal flows are the most degenerate states of fluid motion.

Our conjecture is that all (generic) 2-dimensional flows of an ideal incompressible fluid have a similar asymptotic behavior as $t \to \infty$: every such flow tends to some minimal flow. This hypothesis appears difficult to prove. It sounds close to (but in fact is very far from) "statistical hydrodynamics" in the sense of J. Miller and R. Robert [M], [R].

We do not expect the regularity of minimal flows to be very high; even if $u_0 \in C^\infty$, the corresponding minimal flow $v \in S_{u_0}$ has *a priori* only bounded vorticity, and it is unclear whether its vorticity is at least continuous.

There exist three classes of minimal flows. If a minimal flow belongs to the first class, then every mixing operator applied to its vorticity can only increase the kinetic energy. For a flow from the second class, every mixing of its vorticity decreases its energy. The third class contains only one flow (up to a multiplicative constant), namely the flow with constant vorticity.

Every minimal flow realizes a *global* minimum or maximum (for minimal flows respectively of the first and the second class) of the energy E on its orbit V; for the flow of the third class its orbit consists of one point. But we cannot assert (and it may be false) that this critical point is nondegenerate. In some examples the maximum of energy is attained on a compact set $H \subset V$, containing v (the convexity considerations show that the global minimum of energy is always assumed at a single point). In fact, consider the following example. Suppose that the flow domain M is a circular disk $|x| < 1$, and the velocity field $u_0(x)$ has the form of two small spots of vorticity of opposite sign, such that the total vorticity is zero. If we decrease the size of these spots, keeping the total vorticity (i.e. the circulation along the contour encircling each of the spots) constant, we can make the energy E arbitrarily big, because it grows as logarithm of the spot diameter. We assume that the size of both spots is small enough. The set S_{u_0}, defined above, contains a minimal flow v; if the size of vortices is small enough, this flow looks like two standing vortices of opposite signs (this configuration realizes the global maximum of the energy E on its isovortical manifold V). But this flow is not axisymmetric, and therefore the same minimal value of E is assumed on the whole circle in V, consisting of the flows, obtained from v by rotations. If we add to v the velocity field of a slow rotation of the disk, we obtain a nonsteady solution of the Euler equations in the disk, having the form of two small vortices of opposite signs, slowly rotating in the disk on the backdrop of a small uniform vorticity. (This type of flows is analogous to the precession of the axisymmetric rotating body; see [S1] for more details).

Thus minimal flows are not generally stable.What may be asserted is that they are "compactly unstable". This means that if v is a minimal flow, then there exists a compact $H \subset J_1$, containing v, such that for all initial velocities $w(x, 0)$, close to v, the flow $w(x, t)$ is close to H for all $t > 0$.

3.5. Spectrally stable and Arnold stable flows

We call a steady flow U_0 *spectrally stable* if the linearized equations (2.1), (2.2) have no exponentially growing solutions, i.e. the linearized operator L has no eigenvalues with a positive real part. What is the relation between the classes of spectrally stable and Arnold stable (or minimal) flows? In the case of a channel flow the essential spectrum of the linearized evolution operator e^{tL} lies on the unit circle; applying the theorem 3.1 we conclude that if the flow is spectrally unstable, it is nonlinearly unstable; thus, it is not Arnold stable. For generic steady flows the question is less clear, because unstable eigenvalues of the operator e^{tL} may occur on its essential spectrum. This may happen, if, for example, the flow U_0 contains a thin jet in one part of the flow domain M, and a hyperbolic stagnation point elsewhere in M. In this case the essential spectrum of e^{tL} contains an annulus $e^{-\Lambda t} \leq ||z| \leq e^{\Lambda t}$, and the eigenvalue $e^{\lambda t}$ satisfies $1 < e^{\lambda t} < e^{\Lambda t}$. Theorem 3.1 is no more applicable, and it is now unclear whether the flow U_0 is stable (while it is unlikely that a distant saddle point could stabilize the jet).

In the other direction the answer is clearer: there exist steady flows which are spectrally stable, and are not Arnold stable. Our examples are plane-parallel flows in a channel $-1 < y < 1$, periodic in the x-direction with the period L (so that the flow domain may be regarded as a side surface of a cylinder). In the first example the velocity profile $U(y)$ has the form $U(y) = y + \varepsilon f(y)$ for an arbitrary smooth function $f(y)$ and small $\varepsilon > 0$; in the second example $U(y)$ may be an arbitrary smooth function, but the period L in the x-direction is small (depending on U). In both examples, if ε or respectively L are small enough, there are no unstable eigenvalues. This is proved in the paper of Faddeev [Fa] which is excellent for its clarity.

However these flows, for some profiles U, are not Arnold stable. For example, consider $U(y) = \sin(m\pi y)$ for integer $m > 1$ and sufficiently small period L so that the flow is spectrally stable. Let $U_0 = (U(y), 0)$ denote the flow field and let $\omega_0 = \nabla \times U_0$ be its vorticity. Let us show that this field is neither the point of a local minimum nor the point of a local maximum of the energy E among the fields u isovortical with U_0. To do this, it is sufficient to show that for every $\varepsilon > 0$ there exist two volume preserving diffeomorphisms η and ζ such that $||\omega_0 \circ \eta^{-1} - \omega_0||_{L^2} < \varepsilon$, $||\omega_0 \circ \zeta^{-1} - \omega_0||_{L^2} < \varepsilon$, and $E(\omega_0 \circ \eta^{-1}) < E(\omega_0)$, $E(\omega_0 \circ \zeta^{-1}) > E(\omega_0)$, where $E(\omega)$ denotes the kinetic energy of incompressible flow with vorticity ω. We divide the flow domain $M = \{(x, y)|0 \leq x < L, \ -1 \leq y \leq 1\}$ into small equal cells M_k; for example, they may be equal squares of size δ. Consider permutations of these cells, and for every permutation τ consider the function $\tau \cdot \omega_0 = \omega_0 \circ \tau^{-1}$ (here τ is regarded as a measurable transformation of the flow domain M, preserving the Lebesgue measure). As it is proven in [S2], for every permutation τ there exists a smooth volume preserving diffeomorphism τ', such that $\tau' = \tau$ outside an arbitrarily small neighborhood of $\bigcup_k \partial M_k$. Thus, $||\omega_0 \circ \tau' - \omega_0 \circ \tau||_{L^2}$ may be made arbitrarily small, and we shall consider now only the action of permutations of cells on the vorticity and velocity fields. We define two permutations, τ_1 and τ_2, such that the action of τ_1 on ω_0 increases the energy of the flow, while the action of τ_2 decreases the energy. Note that if ω_0 and ω_0' are two vorticity fields, which are equimeasurable, i.e. $\mathrm{mes}\{\omega_0 < c\} = \mathrm{mes}\{\omega_0' < c\}$ for every c, then there exist a partition of M into sufficiently small equal squares M_k and a permutation τ of these cells such that $||\omega_0' - \omega_0 \circ \tau||_{L^2}$ is arbitrarily small (see [S2]). Furthermore, there exist functions ω_0' which are equimeasurable with ω, $E(\omega_0') > E(\omega_0)$, and $||\omega_0' - \omega_0||_{L^2}$ is as small as we wish. For example, we can inflate a little one period of $\omega_0(x, y) = m \sin my$ in the y-direction while shrinking other periods. Now we can find a permutation τ_1, which approximately transforms ω_0 into ω_0', and then approximate this permutation by a smooth volume preserving diffeomorphism η. Our construction shows that $||\omega_0 \circ \eta - \omega_0||_{L^2}$ may be made arbitrarily small and that $E(\omega_0 \circ \eta) > E(\omega_0)$. Thus ω_0 is not a point of local minimum of E on V.

To show that ω_0 is not a point of local maximum of E on V, observe first that the mean value of ω_0 is zero. This makes the following construction possible. Let us divide M into N equal cells and pick $n << N$ cells by random. Let τ_2 be a random permutation of the chosen n cells. If $N \to \infty$, $n \to \infty$, $n/N \to \delta$, then, with probability 1, $\omega_0 \circ \tau_2$ tends weakly in $L^2(M)$ to $(1 - \varepsilon)\omega_0$; on the other hand, $||\omega_0 \circ \tau_2 - \omega_0||_{L^2} < c\delta$; hence, we can find a diffeomorphism ζ, such that $||\omega_0 \circ \zeta - \omega_0||_{L^2} < c\varepsilon$, while $E(\omega_0 \circ \zeta) < E(\omega_0) - \varepsilon^2$. This shows that ω_0 is not a point of local minimum of E on V. Hence we have the following

Problem. We have shown that there exist steady, plane-parallel flows such that the linearized problem has no unstable eigenvalues, but these flows don't satisfy the conditions of the Arnold stability (or minimality, which is essentially the same). Are they stable in J_1? On one hand, there is no "fast" exponential instability. On the other hand, the vorticity integrals are not constraints that prevent a flow from going far away from U_0, if initially it was close to U_0 in J_1. So, either there exist other constraints of unknown nature, or the above flows are unstable in J_1. However such an instability is quite different from instabilities with which we are familiar.

3.6. Problems and conjectures on the Arnold stable and minimal flows.

We have four remarkable classes of steady flows: Lyapunov stable flows, Arnold stable, minimal flows, and spectrally stable ones (in the sense that the linearized equation has no unstable eigenvalues). What are relations between these classes?

Suppose that the steady flow $U_0(x)$ does not satisfy Arnold's condition of stability. This means that U_0 is a critical point of the energy E restricted on the surface V of isovortical vector fields in the space J_1, but this flow is not the local minimum or maximum of the functional E on V. Then we may anticipate that this flow is unstable, because there is nothing to hold the perturbed flow $u(x, t)$ close to $U_0(x)$; the Energy-Casimir method breaks down. If the flow U_0 is spectrally unstable, then Theorem 3.1 shows that it is nonlinearly unstable. But there is a wide gap between spectrally unstable flows and those which do not satisfy the Arnold's conditions. We conjecture that those flows which are neither Arnold stable nor spectrally unstable are nonlinearly unstable in the space J_1, but the nature of their instability is different from that of linearly unstable flows.

3.7. L^2-instability

In the above theories of nonlinear stability we considered stability of smooth flows with respect to small perturbations, which are small in the J_1 (i.e. in H^1) sense: the vorticity of perturbations should be small in L^2. The theory breaks down if we drop the condition on the vorticity perturbation and consider all (smooth) velocity fields $u(x,0)$ which are close to $U_0(x)$ in L^2 without any conditions on derivatives. Note that this class of perturbations is no less physically significant than the previous one, because it describes perturbations with small energy. Such perturbations may be easily created, for example, by inserting small obstacles in the flow. In this case, the vorticity integrals are completely destroyed and it appears that nothing prevents the flow from going far away from U_0. Hence, the natural conjecture is that every nontrivial flow (i.e.flow having a nonconstant velocity) is unstable with respect to small in L^2 perturbations.

Consider the simplest basic steady flow, namely a parallel flow. Let M be a strip $0 \leq x_2 \leq 1$ in the (x_1, x_2)-plane. We restrict ourselves to the flows having period L along the x_2-axis; this period is the same for all flows that are considered below. Suppose that the velocity field $U_0(x)$ has the form $(U(x_2), 0)$, where U is a given smooth function (the velocity profile). The original question asked, for which profiles U is the flow U_0 is stable. Our first result is the following

Theorem 3.2 *For every nontrivial (i.e. different from constant) velocity profile U the flow U_0 is L^2-unstable. This means that for every function $U(x_2) \neq$ const there exists $C > 0$, such that for every $\varepsilon > 0$ the following is true. There exist $T > 0$ and a smooth force $f(x,t)$, defined in $M \times [0,T]$, such that $\int_0^T \| f(\cdot,t) \|_{L^2} \, dt < \varepsilon$, and f transfers the flow U_0 during the time interval $[0,T]$ into a steady flow u_1, such that $\| U_0 - u_1 \|_{L^2} > C$.*

This theorem is proved in [S3] by an explicit construction, based on the variational method. The next result is much stronger, but here we use a weaker notion of instability (see [S4]). Let X be a Banach space of incompressible vector fields in M, tangent to the boundary. Consider the Euler equations with a nonzero right hand side (i.e. external force):

$$\frac{\partial u}{\partial t} + (u, \nabla)u + \nabla p = f; \tag{3.32}$$

$$\nabla \cdot u = 0. \tag{3.33}$$

Here $f = f(x,t)$ is a smooth in x vector field such that $\nabla \cdot f = 0$ and $f(x,t)|_{\partial M}$ is parallel to ∂M. Consider the behavior of $u(x,t)$ when f is small in the following sense: $\int_0^T \| f(\cdot,t) \|_{L^2} \, dt$ is small, where $[0,T]$ is the time interval (assumed to be long) where the flow is considered. For example, if f has the form $f(x,t) = F(x)\delta(t)$ we return to the initial stability problem.

Definition 3.3. *Suppose that $u(x_1, x_2)$ and $v(x_1, x_2)$ are two steady flows. We say that the force f transfers the flow u into the flow v during the time interval $[0,T]$, if the following is true: if $w(x,t)$ is the solution of the nonhomogeneous Euler equations (3), (4) with the initial condition $w(x,0) = u(x)$, then $w(x,T) = v(x)$.*

We note that if the force f satisfies a stronger condition $\int_0^T \| \omega(\cdot,t) \|_{L^2} \, dt < \varepsilon$, where $\omega = \nabla \times f$ is the vorticity, then for every Arnold stable flow U_0, the resulting perturbation at time t will be small, too. But the following theorem shows that situation in L^2 is quite different.

Theorem 3.3 *Suppose that $U(x_2)$ and $V(x_2)$ are two velocity profiles, such that $\int_0^1 U(x_2)dx_2 = \int_0^1 V(x_2)dx_2$ and $\int_0^1 \frac{1}{2}|U(x_2)|^2dx_2 = \int_0^1 \frac{1}{2}|V(x_2)|^2dx_2$; let $U_0(x_1,x_2) = (U(x_2),0)$, $v_0(x_1,x_2) = (V(x_2),0)$ be corresponding steady parallel flows (having equal momenta and energies). Then for every $\varepsilon > 0$ there exist $T > 0$ and a smooth force $f(x,t)$, such that $\int_0^T \parallel f(\cdot,t) \parallel_{L^2} < \varepsilon$, and f transfers u into v during the time interval $[0,T]$.*

In other words, the flow may be considerably changed by arbitrarily small in L^2 force, provided the time interval is sufficiently long.
This means that the flow of an ideal incompressible fluid is perfectly controllable by arbitrarily small force.

Theorem 3.2 is proven by an explicit construction of the flow. Note first, that if $U_1, U_2, \cdots, U_N$ are velocity profiles, and Theorem 2 is true for every pair (U_i, U_{i+1}) of velocity profiles, then we can pass from U_1 to U_N by simply concatenating the flows connecting U_i and U_{i+1}; thus Theorem 2 is true for the pair (U_1, U_N). Therefore it is enough to construct the sequence of steady flows with profiles $U_1, \cdots, U_N$, and the intermediate nonsteady flows connecting every two successive steady ones. We also note that it is enough to construct a sequence of piecewise-smooth flows, because it is not difficult to smooth them, so that the necessary force will have arbitrarily small norm in $L^1(0,T;L^2(M))$.

As a first step, we change the flow with the profile $U = U_1$ by a piecewise-constant profile U_2 with sufficiently small steps; this may be done by a force with arbitrarily small norm. Thus, $U_2(x_2)$ is a step function, $U_2(x_2) = U_2^{(k)}$ for $x_2^{(k-1)} < x_2 < x_2^{(k)}$, $k = 1, \cdots, K$. Every successive profile U_i is also a stepwise function. We are free to subdivide the steps and change a little the values of velocity, if these changes are small enough.

Every flow u_k is obtained from the previous one u_{k-1} by one of two operations, described in the following theorems.

Theorem 3.4 *Let $U(x_2)$ be a step function, $U(x_2) = U^{(k)}$ for $x_2^{(k-1)} < x_2 < x_2^{(k)}$; let $V(x_2)$ be another step function, obtained by transposition of two adjacent segments $[x_2^{(k-1)}, x_2^{(k)}]$ and $[x_2^{(k)}, x_2^{(k+1)}]$. Let $u(x_1,x_2)$, $v(x_1,x_2)$ be parallel flows with velocity profile $U(x_2), V(x_2)$. Then for every $\varepsilon > 0$ there exist $T > 0$ and a piecewise-smooth force $f(x,t)$, such that $\int_0^T \parallel f(\cdot,t) \parallel_{L^2} < \varepsilon$, and the force f transfers the flow u into the flow v during the time interval $[0,T]$.*

To formulate the next theorem, we recall the law of an elastic collision of two bodies. Suppose that two point masses m_1 and m_2, having velocities u_1 and u_2, collide elastically. Then their velocities after collision will be $v_1 = 2u_0 - u_1$, $v_2 = 2u_0 - u_2$, where $u_0 = (m_1u_1 + m_2u_2)/(m_1 + m_2)$ is the velocity of the center of masses. The transformation $(u_1, u_2) \to (v_1, v_2)$ is called a *transformation of elastic collision.*

Theorem 3.5 *Assume that the profile $U(x_2)$ is like the profile in Theorem 3, and the profile $V(x_2)$ is equal to $U(x_2)$ outside the segment $x_2^{(k-1)} < x_2 < x_2^{(k+1)}$; on the last segment, $V(x_2) = v^{(k)}$, if $x_2^{(k-1)} < x_2 < x_2^{(k)}$, and $V(x_2) = v^{(k+1)}$, if $x_2^{(k)}) < x_2 < x_2^{(k+1)}$, where $(v^{(k)}, v^{(k+1)})$ is obtained from $(u^{(k)}, u^{(k+1)})$ by the transformation of elastic collision, the lengths $x_2^{(k)} - x_2^{(k-1)}$, $x_2^{(k+1)} - x_2^{(k)}$ playing the role of masses m_1, m_1. Let $u(x_1, x_2), v(x_1, x_2)$ be parallel flows with profiles $U(x_2), V(x_2)$. Then for every $\varepsilon > 0$ there exist $T > 0$ and a force $f(x,t)$, such that $\int_0^T \| f(\cdot, t) \|_{L^2} < \varepsilon$, and the force f transfers the flow u into flow v.*

Assume now that $U(x_2)$ and $V(x_2)$ are two velocity profiles, having equal momenta and energies. Then it is not difficult to construct a sequence of step functions $U_2(x_2), U_3(x_2), \cdots, U_N(x_2)$, so that U_2 is L^2-close to $U_1 = U$, U_N is L^2-close to V, and every profile U_k is obtained from U_{k-1} by one of two operations, described in Theorems 3 and 4. Using these theorems and the discussion above, we construct a piecewise-smooth force $f(x,t)$ such that $\int_0^T \| f(\cdot,t) \|_{L^2} \, dt < \varepsilon$ and f transfers U into V during the time interval $[0, T]$.

Theorems 3.4 and 3.5 are proved by explicit construction of the flows. They are true also for circular flows in a disk, with the angular momentum taking the place of momentum in Theorem 3.5. But for generic 2-dimensional domains the situation is not so clear. We don't know whether there is an integral of motion, similar to the angular momentum, in any domain different from the disk. If such integral does not exist, which is most likely, then the natural conjecture is that for any two flows with equal energies the conclusion of Theorem 3.3 is true. But this behavior is paradoxical: just imagine a nearly circular flow in a nearly circular domain (e.g. ellipse), which after some long time changes the sign of the angular velocity. This question requires more study.

3.8. L^2-instability and scattering for the Euler equations

The L^2-instability may be regarded as another side of of the hypothetical picture of an asymptotic behavior of generic flow as $t \to \infty$, developed in section 3.4. According to this hypothesis, the vorticity carried by the flow is mixed more and more until its further mixing becomes impossible because of the energy conservation. Thus the flow $u(x,t)$ tends to some minimal flow $u_+(x)$. This passage from an initial (arbitrary) flow $u_0(x) = u(x,0)$ to the final state (minimal flow) $u_+(x)$ is analogous to the scattering of linear waves on an obstacle (or potential); to make this analogy closer, we can continue the flow back in the direction of negative t; as $t \to \infty$, the flow $u(x,t)$ tends to some minimal flow $u_-(x)$. Consider the passage from u_- to u_+; this is the exact analog of the scattering operator. The vorticity field for $|t|$ very big is a highly oscillating function in the flow domain, which approaches the vorticity of the final flow only in a weak sense. Thus the velocity field

$u(x,t)$ for $t<0$ and $|t|$ very big may be arbitrary close in L^2 to $u_-(x)$. If we take it as the initial condition for the Euler equation, we obtain an example of a small in L^2 perturbation of a minimal flow $u_-(x)$, which grows considerably on a large time interval.

The scattering property was proved by Caglioti and Maffei [CM1] for 1-dimensional Vlasov-Poisson equation, which has some features similar to the 2-dimensional Euler equation but is much simpler. As for the Euler equations themselves, they attempted in [CM2] to construct an asymptotic decomposition of their hypothetical scattering solution as $|t|\to\infty$, having the form of a collection of long and narrow vorticity filaments (this form would be assumed by the vorticity field if it is transported as a passive scalar by a smooth and steady field). However they managed to construct only the first term of this asymptotics. The difficulty lies in the fact that the interaction between the oscillating part of the vorticity field and the mean field decreases very slowly (as $|t|^{-1}$) when $|t|\to\infty$; the vorticity perturbation remains "active" for all t, and in no way can be regarded as a passive scalar. The opposite difficulty was pointed out by Isichenko in his paper [I]; in this smooth picture the transverse motions of fluid parcels decay too rapidly (as $|t|^{-5/2}$) when $|t|\to\infty$, and therefore the state of the minimal flow is out of reach.

Our hypothesis is that the approach to the final state is not that smooth. The flow picture at small scales is being transformed all the time, infinitely many times. Thus, the asymptotic solution in the form assumed in [CM2] simply does not exist. The true picture is much more violent (including infinite series of refolding of vorticity filaments and appearance of new, secondary filaments, which complicates the picture even more). This problem deserves extensive study.

4. Conclusion and further questions

In this paper we did not try to cover all the vast field of the fluid instability; rather we have concentrated on some particular aspects of it. We tried to show that there exist different kinds of instability. In fact the differences are so big that they deserve to be regarded as different phenomena, and not as different kinds of one phenomenon. The difference may be illustrated by the following simple example. Consider a pendulum balanced upside down in the top position. It is certainly unstable; almost any small disturbance will grow in time. Consider, on other hand, a particle moving freely in Euclidean space which is at rest in some point. This equilibrium is also unstable, because every small impulse will result in steady motion, which after a long time will move the particle far away. These two instabilities appear quite different. Technically, the second one is associated with the Jordan cell structure of the governing operator, while in the first case it is an unstable eigenvalue that determines the instability.

In the context of fluid motion, we see the same two sorts of instability but in a much stronger form. Most of the linear and nonlinear instabilities considered in this work belong to the second class; they may be called "slow" instabilities, as opposed to "fast" instabilities, associated with isolated unstable eigenvalues. The reason for the ubiquity of slow instabilities is the fact that steady flows in 2 dimensions themselves have a Jordanian structure at almost every point in the sense that the differential of the flow map (i.e. diffeomorphism produced by the flow during some time) is a Jordan matrix. This fact alone is enough to explain linear growth of perturbations in the smooth norms. On the other hand, instability in the energy (L^2) norm is associated directly with the simple picture of a freely moving particle described above. And finally, the nature of a hypothetical instability in a vorticity norm beyond "Arnold" stability and spectral instability is quite unclear.

If there are unstable discrete eigenvalues in the spectrum of linearized operator, then the flow is definitely unstable. This is "fast" instability with exponential growth of disturbances. Less clear is the situation when the unstable continuous spectrum of the evolution operator is nonempty (e.g., when the basic steady flow has a hyperbolic stagnation point). In this case the spectrum fills an annulus, and for each point of the spectrum we can construct a solution of the linearized equation which grows in time, but not monotonically; rather it has "outbreaks" in some rare time moments, being small most of the time. It is unclear whether we can construct a growing solution to the full nonlinear Euler equations showing similar behavior. May be we have here one more kind of instability, unaccounted for at present in the traditional scheme of instability..

The case of an unstable eigenvalue embedded in the essential spectrum is also unclear. Does a growing solution appear, or will the continuous spectrum "damp" it? What will happen if we change a little the basic steady flow? Does the eigenvalue "dissolve" in the continuous spectrum? Is there something analogous to the Fermi rule?

So we can be optimistic: there is a lot of work ahead of us.

Acknowledgements. This work has been written during our common stay at the Institute des Hautes Études Scientifiques at Bures-sur-Yvette and in the Newton Institute for Mathematical Sciences in Cambridge in August-September 2000. We are very much thankful to these institutions for an exciting scientific atmosphere and excellent working conditions. We are grateful to the organizers of the Program on the Geometry and Topology of Fluid Flows in the Newton Institute for inviting us together to Cambridge. The first author (S. F.) received support from EPSRC grant GR K99015 and NSF grant DMS-9970977.

We are extremely thankful to Victor Yudovich and Misha Vishik for many stimulating discussions.

References

[A1] V. I. Arnold, Sur la topologie des écoulements stationnaires des fluides parfaites. *C.R. Acad. Sci. Paris*, v. 261 (1965), 17–20.

[A2] V. I. Arnold, Notes on the three-dimensional flow pattern of a perfect fluid in the presence of a small perturbation of the initial velocity field (Russian). *Applied Mathematics and Mechanics*, v.36, no. 2 (1972), 236–242.

[A3] V. I. Arnold, Sur la géometrie differentielle des groupes de Lie de dimension infinie et ses applications a l'hydrodynamique des fluides parfaits. *Annales de l'Institut Fourier*, Grenoble, v. 16 (1966), 316–361.

[AK] V. Arnold, B. Khesin, Topological Methods in Hydrodynamics, Springer, 1998.

[BFY] L. Belenkaya, S. Friedlander, and V. Yudovich, The unstable spectrum of oscillating shear flows. *SIAM Journal of Applied Mathematics*, v. 59, no. 5 (1999), 1701–1715.

[BHL] B. J. Bayly, D. D. Holm, A. Lifschitz, 3-D stability of elliptical vortex columns in external strain flows. *Philosophical Transactions of the Royal Society, London, Ser. A*, v. 354, no. 1709 (1996), 61–66.

[CM1] E. Caglioti and C. Maffei, Time asymptotics for solutions of Vlasov-Poisson equation in a circle. *Journal of Statistical Physics*, v. 92, no. 1–2 (1998), 301–323.

[CM2] E. Caglioti and C. Maffei, Scattering theory: a possible approach to the homogenization problem for the Euler equations. *Rendiconti di Matematica e delle sue Applicazioni, Serie VII*, v. 17, no. 3 (1997), 445–475.

[Ch] S. Chandrasekhar, *Hydrodynamic and Hydromagnetic Stability*. Clarendon Press, Oxford, 1961.

[Chi] C. Chicone, A geometric approach to regular perturbation theory with application to hydrodynamics. *Transactions of American Mathematical Society*, v. 347 (1995), no. 12, 4559–4598.

[DFGHMS] T. Dombre, U. Frisch, J. M. Greene, M. Hénon, A. Mehr, and A. M. Soward, Chaotic streamlines in the ABC flows. *Journal of Fluid Mechanics*, v. 167 (1986), 353–391.

[DR] P. G. Drazin and W.H. Reid, *Hydrodynamic Stability*. Cambridge Monographs on Mechanics and Applied Mathematics, Cambridge University Press, Cambridge, 1982.

[E] K. S. Eckhoff, On stability for symmetric hyperbolic systems. *Journal of Differential Equations*, v. 40 (1981), 94–115.

[ES] K. S. Eckhoff and L. Storesletten, On the stability of rotating compressible and inviscid fluids. *Journal of Fluid Mechanics*, v. 99 (1980), 433–448.

[Fa] L. D. Faddeev, On the theory of stability of stationary plane-parallel flows of an ideal fluid (Russian). *Zapiski Nauchnyh Seminarov LOMI*, v. 21 (1971), 164–172.

[F] S. Friedlander, On vortex tube stretching and instabilities in an inviscid fluid. IHES preprint, 2000.

[Fr] F. G. Friedlander, *Sound Pulses*. Cambridge University Press, Cambridge, 1958.

[FGV] S. Friedlander, A. D. Gilbert, and M. M. Vishik, Hydrodynamic instability for certain ABC flows. *Geophysical and Astrophysical Fluid Dynamics*, v. 73 (1993), 97–107.

[FH] S. Friedlander and L. N. Howard, Instability in parallel flow revisited. *Studies in Applied Mathematics*, v. 101 (1998), 1–21.

[FSV] S. Friedlander, W. Strauss, and M. M. Vishik, Nonlinear instability in an ideal fluid. *Annales de l'Institut Henri Poincaré. Analyse Non Linéaire*, v. 14, no. 2 (1997), 187–209.

[FV] S. Friedlander and M. M. Vishik, Instability criteria for steady flows of a perfect fluid. *Chaos*, v.2, no. 3 (1992), 455–460.

[FV1] S. Friedlander and M. M. Vishik, Instability Criteria for the flow of an inviscid incompressible fluid. *Physical Review Letters*, v. 66, no. 17 (Apr. 29, 1991), 2204–2206.

[FVY] S. Friedlander, M. Vishik, V. Yudovich, Unstable eigenvalues associated with inviscid fluid flows. To appear in *Journal of Mathematical Fluid Mechanics*, 2000.

[G] E. Grenier, On the nonlinear instability of the Euler and Prandtl equations. *Communications on Pure and Applied Mathematics*, v. 53, no.1 (2000), 1067–1091.

[I] M. Isichenko, Nonlinear Landau damping in collisionless plasma and inviscid fluid. *Physical Review Letters*, March 24, 1997, v. 78, no. 12, 2369–2372.

[K] H. Koch, Instability for incompressible and inviscid fluids. *Partial differential equations (Praha, 1998)*, 240–247, Chapman & Hall/CRC Res. Notes Math., 406, Chapman & Hall/CRC, Boca Raton, FL, 2000.

[L] D. Ludwig, Exact and asymptotic solutions of the Cauchy problem. *Communications on Pure and Applied Mathematics*, v. 13 (1960), 473–508.

[LH] A. Lifschitz and E. Hameiri, Local stability conditions in fluid dynamics. *Physics of Fluids*, v. A3 (1991), 2644.

[LL] A. J. Lichtenberg and M. A. Lieberman, *Regular and Chaotic Dynamics.* Applied Mathematical Sciences, v. 88, Springer Verlag, 1992.

[M] J. Miller, Statistical mechanics of Euler equations in two dimensions. *Physical Revues Letters*, v. 65 (1990), no. 17, 2137–2140.

[MS] L. Meshalkin and Y. Sinai, Investigation of stability for a system of equations describing the motion of a viscous incompressible fluid (Russian). *Applied Mathematics and Mechanics*, v. 25 (1961), 1140–1143.

[R] R. Robert, A maximum entropy principle for two-dimensional perfect fluid dynamics. *Journal of Statistical Physics*, v. 65 (1991), no. 3–4, 531–553.

[S1] A. Shnirelman, Lattice theory and flows of ideal incompressible fluid. *Russian Journal of Mathematical Physics*, v. 1 (1993), no. 1, 105–114.

[S2] A. Shnirelman, On the geometry of the group of diffeomorphisms and the dynamics of an ideal incompressible fluid.*Math. USSR Sbornik*, v. 56, no. 1, 79–105 (1987).

[S3] A. Shnirelman, On the L^2-instability of fluid flows. Séminaire "Equations aux Derivées Partielles" du Centre de Mathematiques de l'Ecole Polytechnique, 15 fevriere, 2000. http://math.polytechnique.fr/seminaires edp/99-2000/seminaire99-2000.html.

[S4] A. Shnirelman, On the L^2-instability and L^2-contrallability of steady flows of an ideal incompressible fluid. Journees des equations aux derivees partielles, St. Jean de Mont, June 1999.

[Sh] E. E. Shnol, On the instability of plane-parallel flows of perfect fluid. *Journal of Applied Mathematics and Mechanics*, v. 38 (1974), 464–468 (1975).; translated from Prikl. Mat. Meh. 38 (1974), 502–506 (Russian)

[SV] L. Sadun and M. M. Vishik, The spectrum of the second variation for an ideal incompressible fluid. *Physics Letters A*, v. 182 (1993), 394–398.

[V] M. M. Vishik, Spectrum of small oscillations of an ideal fluid and Lyapunov exponents. *Journal de Mathématiques Pures et Appliquées*, v. 75 (1996), 531–557.

[VF] M. M. Vishik and S. Friedlander, Dynamo theory methods for hydrodynamic stability. *Journal de Mathématiques Pures et Appliquées*, v. 72 (1993), 145–180.

[Y1] V. Yudovich, Example of secondary stationary flow or periodic flow appearing while a laminar flow of a viscous incompressible fluid loses its stability. *Applied Mathematics and Mechanics*, v. 29, no. 3 (1965), 453–467.

[Y2] V. Yudovich, The linearization Method in Hydrodynamical Stability Theory. *AMS Translation of Mathematical Monographs*, v. 74, 1989.

[Y3] V. Yudovich, On loss of smoothness for the Euler equations. *Chaos*, v.10, no.3 (2000).

Susan Friedlander
Department of Mathematics
University of Illinois at Chicago
851 S Morgan Street
M/C 249 Chicago, IL 60607–7045
USA
E-mail address: susan@math.nwu.edu

Alexander Shnirelman
School of Mathematical Sciences
Tel–Aviv University
Ramat–Aviv
Tel–Aviv 69978
Israel
E-mail address: shnirelm@math.tau.ac.il

Finite Volume Solution of 2D and 3D Euler and Navier-Stokes Equations

J. Fürst, M. Janda and K. Kozel

Abstract. This contribution deals with the modern finite volume schemes solving the Euler and Navier-Stokes equations for transonic flow problems. We will mention the TVD theory for first order and higher order schemes and some numerical examples obtained by 2D central and upwind schemes for 2D transonic flows in the GAMM channel or through the SE 1050 turbine cascade of Škoda Plzeň.

In the next part two new 2D finite volume schemes are presented. Explicit composite scheme on a structured triangular mesh and implicit scheme realized on a general unstructured mesh. Both schemes are used for the solution of inviscid transonic flows in the GAMM channel and the implicit scheme also for the flows through the SE 1050 turbine cascade using both triangular and quadrilateral meshes. For the case of the flows through the SE 1050 turbine we compare the numerical results with the experiment.

The TVD MacCormack method as well as a finite volume composite scheme are extended to a 3D method for solving flows through channels and turbine cascades.

1. Mathematical model

We consider the system of 2D Navier-Stokes equations for compressible medium in conservative form:

$$\begin{array}{ll} W_t + F_x + G_y = R_x + S_y, & \\ W = [\rho, \rho u, \rho v, e], & p = (\gamma - 1)\left(e - \frac{1}{2}\rho(u^2 + v^2)\right), \\ F = [\rho u, \rho u^2 + p, \rho uv, (e+p)u], & G = [\rho v, \rho uv, \rho v^2 + p, (e+p)v], \\ R = [0, \tau_{11}, \tau_{12}, u\tau_{11} + v\tau_{12} + kT_x], & S = [0, \tau_{21}, \tau_{22}, u\tau_{21} + v\tau_{22} + kT_y], \end{array} \quad (1)$$

where ρ is the density, (u, v) the velocity vector, e the total energy per unit volume, μ the viscosity coefficient, k is the heat conductivity, p is the pressure, γ is the adiabatic coefficient, and the components of the stress tensor τ are

$$\tau_{11} = \mu\left(\frac{4}{3}u_x - \frac{2}{3}v_y\right), \quad \tau_{21} = \tau_{12} = \mu\left(u_y + v_x\right), \quad \tau_{22} = \mu\left(-\frac{2}{3}u_x + \frac{4}{3}v_y\right). \quad (2)$$

The 2D Euler equations are obtained from the Navier-Stokes equations by setting $\mu = k = 0$.

The system of 3D Euler equations is written also in conservative form (here w is the third component of the velocity vector):

$$\begin{aligned} &W_t + F_x + G_y + H_z = 0, && W = [\rho, \rho u, \rho v, \rho w, e], \\ &p = (\gamma - 1)\left(e - \tfrac{1}{2}\rho(u^2 + v^2 + w^2)\right), && F = [\rho u, \rho u^2 + p, \rho uv, \rho uw, (e+p)u], \\ &G = [\rho v, \rho uv, \rho v^2 + p, \rho vw, (e+p)v], && H = [\rho w, \rho uw, \rho vw, \rho w^2 + p, (e+p)w]. \end{aligned} \tag{3}$$

The system of Euler equations is a set of first order PDE of hyperbolic type whereas the system of Navier-Stokes equations is a set of second order PDE of parabolic type.

1.1. Boundary conditions

We assume four types of boundary conditions for the case of Euler equations:

Inlet:: At the inlet we prescribe the direction of the velocity (by the inlet angle for the 2D case and by 2 angles for the 3D case), the value of the stagnation density ρ_0 and the stagnation pressure p_0. We extrapolate the static pressure p from inside and compute the other required quantities using the following relations between the stagnation and the static quantities:

$$p_0 = p\left(1 + \frac{\gamma - 1}{2}M^2\right)^{\frac{\gamma}{\gamma-1}} \qquad \rho_0 = \rho\left(1 + \frac{\gamma - 1}{2}M^2\right)^{\frac{1}{\gamma-1}}$$

where M is the local Mach number defined by $M = \sqrt{u^2 + v^2 + w^2}/a$ where the local speed of sound is $a = \sqrt{\gamma p/\rho}$. For the Navier-Stokes equations we assume $\partial T/\partial \vec{n} = 0$ and W_∞ given.

Outlet:: At the outlet we prescribe the value of the static pressure p and extrapolate the values of the density ρ and of the velocity vector from the flow field. For the viscous flows we assume again $\partial T/\partial \vec{n} = 0$.

Solid wall:: Here we prescribe the non-permeability condition ($\vec{u} \cdot \vec{n} = 0$) for the inviscid case or $\vec{u} = 0$ for the case of viscous flows. Here we assume also the adiabatic walls (i.e. $\partial T/\partial \vec{n} = 0$ where T is the temperature).

Periodicity:: Here we prescribe the periodical condition for all components of the vector of unknowns W.

2. Numerical methods

2.1. TVD schemes for one-dimensional scalar case

The theory for full nonlinear systems is very complicated and we restrict the analysis of the numerical method only to the scalar case. We assume the initial value problem for the one-dimensional scalar equation

$$u_t + f(u)_x = 0 \tag{4}$$

with the initial condition $u(x, 0) = u_0(x)$. This initial value problem is solved for $(x, t) \in R \times R_+$.

However the theoretical results are not straightforward applicable to nonlinear systems and to bounded domains, the simple model case is still important for understanding some properties of the numerical methods and as a hint for constructing good numerical methods for systems.

We approximate the weak solution to (4) by a piecewise constant function $U(x,t) = u_i^n$ for $(i-1)\Delta x < x \leq i\Delta x$ and $(n-1)\Delta t < t \leq n\Delta t$ where Δx and Δt are the mesh spacings in the space and time variables. The initial condition u^0 is computed as

$$u_i^0 = \frac{1}{\Delta x} \int_{(i-1)\Delta x}^{i\Delta x} u_0(x)\, dx. \tag{5}$$

The values u_i^n are computed for $n > 0$ using the following explicit numerical scheme written in so the called conservation form:

$$u_i^{n+1} = u_i^n - \frac{\Delta t}{\Delta x} \left[\tilde{f}(u_{i-p}^n, u_{i-p+1}^n, \ldots, u_{i+q}^n) - \tilde{f}(u_{i-p-1}^n, u_{i-p}^n, \ldots, u_{i+q-1}^n) \right] \tag{6}$$

where $\tilde{f}$ is a continuous function of $p + q + 1$ parameters called the numerical flux function which approximates the physical flux function f in the following sense:

$$\forall v :: \ \tilde{f}(v, v, \ldots, v) = f(v). \tag{7}$$

The analysis of the above mentioned method is complicated even in the one-dimensional scalar case because of the nonlinearity of the flux function f (and consequently $\tilde{f}$). Nevertheless, there exist good theoretical backgrounds for certain subclasses of the general method. Namely, the following facts have been proved:

- the convergence towards the unique (so called viscosity vanishing[1]) weak solution of (4) for the class of *monotone methods*,
- the convergence towards the unique weak solution (4) for the class of *weakly TV bounded methods* (see [3] for details),
- the convergence towards the set of all weak solutions [2] of (4) for the class of *TVD methods* (see [15], [13]).

However the theory of the monotone methods is very strong and is easily extensible to the multidimensional case, the monotone methods are at most of the first order of accuracy. Therefore we prefer the class of TVD methods which is defined as follows:

Definition 2.1. *The numerical method (6) is called* total variation diminishing *(or simply* TVD*), if and only if for each numerical approximation U*

$$TV(u^{n+1}) \leq TV(u^n) = \sum_{i=-\infty}^{+\infty} |u_{i+1}^n - u_i^n|. \tag{8}$$

[1]The viscosity vanishing solution is obtained as a limit of solutions to the problems given by the equation $u_t^\epsilon + f(u^\epsilon)_x = \epsilon u_{xx}^\epsilon$ for $\epsilon \to 0+$.

[2]The uniqueness of the solution follows from Coquel's-LeFloch's theorem for the weakly TV bounded methods.

Let us consider a general one-dimensional method of the form

$$u_i^{n+1} = u_i^n - C_{i-\frac{1}{2}}(u_i^n - u_{i-1}^n) + D_{i+\frac{1}{2}}(u_{i+1}^n - u_i^n), \tag{9}$$

then the following theorem of Harten [15] can be used to check the TVD property.

Lemma 2.2 (Harten,1983). *Let the following conditions be fulfilled* $\forall i \in \mathbf{Z}$*:*

$$C_{i-\frac{1}{2}} \geq 0, \qquad D_{i+\frac{1}{2}} \geq 0, \qquad C_{i+\frac{1}{2}} + D_{i+\frac{1}{2}} \leq 1. \tag{10}$$

Then the numerical method (9) is TVD.

This lemma gives some hints how to correct some classical high order methods in order to be TVD. Unfortunately, this lemma is valid only for the one-dimensional case. In fact, Goodman and LeVeque show in [12], that the TVD property in the multidimensional case implies only a first order accuracy.

In spite of their result, many methods based on one-dimensional high order TVD methods have been constructed for practical multidimensional problems. Although they are not TVD, they remain high order for smooth solutions and usually do not generate oscillations near discontinuities.

For practical computations we use the MacCormack predictor-corrector scheme because of its simple implementation especially for nonlinear systems. The TVD MacCormack scheme has for one-dimensional scalar problem the following form:

$$u_i^{n+\frac{1}{2}} = u_i^n - \frac{\Delta t}{\Delta x}\left(f(u_i^n) - f(u_{i-1}^n)\right), \tag{11}$$

$$\overline{u_i^{n+1}} = \frac{1}{2}\left[u_i^n + u_i^{n+\frac{1}{2}} - \frac{\Delta t}{\Delta x}\left(f(u_{i+1}^{n+\frac{1}{2}}) - f(u_i^{n+\frac{1}{2}})\right)\right], \tag{12}$$

$$\begin{aligned} u_i^{n+1} = {} & \overline{u_i^{n+1}} + \left[G^+(r_i^+) + G^-(r_{i+1}^-)\right](u_{i+1}^n - u_i^n) - \\ & - \left[G^+(r_{i-1}^+) + G^-(r_i^-)\right](u_i^n - u_{i-1}^n) \end{aligned} \tag{13}$$

with $G^\pm$ defined by

$$G^\pm(r_i^\pm) = \frac{|f'(u_i^n)|\Delta t}{2\Delta x}(1 - \frac{|f'(u_i^n)|\Delta t}{\Delta x})\left[1 - \Phi(r_i^\pm)\right] \tag{14}$$

and

$$\Phi(r_i^\pm) = \max\left(0, \min(2r_i^\pm, 1)\right). \tag{15}$$

2.2. Extension to the 2D scalar case

The scalar initial value problem for the 2D problem is given by the equation $u_t + f(u)_x + g(u)_y = 0$ with the initial condition $u(x, y, 0) = u_0(x, y)$. We again solve this problem in the unbounded domain $(x, y, t) \in R \times R \times R_+$.

For a two-dimensional scalar case we analyzed in [9] a class of monotone schemes (especially the so called l^1-contractive schemes) and published in [11] the following lemma:

Lemma 2.3. *Let there exist functions $\tilde{f}_k$ of $2(p+q+1)$ parameters such that*

$$\tilde{f}(u^n_{i-p}, ..., u^n_{i+q}) - \tilde{f}(v^n_{i-p}, ..., v^n_{i+q}) =$$
$$= \sum_{k=-p}^{q} \tilde{f}_k(u^n_{i-p}, ..., u^n_{i+q}, v^n_{i-p}, ..., v^n_{i+q})(u^n_{i+k} - v^n_{i+k}) \tag{16}$$

for arbitrary u and v. If the functions $\tilde{f}_k$ satisfy the conditions

$$\tilde{f}_0(u^n_{i-p}, ..., u^n_{i+q}, v^n_{i-p}, ..., v^n_{i+q}) - \tilde{f}_1(u^n_{i-p-1}, ..., u^n_{i+q-1}, v^n_{i-p-1}, ..., v^n_{i+q-1}) \le \frac{\Delta x}{\Delta t} \tag{17}$$

and for each $k \neq 0$ (we set $\tilde{f}_k = 0$ for $k < -p$ and $k > q$) the relations

$$\tilde{f}_k(u^n_{i-p}, ..., u^n_{i+q}, v^n_{i-p}, ..., v^n_{i+q}) \le \tilde{f}_{k+1}(u^n_{i-p-1}, ..., u^n_{i+q-1}, v^n_{i-p-1}, ..., v^n_{i+q-1}) \tag{18}$$

are valid, then the scheme (6) is monotone (and hence TVD).

This lemma (as well as its multidimensional variant) is proved for example in [9].

A similar lemma is valid also for multidimensional problems. Let us consider only the two-dimensional case. The total variation is defined in the this case by:

$$TV(u^n) = \Delta y \sum_{i,j} |u^n_{i+1,j} - u^n_{i,j}| + \Delta x \sum_{i,j} |u^n_{i,j+1} - u^n_{i,j}|. \tag{19}$$

Let the mesh steps be the same for both coordinates i.e. $\Delta x = \Delta y$. The weak solution is approximated by a piecewise constant function $U(x,y,t) = u^n_{i,j}$ for $(i-1)\Delta x < x \le i\Delta x$, $(j-1)\Delta y < y \le j\Delta y$, and $(n-1)\Delta t < t \le n\Delta t$. The functions $\tilde{f}_k$ and $\tilde{g}_k$ depend in a similar way as in the one-dimensional problem on $2(p+q+1)$ parameters $u^n_{i-p,j}, ..., u^n_{i+q,j}, v^n_{i-p,j}, ..., v^n_{i+q,j}$ for $\tilde{f}_k$ and $u^n_{i,j-p}, ..., u^n_{i,j+q}, v^n_{i,j-p}, ..., v^n_{i,j+q}$ for $\tilde{g}_k$.

Lemma 2.4. *Let there exist functions $\tilde{f}_k$ and $\tilde{g}_k$ of $2(p+q+1)$ parameters such that*

$$\tilde{f}(u^n_{i-p,j}, ..., u^n_{i+q,j}) - \tilde{f}(v^n_{i-p,j}, ..., v^n_{i+q,j})$$
$$= \sum_{k=-p}^{q} \tilde{f}_k(u^n_{i-p,j}, ..., v^n_{i+q,j})(u^n_{i+k,j} - v^n_{i+k,j}) \tag{20}$$

$$\tilde{g}(u^n_{i,j-p}, ..., u^n_{i,j+q}) - \tilde{g}(v^n_{i,j-p,}, ..., v^n_{i,j+q})$$
$$= \sum_{k=-p}^{q} \tilde{g}_k(u^n_{i,j-p}, ..., v^n_{i,j+q})(u^n_{i,j+k} - v^n_{i,j+k}) \tag{21}$$

for each arbitrary u and v.

If the functions $\tilde{f}_k$ and $\tilde{g}_k$ satisfy the conditions

$$\begin{aligned}\tilde{f}_0(u^n_{i-p,j},...,v^n_{i+q,j}) &- \tilde{f}_1(u^n_{i-p-1,j},...,v^n_{i+q-1,j}) + \\ \tilde{g}_0(u^n_{i,j-p},...,v^n_{i,j+q}) &- \tilde{g}_1(u^n_{i,j-p-1},...,v^n_{i,j+q-1}) \leq \frac{\Delta x}{\Delta t}\end{aligned} \tag{22}$$

and for each $k \neq 0$ (we set $\tilde{f}_k = 0$ and $\tilde{f}_k = 0$ for $k < -p$ and $k > q$)

$$\tilde{f}_k(u^n_{i-p,j},...,v^n_{i+q,j}) \leq \tilde{f}_{k+1}(u^n_{i-p-1,j},...,v^n_{i+q-1,j}), \tag{23}$$

$$\tilde{g}_k(u^n_{i,j-p},...,v^n_{i,j+q}) \leq \tilde{g}_{k+1}(u^n_{i,j-p-1},...,v^n_{i,j+q-1}) \tag{24}$$

then the scheme

$$u^{n+1}_{i,j} = u^n_{i,j} - \frac{\Delta t}{\Delta x}\left(\tilde{f}^n_{i+\frac{1}{2},j} - \tilde{f}^n_{i-\frac{1}{2},j}\right) - \frac{\Delta t}{\Delta y}\left(\tilde{g}^n_{i,j+\frac{1}{2}} - \tilde{g}^n_{i,j-\frac{1}{2}}\right), \tag{25}$$

$$\tilde{f}^n_{i+\frac{1}{2},j} = \tilde{f}(u^n_{i-p,j}, u^n_{i-p+1,j},...,u^n_{i+q,j}), \quad \tilde{g}^n_{i,j+\frac{1}{2}} = \tilde{g}(u^n_{i,j-p}, u^n_{i,j-p+1},...,u^n_{i,j+q})$$

is monotone (and hence TVD).

Unfortunately it is not possible to construct a high order TVD scheme for multidimensional problem [12]. The uniform TVD bound is too restrictive and therefore one should consider following approach analyzed for example by Coquel and LeFloch in [4, 3]. Let us assume a two-dimensional conservative scheme of the form:

$$u^{n+1}_{i,j} = u^n_{i,j} - \frac{\Delta t}{\Delta x}\left(\tilde{f}^n_{i+\frac{1}{2},j} - \tilde{f}^n_{i-\frac{1}{2},j}\right) - \frac{\Delta t}{\Delta y}\left(\tilde{g}^n_{i,j+\frac{1}{2}} - \tilde{g}^n_{i,j-\frac{1}{2}}\right), \tag{26}$$

with $\Delta x = \Delta y = const.\Delta t$ and consistent with the equation $u_t + f(u)_x + g(u)_y = 0$.

Theorem 2.5 (Coquel, LeFloch 1991). *Let the numerical fluxes $f^n_{i+\frac{1}{2},j}$ and $g^n_{i,j+\frac{1}{2}}$ can be split onto two parts $f^n_{i+\frac{1}{2},j} = p^n_{i+\frac{1}{2},j} + \frac{\Delta x}{\Delta t}a^n_{i+\frac{1}{2},j}$ and $g^n_{i,j+\frac{1}{2}} = q^n_{i,j+\frac{1}{2}} + \frac{\Delta y}{\Delta t}b^n_{i,j+\frac{1}{2}}$ with p and q being monotone fluxes (e.g. first order Lax-Friedrichs type) and a and b being antidiffusive high order corrections. If there exist for a finite time $T > 0$ constants $C > 0$ and $M > 0$ independent of Δt (and consequently independent of Δx and Δy) such that*

$$||u^n_{i,j}||_\infty < C, \qquad |a^n_{i+\frac{1}{2},j}| < M\Delta x^\alpha, \qquad |b^n_{i,j+\frac{1}{2}}| < M\Delta y^\alpha, \tag{27}$$

with $n\Delta t < T$ and $\alpha \in (\frac{2}{3}, 1)$, then the piecewise constant function with the values given by $u^n_{i,j}$ converges to the unique entropy solution when $\Delta t \to 0$.

Example 2.6 (2D version of the scheme proposed by Davis).
Let us suppose the 2D scheme for the two-dimensional linear equation proposed by

Davis $u_t + cu_x + du_y = 0$:

$$\begin{aligned} u_{i,j}^{n+1} &= u_{i,j}^n - c\frac{\Delta t}{2\Delta x}(u_{i+1,j}^n - u_{i-1,j}^n) + c^2\frac{\Delta t^2}{2\Delta x^2}(u_{i+1,j}^n - 2u_{i,j}^n + u_{i-1,j}^n) + \\ &+ K(r_{i,j}^+, r_{i+1,j}^-)(u_{i+1,j}^n - u_{i,j}^n) - K(r_{i-1,j}^+, r_{i,j}^-)(u_{i,j}^n - u_{i-1,j}^n) - \qquad (28) \\ &- d\frac{\Delta t}{2\Delta y}(u_{i,j+1}^n - u_{i,j-1}^n) + d^2\frac{\Delta t^2}{2\Delta y^2}(u_{i,j+1}^n - 2u_{i,j}^n + u_{i,j-1}^n) + \\ &+ K(s_{i,j}^+, s_{i,j+1}^-)(u_{i,j+1}^n - u_{i,j}^n) - K(s_{i,j-1}^+, s_{i,j}^-)(u_{i,j}^n - u_{i,j-1}^n) \qquad (29) \end{aligned}$$

where

$$K(r_{i,j}^+, r_{i+1,j}^-) = \frac{|c|\Delta t}{2\Delta x}(1 - \frac{|c|\Delta t}{\Delta x})\left[1 - \tilde{\Phi}(r_{i,j}^+, r_{i+1,j}^-)\right] \qquad (30)$$

$$K(s_{i,j}^+, s_{i,j+1}^-) = \frac{|c|\Delta t}{2\Delta y}(1 - \frac{|c|\Delta t}{\Delta y})\left[1 - \tilde{\Phi}(s_{i,j}^+, s_{i,j+1}^-)\right] \qquad (31)$$

$$\tilde{\Phi}(r_{i,j}^+, r_{i+1,j}^-) = \max\left(0, \min(2r_{i,j}^+, r_{i+1,j}^-, 1), \min(r_{i,j}^+, 2r_{i+1,j}^-, 1)\right) \qquad (32)$$

$$\tilde{\Phi}(s_{i,j}^+, s_{i,j+1}^-) = \max\left(0, \min(2s_{i,j}^+, s_{i,j+1}^-, 1), \min(p_{i,j}^+, 2p_{i,j+1}^-, 1)\right) \qquad (33)$$

$$r_{i,j}^+ = (u_{i,j}^n - u_{i-1,j}^n)/(u_{i+1,j}^n - u_{i,j}^n) \qquad (34)$$

$$r_{i,j}^- = (u_{i+1,j}^n - u_{i,j}^n)/(u_{i,j}^n - u_{i-1,j}^n) \qquad (35)$$

$$s_{i,j}^+ = (u_{i,j}^n - u_{i,j-1}^n)/(u_{i,j+1}^n - u_{i,j}^n) \qquad (36)$$

$$s_{i,j}^- = (u_{i,j+1}^n - u_{i,j}^n)/(u_{i,j}^n - u_{i,j-1}^n). \qquad (37)$$

Let us replace $\tilde{\Phi}$ *by the new limiter:*

$$\bar{\tilde{\Phi}}(r_{i,j}^+, r_{i+1,j}^-) = \min\left(\tilde{\Phi}(r_{i,j}^+, r_{i+1,j}^-), \frac{M'\Delta x^\alpha}{|u_{i+1,j}^n - u_{i,j}^n|}\right) \qquad (38)$$

$$\bar{\tilde{\Phi}}(s_{i,j}^+, s_{i,j+1}^-) = \min\left(\tilde{\Phi}(s_{i,j}^+, s_{i,j+1}^-), \frac{M'\Delta y^\alpha}{|u_{i,j+1}^n - u_{i,j}^n|}\right). \qquad (39)$$

Then the antidiffusive fluxes a and b are bounded by $M'\Delta x^\alpha$ *and* $M'\Delta y^\alpha$ *respectively and the modification of the two-dimensional scheme is convergent.*

Similar theorem is valid also for three-dimensional case.

2.3. TVD schemes for hyperbolic systems

Let us consider a linear system

$$W_t + AW_x = 0. \qquad (40)$$

The solution is now a vector-valued function $W : \mathbf{R} \times \mathbf{R}^+ \to \mathbf{R}^m$ (m is the number of equations) and A is an $m \times m$ constant matrix. The system (40) is called *hyperbolic* if the matrix A has m real eigenvalues and m linearly independent right eigenvectors. In that case, we can decompose the matrix A into

$$A = R\Lambda R^{-1} \qquad (41)$$

where R is the regular matrix composed of the eigenvectors of A and Λ is the diagonal matrix containing the eigenvalues of A ($\Lambda = \mathrm{diag}(a^{(1)}, ..., a^{(m)})$).

Let us define a new set of variables by $V = R^{-1}W$ (we call it the *characteristic variables*). Multiplying the original system (40) by R^{-1} one gets

$$R^{-1}W_t + R^{-1}AW_x = 0 \tag{42}$$

and hence, using the characteristic variables,

$$V_t + \Lambda V_x = 0 \tag{43}$$

which is a set of m independent scalar problems. For each component we can use the TVD MacCormack scheme defined in the previous section. In order to get the TVD MacCormack scheme for the original variables W, we multiply the scheme written for the characteristic variables V by the matrix R. Finally, we get

$$W_i^{n+\frac{1}{2}} = W_i^n - \frac{\Delta t}{\Delta x}\left(AW_i^n - AW_{i-1}^n\right), \tag{44}$$

$$\overline{W_i^{n+1}} = \frac{1}{2}\left[W_i^n + W_i^{n+\frac{1}{2}} - \frac{\Delta t}{\Delta x}\left(AW_{i+1}^{n+\frac{1}{2}} - AW_i^{n+\frac{1}{2}}\right)\right], \tag{45}$$

$$\begin{aligned} W_i^{n+1} = {} & \overline{W_i^{n+1}} + R\left[\tilde{G}^+(\tilde{r}_i^+) + \tilde{G}^-(\tilde{r}_{i+1}^-)\right] R^{-1}(W_{i+1}^n - W_i^n) - \\ & -R\left[\tilde{G}^+(\tilde{r}_{i-1}^+) + \tilde{G}^-(\tilde{r}_i^-)\right] R^{-1}(W_i^n - W_{i-1}^n). \end{aligned} \tag{46}$$

Here $\tilde{r}_i^{\pm}$ are vectors with m components

$$(\tilde{r}_i^+)^{(l)} = \left(R^{-1}(W_i^n - W_{i-1}^n)\right)^{(l)} / \left(R^{-1}(W_{i+1}^n - W_i^n)\right)^{(l)}, \tag{47}$$

$$(\tilde{r}_i^-)^{(l)} = \left(R^{-1}(W_{i+1}^n - W_i^n)\right)^{(l)} / \left(R^{-1}(W_i^n - W_{i-1}^n)\right)^{(l)}, \tag{48}$$

where $r^{(l)}$ denotes the l-th component of the vector r.

The viscosity coefficients $\tilde{G}$ are $m \times m$ diagonal matrices with the elements given by

$$\tilde{G}^{\pm}(\tilde{r}_i^{\pm})^{(l,l)} = \frac{1}{2}\frac{|a^{(l)}|\Delta t}{\Delta x}(1 - \frac{|a^{(l)}|\Delta t}{\Delta x})[1 - \Phi((\tilde{r}_i^{\pm})^{(l)})]. \tag{49}$$

In order to avoid the evaluation of eigenvectors of the Jacobian matrix A we use the so called simplified TVD scheme proposed by D. M. Causon in [2] which is for the case of the one-dimensional nonlinear system written as:

$$W_i^{n+\frac{1}{2}} = W_i^n - \frac{\Delta t}{\Delta x}\left(F(W_i^n) - F(W_{i-1}^n)\right), \tag{50}$$

$$\overline{W_i^{n+1}} = \frac{1}{2}\left[W_i^n + W_i^{n+\frac{1}{2}} - \frac{\Delta t}{\Delta x}\left(F(W_{i+1}^{n+\frac{1}{2}}) - F(W_i^{n+\frac{1}{2}})\right)\right], \tag{51}$$

$$\begin{aligned} W_i^{n+1} = {} & \overline{W_i^{n+1}} + \left[\overline{G}^+(\overline{r}_i^+) + \overline{G}^-(\overline{r}_{i+1}^-)\right](W_{i+1}^n - W_i^n) - \\ & - \left[\overline{G}^+(\overline{r}_{i-1}^+) + \overline{G}^-(\overline{r}_i^-)\right](W_i^n - W_{i-1}^n) \end{aligned} \tag{52}$$

with $\overline{G}^{\pm}$ and $\overline{r}^{\pm}$ given by the formulas

$$\begin{aligned} \overline{r}_i^+ &= \frac{<W_i^n - W_{i-1}^n, W_{i+1}^n - W_i^n>}{<W_{i+1}^n - W_i^n, W_{i+1}^n - W_i^n>}, & \overline{r}_i^- &= \frac{<W_i^n - W_{i-1}^n, W_{i+1}^n - W_i^n>}{<W_i^n - W_{i-1}^n, W_i^n - W_{i-1}^n>}, \\ \overline{G}^{\pm}(\overline{r}_i^{\pm}) &= \tfrac{1}{2} C(\nu_i)\left[1 - \Phi(\overline{r}_i^{\pm})\right], & C(\nu_i) &= \begin{cases} \nu_i(1-\nu_i), & \nu_i \le 0.5 \\ 0.25, & \nu_i > 0.5, \end{cases} \\ \nu_i &= \rho_{A_i} \tfrac{\Delta t}{\Delta x}. \end{aligned} \tag{53}$$

Here $< .,. >$ denotes the standard inner (scalar) product in $\mathbf{R}^m$ and ρ_{A_i} is the spectral radius of the Jacobi matrix $\partial F / \partial W$ at the point W_i (for the case of the Euler equations, $\rho_{A_i} = |u_i| + c_i$ where u_i is the local flow speed and c_i is the local sound speed.

3. Composite schemes

In the simplest 1D case there exist two forms of Lax-Friedrichs (LF) scheme:

1. standard one-step version (non-staggered scheme):

$$W_i^{n+1} = \frac{1}{2}(W_{i+1}^n + W_{i-1}^n) - \frac{\Delta t}{2\Delta x}(F_{i+1}^n - F_{i-1}^n), \tag{54}$$

2. two-step version (staggered scheme):

$$W_{i+1/2}^* = \frac{1}{2}(W_{i+1}^n + W_i^n) - \frac{\Delta t}{2\Delta x}(F_{i+1}^n - F_i^n), \tag{55}$$

$$W_i^{n+1} = \frac{1}{2}(W_{i+1/2}^* + W_{i-1/2}^*) - \frac{\Delta t}{2\Delta x}(F_{i+1/2}^* - F_{i-1/2}^*). \tag{56}$$

The two-step Lax-Wendroff (LW) scheme can be written as:

1. non-staggered scheme:

$$W_i^* = W_i^n - \frac{\Delta t}{4\Delta x}(F_{i+1}^n - F_{i-1}^n) + \epsilon(W_{i+1}^n - 2W_i^n + W_{i-1}^n), \tag{57}$$

$$W_i^{n+1} = W_i^n - \frac{\Delta t}{2\Delta x}(F_{i+1}^* - F_{i-1}^*), \tag{58}$$

2. or staggered scheme:

$$W_{i+1/2}^* = \frac{1}{2}(W_{i+1}^n + W_i^n) - \frac{\Delta t}{2\Delta x}(F_{i+1}^n - F_i^n), \tag{59}$$

$$W_i^{n+1} = W_{i+1}^n - \frac{\Delta t}{\Delta x}(F_{i+1/2}^* - F_{i-1/2}^*). \tag{60}$$

The one-dimensional composite scheme LWLFN is given by (N-1) LW steps followed by one LF step:

$$W^{n+N} = LF \circ \underbrace{LW \circ LW \circ \ldots \circ LW}_{N-1} W^n. \tag{61}$$

The extension for two-dimensional case is done using the finite volume approach. Let us assume a triangular mesh with triangles denoted by T_i. The cell-centered LF scheme corresponding to the one-dimensional LF non-staggered scheme is given by:

$$W_i^{n+1} = \frac{1}{3}\sum_{k=1}^{3} W_k^n - \frac{\Delta t}{\mu_i}\sum_{k=1}^{3}(F_{k,i}^n \Delta y_k - G_{k,i}^n \Delta x_k) \tag{62}$$

where $\mu_i = \int\int_{T_i} dx\,dy$, $F_{k,i}^n = (F_k^n + F_i^n)/2$, $G_{k,i}^n = (G_k^n + G_i^n)/2$ and W_k^n is the value of W at the one of three neighboring triangles T_k of the triangle T_i.

For the two-step LW scheme one has:

$$W_i^{n+1/2} = \frac{1}{3}\sum_{k=1}^{3} W_k^n - \frac{\Delta t}{2\mu_i}\sum_{k=1}^{3}(F_{k,i}^n \Delta y_k - G_{k,i}^n \Delta x_k), \tag{63}$$

$$W_i^{n+1} = W_i^n - \frac{\Delta t}{\mu_i}\sum_{k=1}^{3}(F_{k,i}^{n+1/2} \Delta y_k - G_{k,i}^{n+1/2} \Delta x_k). \tag{64}$$

The cell-vertex corresponding to the one-dimensional staggered version has the following form (assuming for simplicity that six triangles meet at each vertex):
LF:

$$W_i^* = \frac{1}{3}\sum_{k=1}^{3} W_{0,k}^n - \frac{\Delta t}{2\mu_i}\sum_{k=1}^{3}(\tilde{F}_k \Delta y_k - \tilde{G}_k \Delta x_k), \tag{65}$$

$$\tilde{F}_k = \frac{1}{2}(F_{0,k+1} - F_{0,k}),\ \tilde{G}_k = \frac{1}{2}(G_{0,k+1} - G_{0,k}),$$

$$W_{0,k}^{n+1} = \frac{1}{6}\sum_{i=1}^{6} W_i^* - \frac{\Delta t}{2\mu_{0,k}}\sum_{i=1}^{6}(\tilde{F}_i^* \Delta y_i - \tilde{G}_i^* \Delta x_y), \tag{66}$$

$$\tilde{F}_i^* = \frac{1}{2}(F_{i+1}^* - F_i^*),\ \tilde{G}_i^* = \frac{1}{2}(G_{i+1}^* - G_i^*).$$

LW scheme uses the same first step as the above described LF scheme and the second step is of the form:

$$W_{0,k}^{n+1} = W_{0,k}^n - \frac{\Delta t}{2\mu_{0,k}}\sum_{i=1}^{6}(\tilde{F}_i^* \Delta y_i - \tilde{G}_i^* \Delta x_y). \tag{67}$$

Here $W_{0,k}$ are located at the vertices $V_{0,k}$ of the triangle T_i and W_i^* are located at the centers of gravity P_i of triangles T_i.

The composite scheme consists again as $N-1$ steps of LW scheme followed by one LF step. The value of N is determined by numerical testing. The cell vertex scheme was also extended to three-dimensional case.

Remark: Composite schemes were introduced by R. Liska and B. Wendroff for a 2D nonlinear hyperbolic problem solved by finite difference methods on an orthogonal grid. The idea is following: in many high order schemes one has to add some artificial viscosity terms in order to achieve solution without oscillations near shock waves or steep gradients. Composite schemes use certain number of a high order scheme (with low value of artificial viscosity) and the arising oscillations are then smeared by one step of a more dissipative scheme (for example Lax-Friedrichs scheme).

We extended the original finite difference schemes to 2D and 3D finite volume method of the cell centered or cell vertex form [14].

4. Implicit finite volume method for 2D inviscid flows

The numerical solution is again obtained by the finite volume approach: The domain Ω is approximated by a polygonal domain Ω_h and this polygonal domain is divided into m polygonal convex cells[3] C_i possess the following property:

$$\overline{\Omega_h} = \bigcup_{i=1}^{m} \overline{C_i} \text{ and } C_i \cap C_j = \emptyset \text{ for } i \neq j \; .$$

Figure 1 shows a sample of such a domain divided into 13 triangular, quadrilateral, and pentagonal cells. Let m_i denote the number of cells adjacent to C_i (i.e. number of cells that share an edge with the cell C_i) and let the set $N_i = \{i_1, i_2, ..., i_{m_i}\}$ contain their indices (see Fig. 1 where $m_i = 5$). Next, let us denote by B_i^{inlet} the set of edges shared by the cell C_i and the inlet boundary Γ_h^{inlet} of Ω_h; similarly for B_i^{outlet} and B_i^{wall}.

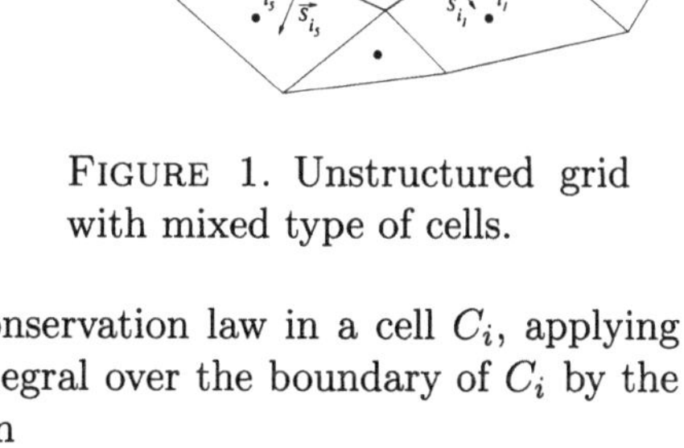

FIGURE 1. Unstructured grid with mixed type of cells.

The basic finite volume scheme is obtained in the usual way: integrating the conservation law in a cell C_i, applying Green's theorem and approximating the integral over the boundary of C_i by the numerical flux functions. The scheme is then

$$W_i^{n+1} = W_i^n - \Delta t R^1(W^n)_i. \tag{68}$$

[3]Since the structured grid can be viewed from the mathematical point of view as a special case of the unstructured grid, we present only the scheme for unstructured meshes.

Here W_i^n stands for the approximation of the solution in the cell C_i at a time $t = n\Delta t$ and $R^1(W^n)_i$ is the component of the residual vector computed as

$$R^1(W^n)_i = \frac{1}{\mu(C_i)}\left[\sum_{j\in N_i} H(W_i^n, W_j^n, \vec{S}_{i,j}) + \sum_b \sum_{e\in B_i^b} H^b(W_i^n, \vec{S}_e)\right]. \tag{69}$$

The superscript 1 denotes the first order approximation, $\mu(C_i)$ is the volume of the cell C_i, $\vec{S}_{i,j}$ denotes the outer normal vector to the common edge between C_i and C_j, the function H is the numerical flux, b denotes the type of the boundary conditions and belongs to the set $b \in \{inlet, outlet, wall\}$, H^b is the numerical flux through the boundary and $\vec{S}_e$ denotes the outer normal vector to the boundary edge e. Both vectors $\vec{S}_{i,j}$ and $\vec{S}_e$ have the length equal to the length of the corresponding edge. The numerical flux $H(W_i^n, W_j^n, \vec{S}_{i,j})$ in the previous formula is the numerical approximation of the integral of the physical flux function over the common edge $e_{i,j}$ shared between C_i and C_j:

$$H(W_i^n, W_j^n, \vec{S}_{i,j}) \approx \int_{e_{i,j}} \left(\begin{pmatrix} \rho u \\ \rho u^2 + p \\ \rho uv \\ (e+p)u \end{pmatrix} n_x + \begin{pmatrix} \rho v \\ \rho uv \\ \rho v^2 + p \\ (e+p)v \end{pmatrix} n_y \right) dS \tag{70}$$

where n_x and n_y are the components of the unit normal vector to the edge $e_{i,j}$ oriented as the outer normal for the cell C_i. Analogous, $H^b(W_i^n, \vec{S}_e)$ is the approximation of the flux through the edge on the boundary.

4.1. First order implicit scheme

As a building block for the implicit scheme we choose the first order finite volume scheme based on Osher's flux and the related approximate Riemann solver (see [16]). The advantage of the Osher flux is that one can evaluate simply the Jacobians of the numerical flux function which are needed for the implicit scheme.

The usual explicit first order scheme is then

$$W^{n+1} = W^n - \Delta t R^1(W^n) \tag{71}$$

and the implicit scheme is obtained from the explicit version (68) by replacing $R^1(W^n)$ by $R^1(W^{n+1})$:

$$W_i^{n+1} = W_i^n - \Delta t R^1(W^{n+1})_i. \tag{72}$$

The operator R is nonlinear, so we can't solve this equation directly. Therefore we linearize the equation at the point W^n:

$$W^{n+1} = W^n - \Delta t \left(R^1(W^n) + \frac{\partial R^1}{\partial W}\left(W^{n+1} - W^n\right)\right), \tag{73}$$

hence,

$$\left(\frac{\mathbf{I}}{\Delta t} + \frac{\partial R^1}{\partial W}\right)\left(W^{n+1} - W^n\right) = -R^1(W^n). \tag{74}$$

The matrix $\frac{\partial R^1}{\partial W}$ is evaluated at the point W^n using the expressions for the Jacobian matrices of Osher's flux functions $\frac{\partial H(W_L,W_R,\vec{S})}{\partial W_L}$ and $\frac{\partial H(W_L,W_R,\vec{S})}{\partial W_R}$ and the appropriate Jacobian matrices of boundary fluxes.

The resulting system of linearized equations is solved using a GMRES method preconditioned with the ILU decomposition.

4.2. Second order semi-implicit scheme

In order to improve the accuracy of the basic first order scheme we use a piecewise linear reconstruction of the solution (for details see [10]). The second order semi-implicit scheme uses the high order residual R^2 on the right hand side (explicit part) and the matrix on the left hand side is computed from the low order residual:

$$\left(\frac{\mathbf{I}}{\Delta t}+\frac{\partial R^1}{\partial W}\right)\left(W^{n+1}-W^n\right)=-R^2(W^n). \tag{75}$$

The second order residual R^2 is computed in the following way: we compute first the approximation of the gradient of the solution $\overline{\text{grad}}W^n$ in each cell [4] and then using this gradient we define the second order residual vector by

$$\begin{aligned} W^L_{i,j} &= W^n_i+(\vec{x}_{i,j}-\vec{x}_i)\cdot\overline{\text{grad}}W^n_i, \qquad (76)\\ W^R_{i,j} &= W^n_j+(\vec{x}_{j,i}-\vec{x}_j)\cdot\overline{\text{grad}}W^n_j,\\ R^2(W^n)_i &= \frac{1}{\mu(C_i)}\left[\sum_{j\in N_i}H(W^L_{i,j},W^R_{i,j},\vec{S}_{i,j})+\sum_b\sum_{e\in B^b_i}H^b(W^L_{i,j},\vec{S}_e)\right], \end{aligned}$$

where $\vec{x}_i$ is the center of gravity of the cell i and $\vec{x}_{i,j}$ is the center of the common edge between the cells i and j. The numerical fluxes H and H^b are computed in the same way as for the first order scheme (i.e. using Osher's Riemann solver) but instead of W_i and W_j we use the interpolated values $W^R_{i,j}$ and $W^L_{i,j}$.

In the steady case one has $R^2(W^n)=0$ and therefore the scheme is high order. Moreover, as the second order residual is of ENO type, the stationary solution is essentially non-oscillatory. In the unsteady case the scheme is only low order due to replacing R^2 by R^1 in the implicit part.

5. 2D transonic inviscid flow through a channel and a turbine cascade

5.1. Transonic flow through the 2D test channel with a bump

As a first test case we choose the transonic flow through the two-dimensional test channel with a bump, i.e. the so-called Ron-Ho-Ni channel. This is a well-known test case and it was solved by many researchers. See for example [5], [6].

We use the structured mesh with 120×30 quadrilateral cells for the TVD MacCormack scheme, an unstructured triangular mesh with 4424 triangles (with

[4]The evaluation of the gradients is described in in detail [10].

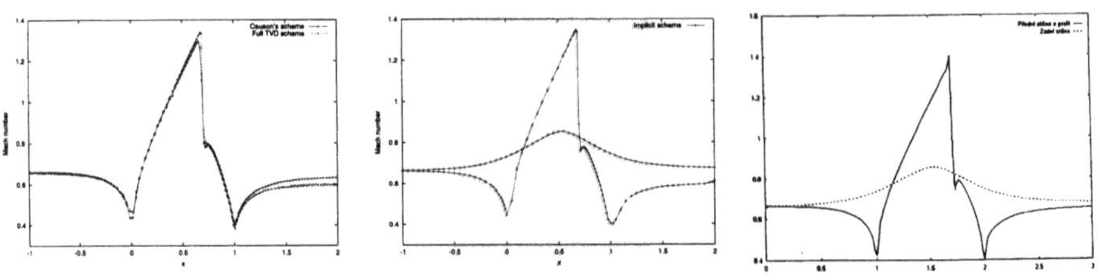

(a) TVD MacCormack scheme (b) Implicit scheme (c) Composite scheme

FIGURE 2. Distribution of the Mach number along the lower and upper walls for the 2D channel.

refinement in the vicinity of the shock wave) for the implicit scheme and a structured triangular mesh with 10800 triangles for the composite scheme. At the inlet ($x = -1$) we prescribe the stagnation pressure $p_0 = 1$, the stagnation density $\rho_0 = 1$ and the inlet angle $\alpha_1 = 0$. At the outlet we keep the pressure $p_2 = 0.737$. The upper ($y = 1$) and lower part are solid walls.

Figure 2(a) shows the distribution of the Mach number along the upper and the lower wall after 30000 iterations of two above mentioned variants of the MacCormack scheme with $CFL = 0.5$ while figure 2(b) shows the results obtained by the implicit scheme and figure 2(c) by the composite scheme.

We can see that the results obtained by the full TVD MacCormack scheme are quite good. Causon's simplified scheme uses too much artificial dissipation.

5.2. Transonic flow through the 2D turbine cascade SE 1050

Next we solve the transonic flow through the 2D turbine cascade SE 1050 given by Škoda Plzeň. Figure 3 shows the results of the interferometric measurement obtained for the inlet Mach number $M_1 = 0.395$ [17]. One can see the characteristic structure of the shock waves emitted from the outlet edge and the reflected shock waves. Moreover, one can notice the recompression zone on the suction side of the blade. Our computation was performed on a structured mesh with 200×40 quadrilateral cells for the full TVD MacCormack scheme (see Fig. 4(a)) and on an unstructured mesh with 7892 triangles for the implicit scheme (Fig. 4(b)).

5.3. Laminar viscous flow through a 2D turbine cascade

Next we solve the transonic viscous flow through the DCA 8% cascade. We consider the flow with the non-dimensional viscosity $\mu = 10^{-4}$ which gives, for the inlet conditions $p_0 = 1$, $\rho_0 = 1$, $\alpha_1 = 2^o$ and outlet pressure $p_2 = 0.48$, the value of the Reynolds number $Re = 6450$, inlet Mach number $M_1 = 0.76$ and outlet Mach number $M_2 = 1.03$.

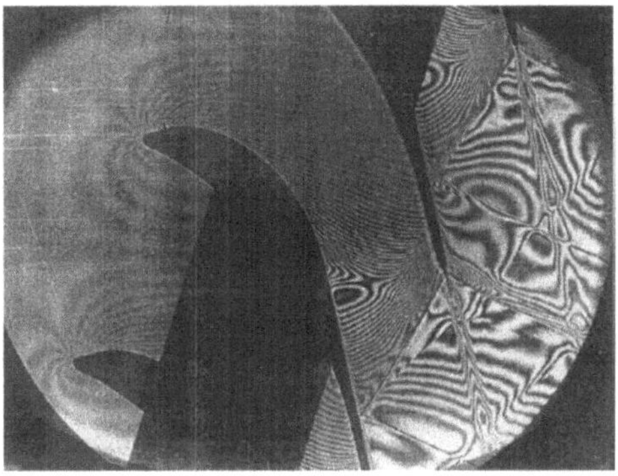

FIGURE 3. Interferometric measurement of SE 1050

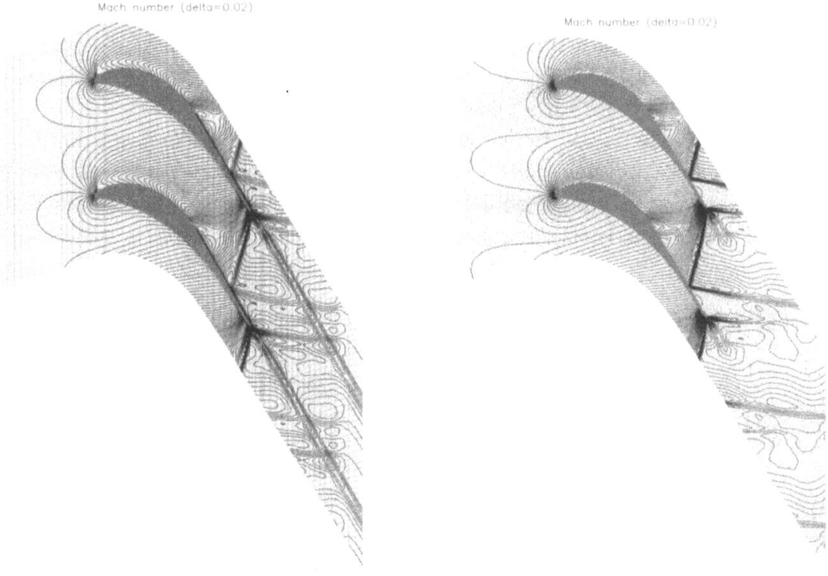

(a) TVD MacCormack scheme, structured grid

(b) Implicit scheme, unstructured grid

FIGURE 4. Distribution of the Mach number in a 2D turbine cascade

We use a simple structured mesh with 90×50 quadrilateral cells refined in the vicinity of profiles. Figure 5 shows the results obtained by an improved version of Causon's scheme [5] after 50000 and 50200 iterations. We can see that the solution is non-stationary (see the changes in the shape of the wake). Similar non-stationary solution was obtained also by using a finite difference ENO scheme [1]. Let us mention that these flow conditions (it means relatively low Reynolds

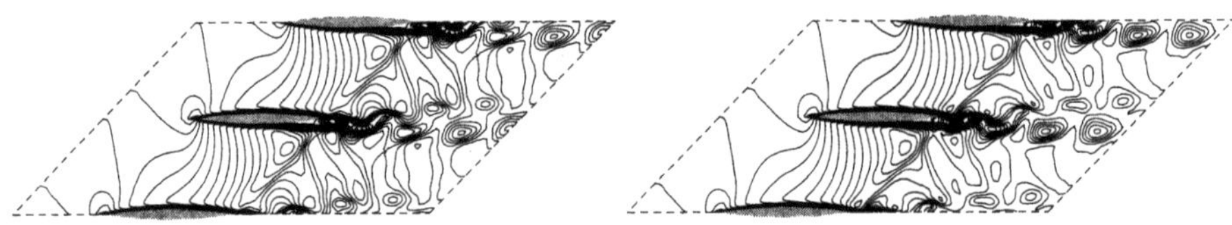

(a) 50000 time steps (b) 50200 time steps

FIGURE 5. Isolines of Mach number for the non-stationary laminar transonic flow through the DCA 8% cascade.

number and high Mach number) are not interesting for practical applications. We have done this computation in order to show that the effects of artificial viscosity can be very important for a viscous flow calculation. As a matter of fact, a similar case was formerly solved by M. Huněk and K. Kozel [8] using an implicit residual averaging method. Their method gave stationary results whereas our computation leads to a non-stationary solution. This is probably due to the fact, that the residual smoothing method uses too much artificial viscosity.

6. 3D inviscid transonic flows

6.1. 3D transonic flow through a channel

The 2D composite scheme of cell-vertex type was extended to 3D case using the following finite volume mesh given by fig. 6. Figure 6(a) shows basic finite volume, fig. 6(b) shows the dual finite volume and fig. 6(c) shows the global 3D basic mesh. We used $180 \times 30 \times 12$ mesh and 3D channel was changed by bump with the height equal to 0.18 at $z = 0$ decreasing to 0.10 at $z = 1$. Figure 7 shows Mach number distribution in the planes $z = const.$. Figure 7(a) shows the solution at the wall with the bump (with the shock wave) and smooth solution at the upper wall. Figure 7(b) shows the change of the strength of the shock wave for z between 0 and 1. Here we use 25 steps of LW scheme followed by 1 LF step.

6.2. 3D transonic flow through a turbine cascade

The three-dimensional Causon's scheme and its improved variant are used for the computation of the transonic flow through the stator stage of the real 3D turbine given by Škoda Plzeň company.

[5] The so called modified scheme which we published for example in [11].

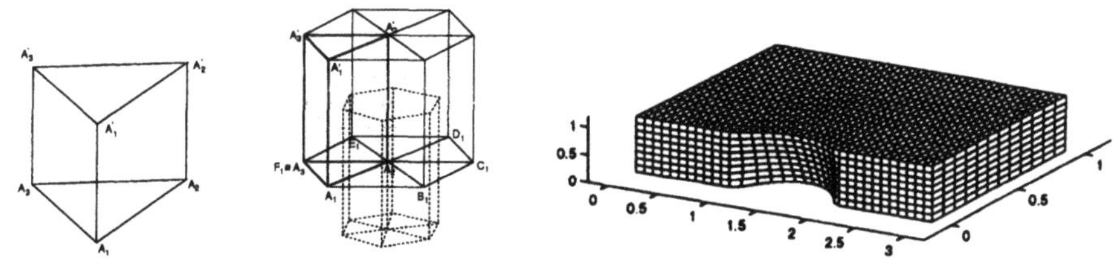

(a) Basic cell (b) Dual cell (c) Global mesh

FIGURE 6. 3D finite volume mesh for composite schemes.

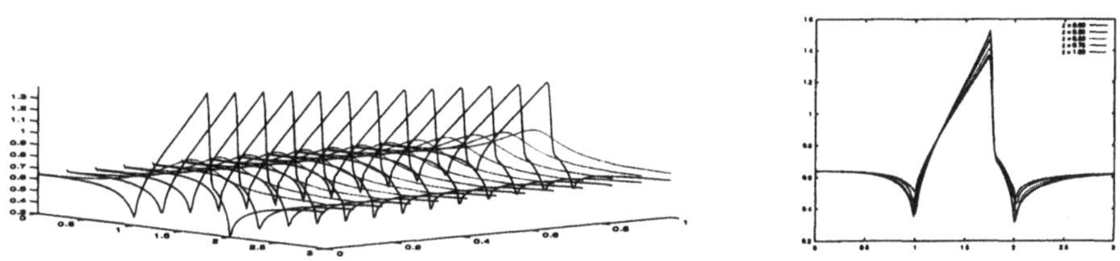

(a) Mach number along the walls (b) Mach number along the lower wall (projections)

FIGURE 7. Mach number distribution in 3D channel

At the inlet we prescribe the stagnation pressure $p_0(r) = 0.38274$, stagnation density $\rho_0(r) = 1$. The direction of the velocity at the inlet is given by two angles $\alpha_1(r)$ and $\mu_1(r)$.

We use a structured mesh with $90 \times 24 \times 17$ hexahedral cells. Figure 8 shows distribution of the Mach number obtained by the above mentioned modification of 3D Causon's scheme. Figures 9(a)-10(b) show the distribution of the Mach number for different section and on the pressure and suction side of the blade. Similar results were obtained by J. Fořt and J. Halama [7] by using a cell vertex scheme of Ni.

Acknowledgment
This work was supported by grants No. 101/98/K001, 201/99/0267 of GAČR and by the Research Plan MSM 98/210000010.

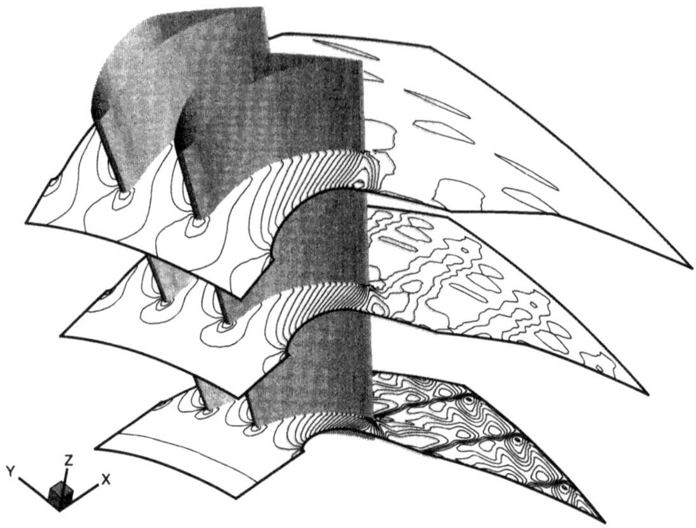

FIGURE 8. Mach number distribution in the 3D turbine.

References

[1] Philippe Angot, Jiří Fürst, and Karel Kozel. TVD and ENO schemes for multidimensional steady and unsteady flows. a comparative analysis. In Fayssal Benkhaldoun and Roland Vilsmeier, editors, *Finite Volumes for Complex Applications. Problems and Perspectives*, pages 283–290. Hermes, july 1996.

[2] D. M. Causon. High resolution finite volume schemes and computational aerodynamics. In Josef Ballmann and Rolf Jeltsch, editors, *Nonlinear Hyperbolic Equations - Theory, Computation Methods and Applications*, volume 24 of *Notes on Numerical Fluid Mechanics*, pages 63–74, Braunschweig, March 1989. Vieweg.

[3] Frédéric Coquel and Philippe Le Floch. Convergence of finite difference schemes for conservation laws in several space dimensions: the corrected antidiffusive flux approach. *Mathematics of computation*, 57(195):169–210, july 1991.

[4] Frédéric Coquel and Philippe Le Floch. Convergence of finite difference schemes for conservation laws in several space dimensions: a general theory. *SIAM J. Numer. Anal.*, 30(3):675–700, June 1993.

[5] Vít Dolejší. *Sur des méthodes combinant des volumes finis et des éléments finis pour le calcul d'ecoulements compressibles sur des maillages non structurés*. PhD thesis, L'Université Méditerranée Marseille et Univerzita Karlova Praha, 1998.

[6] Miloslav Feistauer, Jiří Felcman, and Mária Lukáčová-Medviďová. Combined finite element-finite volume solution of compressible flow. *Journal of Computational and Applied Mathemetics*, (63):179–199, 1995.

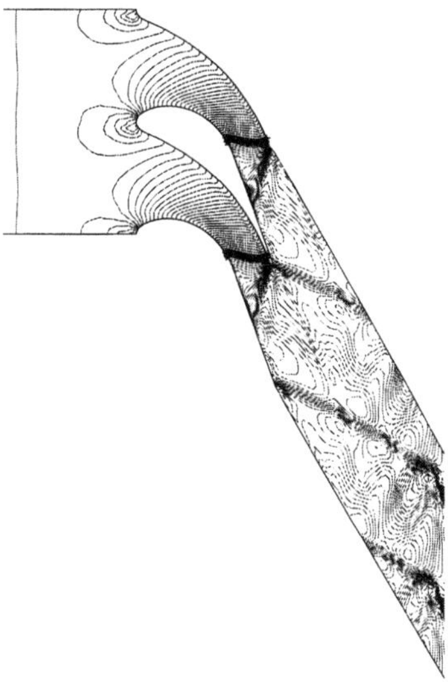

(a) Hub ($k = 1$)

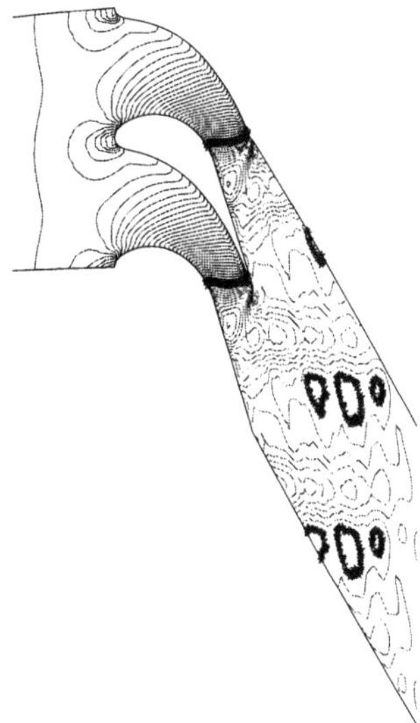

(b) Middle ($k = 9$)

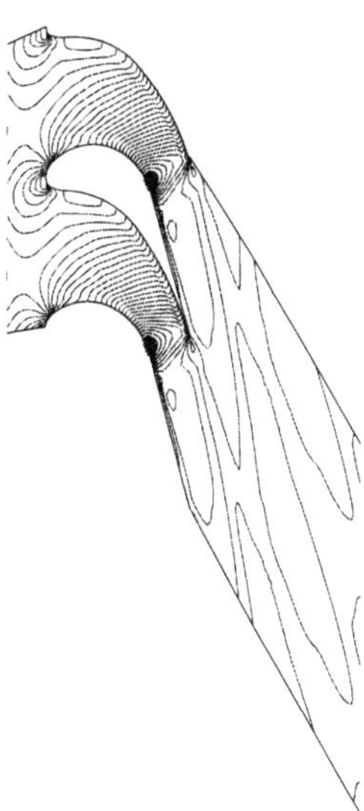

(c) Tip ($k = 18$)

FIGURE 9. Distribution of Mach number for the sections $k = const.$

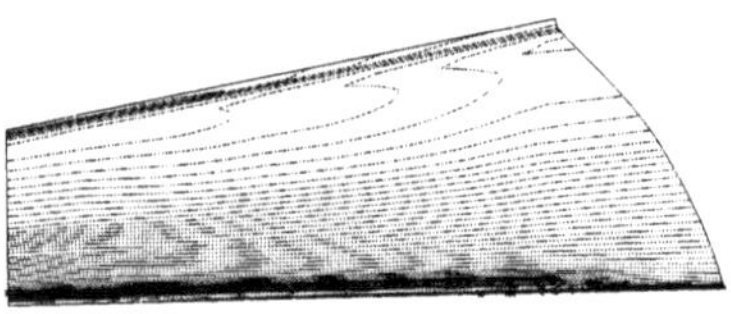

(a) Mach number distribution on the pressure side of the blade

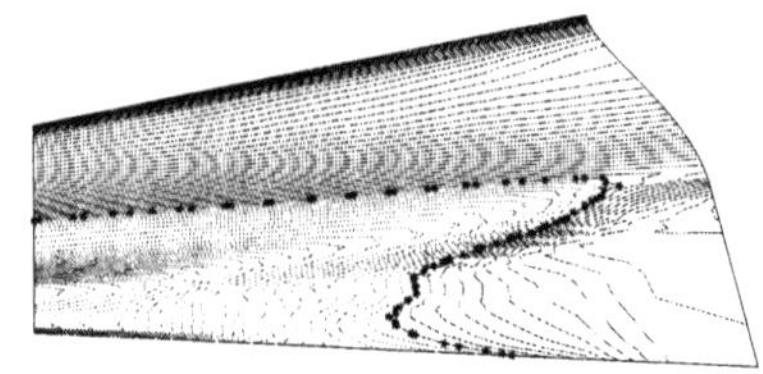

(b) Mach number distribution on the suction side of the blade

FIGURE 10. Distribution of Mach number on the blade

[7] J. Fořt, J. Halama, A. Jirásek, M. Kladrubský, and K. Kozel. Numerical solution of several 2d and 3d internal and external flow problems. In R. Rannacher M. Feistauer and K. Kozel, editors, *Numerical Modelling in Continuum Mechanics*, pages 283–291, September 1997.

[8] Jaroslav Fořt, Miloš Huněk, Karel Kozel, J. Lain, Miroslav Šejna, and Miroslava Vavřincová. Numerical simulation of steady and unsteady flows through plane cascades. In S. M. Deshpande, S. S. Desai, and R. Narasimha, editors, *Fourteenth International Conference on Numerical Methods in Fluid Dynamics*, Lecture Notes in Physics, pages 461–465. Springer, 1994.

[9] Jiří Fürst. Modern difference schemes for solving the system of Euler equations. Diploma thesis, Faculty of Nuclear Science and Physical Engineering, CTU Prague, 1994. (in czech).

[10] Jiří Fürst. *Numerical modeling of the transonic flows using TVD and ENO schemes.* PhD thesis, ČVUT v Praze and l'Université de la Méditerrané, Marseille, 2000. in preparation.

[11] Jiří Fürst and Karel Kozel. Using TVD and ENO schemes for numerical solution of the multidimensional system of Euler and Navier-Stokes equations. In *Pitman Research Notes*, number 388 in Mathematics Series, 1997. Conference on Navier-Stokes equations, Varenna 1997.

[12] J.B. Goodman and R.J. LeVeque. On the accuracy of stable schemes for 2D scalar conservation laws. *Math. Comp.*, 45:503–520, 1988.

[13] Ami Harten. High resolution schemes for hyperbolic conservation laws. *Journal of Computational Physics*, 49:357–393, 1983.

[14] Michal Janda, Karel Kozel, and Richard Liska. Composite schemes on triangular meshes. In *Proceedings of HYP 2000*, Magdeburg, March 2000. to appear.

[15] Randall J. Le Veque. *Numerical Methods for Conservation Laws.* Birkhäuser Verlag, Basel, 1990.

[16] Stanley Osher and Sukumar Chakravarthy. Upwind schemes and boundary conditions with applications to Euler equations in general geometries. *J. Comp. Phys.*, (50):447–481, 1983.

[17] M. Šťastný and P. Šafařík. Experimental analysis data on the transonic flow past a plane turbine cascade. *ASME Paper*, (91-GT-313), 1990.

Department of Technical Mathematics,
Faculty of Mechanical Engineering,
Czech Tecnical University,
Karlovo nám. 13
121 35, Praha, Czech Republic
E-mail address: furst@marian.fsik.cvut.cz

On a Conjecture Concerning the Stokes Problem in Nonsmooth Domains

John G. Heywood

Abstract. This paper continues the efforts of Xie [4], [5], [6], [7], [8] and the present author [2] to prove an inequality of the form $\sup_\Omega |\mathbf{u}|^2 \leq c \|\nabla \mathbf{u}\| \|\tilde{\Delta}\mathbf{u}\|$ for solutions of the Stokes equations in an arbitrary three-dimensional domain Ω, with a constant c independent of the domain. Here the norms $\|\cdot\|$ are L^2-norms over Ω, and $\tilde{\Delta}$ denotes the Stokes operator. The function $\mathbf{u}$ is assumed to be solenoidal, and to vanish on the boundary and at spatial infinity. Xie [6] proved the inequality modulo one missing point that he left as a conjecture. Recently, in [2], we showed that the desired inequality would also follow from another conjecture that seems to us more approachable. The present paper offers some partial results and observations from our efforts to prove this new conjecture.

1. Introduction

We have described at length, in [2] and [3], the potential importance to the Navier-Stokes theory of an inequality of the form

$$\sup_\Omega |\mathbf{u}|^2 \leq c \|\nabla \mathbf{u}\| \left\|\tilde{\Delta}\mathbf{u}\right\|, \tag{1}$$

valid for an arbitrary three-dimensional domain Ω, with a constant independent of the domain. Here $\mathbf{u}$ is a solenoidal vector field that vanishes on $\partial\Omega$ and tends to zero at spatial infinity if Ω is unbounded. The norms are L^2-norms over Ω, and $\tilde{\Delta}$ is the Stokes operator defined in Ω. To date, the most general domains for which this inequality has been proven are Lipschitz domains, with a constant dependent on the domain, which is a result of Brown and Shen [1]. However, we have long believed that assumptions about the regularity of the boundary should not be necessary for this inequality, as explained in [2], and that the constant should be independent of the domain. In this direction, Xie [6] has shown that once the inequality is proven for smoothly bounded domains, with a constant independent of the domain, then one can obtain the inequality for any open set Ω by passing to a limit.

Furthermore, as a model problem, Xie [4], [5], [7] proved an analogue of (1) for the Laplacian, with the constant $c = (2\pi)^{-1}$, valid for any open set $\Omega \subset R^3$. He also showed that this constant is the best possible constant, in any domain. His methods for the Laplacian are all applicable to the Stokes operator, except that at one point he used the maximum principle. The conclusion that he drew from the maximum principle appears likely to be true also for the Stokes operator, and he conjectured it to be so in [6]. There, he proved that if his conjecture is true, then so is the inequality (1) for the Stokes operator, for arbitrary domains, with the constant $c = (3\pi)^{-1}$. He also proved that if an inequality of the form (1) is valid at all, then the constant can be no smaller than $c = (3\pi)^{-1}$.

Recently, in [2], we made a new conjecture that seems to us more approachable than Xie's. The main result of [2] was a proof that this new conjecture, if proven, provides an alternative means of completing Xie's argument for the inequality (1), again with the constant $c = (3\pi)^{-1}$. The purpose of the present paper is to present some partial results towards the proof of this new conjecture.

For the Laplacian, the analogue of Xie's conjecture follows from the maximum principle. For the Laplacian, the analogue of our conjecture is still unproven, and would seem to be best approached by variational methods. A proof by such methods could probably be transferred to the Stokes operator. Therefore, it seems best to approach our conjecture first in the context of the Laplacian, as a model problem. Through the remainder of this paper, we shall deal only with the case of the Laplacian, using methods that we think can be generalized to the Stokes operator.

Our conjecture is that a term μ_m appearing in Xie's reasoning tends to infinity as $m \to \infty$. To define μ_m, and to make the present paper reasonably self contained, it is necessary to repeat part of Xie's argument from [4],[5]. We should mention that Xie later gave a second proof of his inequality for the Laplacian [7], which had the advantage of also yielding a related two-dimensional inequality. But our conjecture concerns only his first proof, which was extended to the Stokes operator modulo his conjecture in [6], and modulo our conjecture in [2].

2. Xie's argument for the Laplacian

In Xie's theorem, the domain Ω is an arbitrary open set in R^3, and the homogeneous Sobolev space $\hat{H}_0^1(\Omega)$ is defined to be the completion of $C_0^\infty(\Omega)$ in the Dirichlet norm

$$\|\nabla u\| = \left(\int_\Omega |\nabla u|^2 \, dx\right)^{1/2}.$$

The assumption that $\Delta u \in L^2(\Omega)$ is interpreted to mean that there exists a function $f \in L^2(\Omega)$ such that

$$\int_\Omega \nabla u \cdot \nabla \phi \, dx = -\int_\Omega f \phi \, dx , \quad \text{for all } \phi \in C_0^\infty(\Omega),$$

and Δu is defined to be this function f.

Theorem 2.1. *(Xie 1991 [4]) Let $\Omega \subset R^3$ and $u \in \hat{H}_0^1(\Omega)$, with $\Delta u \in L^2(\Omega)$. Then*

$$\sup_{\Omega} |u|^2 \leq \frac{1}{2\pi} \|\nabla u\| \, \|\Delta u\| \,, \tag{2}$$

and the constant $(2\pi)^{-1}$ is the best possible, in any domain.

The central part of the proof is to obtain the inequality (2) for a bounded domain with smooth boundary, and for functions u which are finite combinations of the eigenfunctions of the Laplacian.

Let these eigenvalues and eigenfunctions, ordered and orthonormal as usual, be denoted by λ_n and ϕ_n. Thus,

$$-\Delta \phi_n = \lambda_n \phi_n \,, \quad \phi_n|_{\partial\Omega} = 0 \,, \quad (\phi_i, \phi_j) = \delta_{i,j} \,.$$

Following Xie, choose an arbitrary point $\mathbf{y} \in \Omega$, and consider it fixed in all that follows. Also, let m be an arbitrary positive integer, which will be fixed through the first part of the following argument. We will show that

$$|u(\mathbf{y})|^2 \leq \frac{1}{2\pi} \|\nabla u\| \, \|\Delta u\| \,, \tag{3}$$

for all functions of the form

$$u = \sum_{n=1}^{m} c_n \phi_n \,.$$

With m fixed, we may vary the coefficients $(c_1, c_2, \cdots, c_n) \in R^m \setminus \{0\}$ to maximize

$$\frac{u^2(\mathbf{y})}{\|\nabla u\| \, \|\Delta u\|} = \frac{\left(\sum_{n=1}^{m} c_n \phi_n(\mathbf{y})\right)^2}{\left(\sum_{n=1}^{m} \lambda_n c_n^2\right)^{1/2} \left(\sum_{n=1}^{m} \lambda_n^2 c_n^2\right)^{1/2}} \,. \tag{4}$$

This quotient certainly attains a maximum, since it is constant on rays emanating from the origin in the coefficient space R^m, and therefore takes all of its values on the compact unit sphere in R^m. Choose $(\tilde{c}_1, \tilde{c}_2, \cdots, \tilde{c}_n)$ to be a point where the maximum is attained, and set

$$\tilde{u}_m = \sum_{n=1}^{m} \tilde{c}_n \phi_n \,, \qquad \tilde{\mu}_m = \|\Delta \tilde{u}_m\|^2 / \|\nabla \tilde{u}_m\|^2 \,.$$

Of course, the derivatives of the quotient (4) with respect to the variables c_n must vanish at this point. Setting the derivative of the logarithm of (4) equal to zero at the critical point, one obtains

$$\frac{2\phi_n(\mathbf{y})}{\tilde{u}_m(\mathbf{y})} - \frac{\lambda_n \tilde{c}_n}{\|\nabla \tilde{u}_m\|^2} - \frac{\lambda_n^2 \tilde{c}_n}{\|\Delta \tilde{u}_m\|^2} = 0 \,,$$

for $n = 1, \cdots, m$, which can be rewritten as

$$\frac{2\phi_n(\mathbf{y})}{\tilde{\mu}_m + \lambda_n} = \frac{\tilde{u}_m(\mathbf{y})}{\|\Delta \tilde{u}_m\|^2} \lambda_n \tilde{c}_n \,.$$

Squaring this, summing n from 1 to m, and multiplying by $\sqrt{\tilde{\mu}_m}$, one obtains

$$\frac{\tilde{u}_m^2(\mathbf{y})}{\|\nabla \tilde{u}_m\| \, \|\Delta \tilde{u}_m\|} = 4\sqrt{\tilde{\mu}_m} \sum_{n=1}^{m} \left(\frac{\phi_n(\mathbf{y})}{\tilde{\mu}_m + \lambda_n} \right)^2 . \tag{5}$$

It is remarkable that Xie's argument has no need for any further characterization of the critical point $(\tilde{c}_1, \tilde{c}_2, \cdots, \tilde{c}_n)$, or the function $\tilde{u}_m$, or the value of $\tilde{\mu}_m$. Our conjecture is that $\tilde{\mu}_m \to \infty$, as $m \to \infty$, the usefulness of which will be explained later. Though this part of the argument, we continue to follow Xie in identifying the right side of (5) in terms of the Green's function $G_\mu(\mathbf{x}; \mathbf{y})$ for the Helmholtz operator $-\Delta + \mu$ with Dirichlet boundary conditions. This Green's function is easily seen to satisfy

$$\int_\Omega G_\mu^2(\mathbf{x}; \mathbf{y}) \, dx = \sum_{n=1}^{\infty} \left(\int_\Omega G_\mu(\mathbf{x}; \mathbf{y}) \phi_n(\mathbf{x}) \, dx \right)^2 = \sum_{n=1}^{\infty} \left(\frac{\phi_n(\mathbf{y})}{\mu + \lambda_n} \right)^2 . \tag{6}$$

Also, the fundamental singularity for the Helmholtz operator is

$$g_\mu(\mathbf{x}; \mathbf{y}) = \frac{e^{-\sqrt{\mu}|\mathbf{x}-\mathbf{y}|}}{4\pi |\mathbf{x} - \mathbf{y}|} \,. \tag{7}$$

At this point, Xie inferred from the maximum principle that

$$0 \le G_\mu(\mathbf{x}; \mathbf{y}) \le g_\mu(\mathbf{x}; \mathbf{y}) \,, \tag{8}$$

and hence, setting $r = |\mathbf{x} - \mathbf{y}|$, that

$$\int_\Omega G_\mu^2(\mathbf{x}; \mathbf{y}) \, dx \le \int_{R^3} g_\mu^2(\mathbf{x}; \mathbf{y}) \, dx = \int_0^\infty \left(\frac{e^{-\sqrt{\mu} r}}{4\pi r} \right)^2 4\pi r^2 \, dr = \frac{1}{8\pi\sqrt{\mu}} \,. \tag{9}$$

Finally, setting $\mu = \tilde{\mu}_m$ in (6) and (9), and combining them with (5), he obtained

$$\frac{\tilde{u}_m^2(\mathbf{y})}{\|\nabla \tilde{u}_m\| \, \|\Delta \tilde{u}_m\|} \le \frac{1}{2\pi} \,,$$

i.e. (3), which is the main step in proving Theorem 1.

In [6], Xie carried over the entire proof of Theorem 1 to the Stokes equations, except for the use of the maximum principle in proving (9). At this point of the argument he conjectured that the analogue of the inequality

$$\int_\Omega G_\mu^2(\mathbf{x}; \mathbf{y}) \, dx \le \int_{R^3} g_\mu^2(\mathbf{x}; \mathbf{y}) \, dx \tag{10}$$

also holds for the spectral Stokes operator $-\tilde{\Delta} + \mu$, despite the fact that the maximum principle does not hold.

3. Alternative conjecture

While the maximum of the quotient (4) depends only on m, the maximizing function $\tilde{u}_m$ is not unique, and the value of the quotient $\tilde{\mu}_m$ might depend upon the choice of the maximizing function $\tilde{u}_m$. Therefore, we shall assume that $\tilde{u}_m$ is chosen from among the functions that maximize (4), to be one that also maximizes the quotient $\tilde{\mu}_m$. Then we denote this choice of $\tilde{u}_m$ and $\tilde{\mu}_m$ simply by u_m and μ_m.

In place of Xie's conjecture (10), we proposed the following conjecture in [2]:

Conjecture 3.1.

$$\mu_m \to \infty\,, \qquad as\ m \to \infty.$$

The main result of [2] was a proof of the following lemma (in the context of the Laplacian) and its analogue for the Stokes operator:

Lemma 3.2.

$$\frac{\int_\Omega G_\mu^2(\mathbf{x};\mathbf{y})\,dx}{\int_{R^3} g_\mu^2(\mathbf{x};\mathbf{y})\,dx} \to 1\,, \qquad as\ \mu \to \infty. \tag{11}$$

From this lemma and the identity

$$8\pi\sqrt{\mu}\int_{R^3} g_\mu^2(\mathbf{x};\mathbf{y})\,dx = 1$$

proved in (9), it follows that

$$\sqrt{\mu}\int_\Omega G_\mu^2(\mathbf{x};\mathbf{y})\,dx = \frac{\sqrt{\mu}\int_\Omega G_\mu^2(\mathbf{x};\mathbf{y})\,dx}{8\pi\sqrt{\mu}\int_{R^3} g_\mu^2(\mathbf{x};\mathbf{y})\,dx} \to \frac{1}{8\pi}\,, \qquad \text{as } \mu \to \infty\,. \tag{12}$$

Combining this with (5) and (6), we have, for any integer m, and for all $\hat{m} \geq m$,

$$\frac{u_m^2(\mathbf{y})}{\|\nabla u_m\|\,\|\triangle u_m\|} \leq \frac{u_{\hat{m}}^2(\mathbf{y})}{\|\nabla u_{\hat{m}}\|\,\|\triangle u_{\hat{m}}\|} \leq 4\sqrt{\mu_{\hat{m}}}\int_\Omega G_{\mu_{\hat{m}}}^2(\mathbf{x};\mathbf{y})\,dx \to \frac{1}{2\pi} \tag{13}$$

as $\hat{m} \to \infty$, provided that $\mu_{\hat{m}} \to \infty$, as $\hat{m} \to \infty$, which is our new conjecture. Of course, (13) implies (3). Before embarking on an attempt to prove the new conjecture, it may be well to repeat some heuristic reasoning for it from [2], based on Xie's proof that his constant $c = (2\pi)^{-1}$ is optimal.

4. Heuristic support for the new conjecture

If Ω is all of R^3, the function

$$u(\mathbf{x}) = \begin{cases} 1 & , \quad \mathbf{x} = \mathbf{0} \\ \dfrac{1 - e^{-|\mathbf{x}|}}{|\mathbf{x}|} & , \quad \mathbf{x} \neq \mathbf{0} \end{cases} \tag{14}$$

satisfies the hypotheses of Theorem 1, and also its conclusion, with equality, i.e.,

$$\sup_{R^3} |u|^2 = \frac{1}{2\pi}\,\|\nabla u\|\,\|\triangle u\|\,, \tag{15}$$

since

$$\int_{R^3} |\Delta u|^2 \, dx = 2\pi \, , \quad \int_{R^3} |\nabla u|^2 \, dx = 2\pi \, , \quad \sup |u| = 1 \, .$$

Moreover, the same is true of the scaled functions

$$v_\alpha(\mathbf{x}) \equiv u(\alpha \mathbf{x}) \, ,$$

since

$$\int_{R^3} |\Delta v_\alpha|^2 \, dx = 2\pi\alpha \, , \quad \int_{R^3} |\nabla v_\alpha|^2 \, dx = 2\pi \frac{1}{\alpha} \, , \quad \sup |v_\alpha| = \sup |u| = 1.$$

Thus, we have a one parameter family of functions for which equality holds in (2), while the ratio $\mu_\alpha \equiv \|\Delta v_\alpha\|^2 / \|\nabla v_\alpha\|^2 = \alpha^2$ takes on all positive values. The functions v_α become very singular as $\alpha \to \infty$, but very flat as $\alpha \to 0$.

Xie proved that the constant $(2\pi)^{-1}$ cannot be improved in any domain Ω by choosing a point $\mathbf{y} \in \Omega$, and then smoothly truncating the functions $v_\alpha(|\mathbf{x} - \mathbf{y}|)$ to have supports in Ω. As $\alpha \to \infty$, the functions $v_\alpha(|\mathbf{x} - \mathbf{y}|)$ become very small outside of a small neighborhood of $\mathbf{y}$, and hence can be truncated with very little change in the norms $\|\Delta v_\alpha\|$ and $\|\nabla v_\alpha\|$. Thus, the ratios $v_\alpha^2(\mathbf{y}) / \|\nabla v_\alpha\| \, \|\Delta v_\alpha\| = (2\pi)^{-1}$and $\mu_\alpha = \|\Delta v_\alpha\|^2 / \|\nabla v_\alpha\|^2$ are nearly preserved by the truncation, and the latter tend to infinity. Thus, in proving that $(2\pi)^{-1}$ is the best possible constant in his inequality for the Laplacian (and similarly $(3\pi)^{-1}$ for the Stokes operator) Xie constructed a sequence of functions for which the ratio $\mu = \|\Delta v\|^2 / \|\nabla v\|^2$ tends to infinity.

5. An identity for the determination of μ_m

Because of the homogeneity of (4), there exists a maximum $(c_1, \cdots, c_n)$ satisfying the constraint

$$\sum_{n=1}^{m} c_n \phi_n(\mathbf{y}) = 1.$$

We shall obtain an identity for the determination of μ_m by reconsidering the maximization of the quotient (4) subject to this constraint. We continue, of course, to regard $\mathbf{y}$ as fixed.

For each positive integer m, there might still be several possible choices of the maximizing coefficients, even after imposing the constraint. We choose one, denote it by $\{c_{m,n}\}$, $n = 1, \cdots, m$, and set

$$A_m = \sum_{n=1}^{m} \lambda_n c_{m,n}^2 \, , \qquad B_m = \sum_{n=1}^{m} \lambda_n^2 c_{m,n}^2 \, .$$

Lemma 5.1.

$$c_{m,n} = \frac{2\phi_n(\mathbf{y}) A_m B_m}{B_m \lambda_n + A_m \lambda_n^2} \tag{16}$$

Proof: In effect, the $\{c_{m,n}\}$ minimize

$$f(c_1, \cdots, c_n) \equiv \left(\sum_{n=1}^{m} \lambda_n c_n^2 \right) \left(\sum_{n=1}^{m} \lambda_n^2 c_n^2 \right) \tag{17}$$

subject to the constraint

$$g(c_1, \cdots, c_n) \equiv \sum_{n=1}^{m} c_n \phi_n(\mathbf{y}) = 1. \tag{18}$$

Thus, there exists a Lagrangian multiplier λ such that

$$c_{m,n} \left[\lambda_n \sum_{k=1}^{m} \lambda_k^2 c_{m,k}^2 + \lambda_n^2 \sum_{k=1}^{m} \lambda_k c_{m,k}^2 \right] = \lambda \phi_n(\mathbf{y}) \ , \tag{19}$$

for $n = 1, \cdots, m$.

Multiplying (19) by $c_{m,n}$, summing over n, and using the constraint (18), we obtain

$$2f(c_{m,1}, \cdots, c_{m,n}) = \sum_{n=1}^{m} \lambda c_{m,n} \phi_n(\mathbf{y}) = \lambda \ .$$

To eliminate λ, multiply this by $\phi_n(\mathbf{y})$ and substitute the left side of (19) for $\lambda \phi_n(\mathbf{y})$ to get

$$2\phi_n(\mathbf{y}) f(c_{m,1}, \cdots, c_{m,n}) = c_{m,n} \left[\lambda_n \sum_{k=1}^{m} \lambda_k^2 c_{m,k}^2 + \lambda_n^2 \sum_{k=1}^{m} \lambda_k c_{m,k}^2 \right] ,$$

which simplifies to the desired identity (16).

Lemma 5.2. *The ratio $\mu_m = B_m/A_m$ is a root of the equation*

$$f_m(\mu) \equiv \sum_{n=1}^{m} \frac{\phi_n^2(\mathbf{y})}{(\mu + \lambda_n)^2} \left[\frac{\mu}{\lambda_n} - 1 \right] = 0 \ . \tag{20}$$

Proof: Square (16), multiply the result by λ_n, sum, and simplify to obtain

$$A_m = 4A_m^2 B_m^2 \sum_{n=1}^{m} \frac{\phi_n^2(\mathbf{y}) \lambda_n}{(B_m \lambda_n + A_m \lambda_n^2)^2} = 4A_m B_m \sum_{n=1}^{m} \frac{\phi_n^2(\mathbf{y})}{(\mu_m + \lambda_n)^2} \left(\frac{\mu_m}{\lambda_n} \right) \ . \tag{21}$$

Similarly, square (16), multiply the result by λ_n^2, sum, and simplify to obtain

$$B_m = 4A_m^2 B_m^2 \sum_{n=1}^{m} \frac{\phi_n^2(\mathbf{y}) \lambda_n^2}{(B_m \lambda_n + A_m \lambda_n^2)^2} = 4B_m^2 \sum_{n=1}^{m} \frac{\phi_n^2(\mathbf{y})}{(\mu_m + \lambda_n)^2} \ . \tag{22}$$

Multiplying (22) by A_m/B_m and combining with (21) gives $f_m(\mu_m) = 0$, as claimed. The following is immediately apparent from (20).

Lemma 5.3. *Each function $f_m(\mu)$ is negative on the interval $0 \le \mu < \lambda_1$, and positive on the interval $\lambda_m < \mu < \infty$. Hence it has at least one root in the interval $\lambda_1 \le \mu \le \lambda_m$, and no roots outside of this interval.*

It would seem very useful to know that each function $f_m(\mu)$ has only one zero. This may be true, but it does not follow from the general form of the expression (20) without further information about the values of the λ_k and $\phi_k^2(\mathbf{y})$. In fact, one can assign hypothetical values for the λ_k and $\phi_k^2(\mathbf{y})$ so that $f_m(\mu)$ has $m-1$ zeros.

The following lemma is likely to be important as a substitute for not knowing that the zeros of $f_m(\mu)$ are unique. It is an easy consequence of the identity (6), or more simply

$$\sum_{n=1}^{\infty} \frac{\phi_n^2(\mathbf{y})}{\lambda_n^2} = \int_\Omega G^2(\mathbf{x};\mathbf{y})\, dx ,$$

where $G(\mathbf{x},\mathbf{y})$ is the Green's function for the Laplacian.

Lemma 5.4. *The infinite series*

$$f_\infty(\mu) \equiv \sum_{n=1}^{\infty} \frac{\phi_n^2(\mathbf{y})}{(\mu+\lambda_n)^2}\left[\frac{\mu}{\lambda_n} - 1\right] \tag{23}$$

is uniformly and absolutely convergent on bounded subintervals of $\mu \in [0,\infty)$. Moreover, if μ is regarded as a complex variable, the series is similarly convergent in bounded subsets of the right half-plane, and hence represents an analytic function there, which can be differentiated term by term.

Our main hope now is to prove that $f_\infty(\mu) < 0$ for all $\mu \ge 0$. Combined with the first part of the last lemma, this would easily imply that the roots μ_m of (20) must tend to infinity, as $m \to \infty$.

6. On the average values of $\phi_n^2(\mathbf{y})$, as $n \to \infty$

In order that $f_\infty(\mu) < 0$, it is necessary that the positive terms of the series (23), i.e., those for which $\lambda_n < \mu$, are dominated be the subsequent negative terms. This clearly depends upon the pointwise values of the squares of the eigenfunctions, $\phi_n^2(\mathbf{y})$. We seem to need some kind of information that the averages of their values should not decay as $n \to \infty$. In the case of a cube, knowing the eigenfunctions explicitly, it is easily shown that the average of the first N of them tends to a positive constant as $N \to \infty$. In general, the constant depends on the point $\mathbf{y}$, but it is the same for all points with irrational coordinates, the common value of the constant for these points being $1/V$, where V is the volume of Ω. This suggests the kind of result we seem to need, but the proof is not applicable to arbitrary domains.

However, as a substitute for this, there is something we know about the averages of large numbers of the $\phi_n^2(\mathbf{y})$ in an arbitrary smoothly bounded domain. From (6) and (12) we have

$$\sum_{n=1}^{\infty} \frac{\phi_n^2(\mathbf{y})}{\left(\mu^{3/4}+\mu^{-1/4}\lambda_n\right)^2} = \sqrt{\mu}\int_\Omega G_\mu^2(\mathbf{x};\mathbf{y})\,dx \to \frac{1}{8\pi}, \quad \text{as } \mu \to \infty. \tag{24}$$

Since

$$\lambda_n \sim \left(\frac{6\pi^2}{V}\right)^{2/3} n^{2/3} \equiv \tilde{c}n^{2/3}, \quad \text{as } n \to \infty,$$

and since the integral

$$\int_0^\infty \frac{dx}{\left(\mu^{3/4}+\mu^{-1/4}\tilde{c}x^{2/3}\right)^2} = \frac{3\pi}{4\tilde{c}^{3/2}} = \frac{V}{8\pi}$$

is independent of μ, we can regard the left side of (24) as a weighted average of the $\phi_n^2(\mathbf{y})$, with the approximate weight function

$$f(x;\mu) = \frac{1}{\left(\mu^{3/4}+\mu^{-1/4}\tilde{c}x^{2/3}\right)^2}.$$

These weight functions are monotonically decreasing with respect to both x and μ. Hence in these averages, the weights given to great numbers of the $\phi_n^2(\mathbf{y})$ become more nearly equal with increasing μ. Therefore, we can interpret (24) as meaning that the average value of great numbers of the $\phi_n^2(\mathbf{y})$ tends to $1/V$. Thus (24) may well embody the information that we expect to need concerning the values of the $\phi_n^2(\mathbf{y})$, in trying to prove that $f_\infty(\mu) < 0$ for all $\mu \geq 0$.

7. Heuristic indications that $f_\infty(\mu) < 0$, for all $\mu \geq 0$

Let us normalize the volume to $V = 6\pi^2$, so that $\lambda_n \sim n^{2/3}$, and $\tilde{c} = 1$. Further, for the sake of a heuristic argument, let us replace λ_n by $n^{2/3}$, and $\phi_n^2(\mathbf{y})$ by $1/6\pi^2$ in (23). The analogue of our desired result then becomes

$$\hat{f}_\infty(\mu) \equiv \sum_{n=1}^{\infty} \frac{1}{\left(\mu+n^{2/3}\right)^2}\left[\frac{\mu}{n^{2/3}}-1\right] < 0, \quad \text{for all } \mu \geq 0. \tag{25}$$

In fact, we can prove this by an integral comparison test. One may verify that

$$\int_a^\infty \frac{1}{\left(\mu+x^{2/3}\right)^2}\left[\frac{\mu}{x^{2/3}}-1\right] dx = -\frac{3a^{1/3}}{\mu+a^{2/3}}, \quad \text{for all } \mu > 0,$$

and hence that

$$\int_0^\infty \frac{1}{\left(\mu+x^{2/3}\right)^2}\left[\frac{\mu}{x^{2/3}}-1\right] dx = 0, \quad \text{for all } \mu > 0. \tag{26}$$

For $\mu > 0$, the integrand decreases on the interval $0 < x < \left(1+\sqrt{2}\right)^{3/2} \mu^{3/2}$, tending to $+\infty$, as $x \to 0^+$, and passing through zero at $x = \mu^{3/2}$. For $x > \left(1+\sqrt{2}\right)^{3/2} \mu^{3/2}$, it remains negative, but increases toward zero. If one plots the graph of this, and also the terms of the series on the left of (25) as a step function defined on $[0, \infty)$, it will be clear that the series is less than the integral, proving (25).

In fact, the integral, which equals zero, will exceed the series by approximately the integral over interval $(0, 1)$, minus the first term of the series, plus half of the first term of the series. Thus

$$\begin{aligned} \sum_{n=1}^{\infty} \frac{1}{\left(\mu + n^{2/3}\right)^2} \left[\frac{\mu}{n^{2/3}} - 1\right] &\sim - \left[\frac{3}{\mu+1} - \frac{\mu - 1}{2\left(\mu+1\right)^2}\right] \\ &= -\frac{5\mu + 7}{2\left(\mu+1\right)^2}, \quad \text{as } \mu \to \infty. \end{aligned} \tag{27}$$

This seems to provide another strong heuristic argument for the correctness of our conjecture. However, it is so very delicate, with the integral (26) being identically zero, that there seems to be little room for estimates to get back to the precise values of λ_n and $\phi_n^2(\mathbf{y})$, from their asymptotic and average values used in the heuristic calculation (27). It appears that if the relation (24) is to be used, it will probably have to fit directly into the final argument, without a string of intermediate estimates.

Acknowledgments

This work has been supported by the Natural Sciences and Engineering Research Council of Canada.

References

[1] Brown, R.M. & Z. Shen, Estimates for the Stokes operator in Lipschitz domains, Indiana Univ. Math. J., 44 (1995), 1183–1206.

[2] Heywood, J.G., An Alternative to Xie's Conjecture Concerning the Stokes Problem in Nonsmooth Domains, Annali dell' Universita' di Ferrara set VII, n. 46, 2000, 267–284.

[3] Heywood, J.G. & W. Xie, Smooth solutions of the vector Burgers equation in nonsmooth domains, Diff. and Int. Equ. 10 (1997), 961–974.

[4] Xie, W., A sharp inequality for the Poisson's equation in arbitrary domains and its application to Burgers equation, Ph.D. Thesis, University of British Columbia, 1991.

[5] Xie, W., A sharp pointwise bound for functions with L^2 Laplacians and zero boundary values on arbitrary three-dimensional domains, Indiana Univ. Math. J., 40 (1991), 1185–1192.

[6] Xie, W., On a three-norm inequality for the Stokes operator in nonsmooth domains, in The Navier-Stokes Equations II: Theory and Numerical Methods, Springer-Verlag Lecture Notes in Math. 1530 (1992), 310–315.

[7] Xie, W., Integral representations and L^∞ bounds for solutions of the Helmholtz equation on arbitrary open sets in R^2 and R^3, Diff. and Int. Equ. 8 (1995), 689–698.

[8] Xie., W., Sharp Sobolev interpolation inequalities for the Stokes operator, Diff. and Int. Eq., 10 (1997), 393–399.

Department of Mathematics
University of British Columbia
1984 Mathematics Road
Vancouver, B.C. V6T 1Z2, Canada
E-mail address: heywood@math.ubc.ca

On Well–Posedness of the Navier–Stokes Equations

Hideo Kozono

Abstract. We discuss uniqueness, regularity and stability of weak and strong solutions to the Navier-Stokes equations. We first introduce the class $L^s(0,T;$ $L^r(\mathbb{R}^n))$ of Serrin and give a brief survey of well–posedness. Then we devote ourselves to various estimates in the BMO–Hardy spaces and to the Sobolev embedding in the Besov space in the critical case. Making use of the technique in the harmonic analysis, we prove a criterion on regularity and break-down of the solutions. Finally, we show the stability of smooth solutions with definite convergence rates.

Introduction

The purpose of this article is to give a survey on the recent development of well-posedness on the Navier-Stokes equations. We are mainly concerned with the results given by the author. Consider the Navier-Stokes equations in $\mathbb{R}^n (n \geq 2)$:

$$\text{(N-S)} \qquad \begin{cases} \dfrac{\partial u}{\partial t} - \Delta u + u \cdot \nabla u + \nabla p = 0, & x \in \mathbb{R}^n, t \in (0,T), \\ \text{div } u = 0 \quad x \in \mathbb{R}^n, t \in (0,T), \\ u\,|_{t=0} = a, \end{cases}$$

where $u = u(x,t) = (u^1(x,t), \cdots, u^n(x,t))$ and $p = p(x,t)$ denote the unknown velocity vector and the pressure of the fluid at the point $(x,t) \in \mathbb{R}^n \times (0,T)$, respectively, while $a = a(x) = (a^1(x), \cdots, a^n(x))$ is the given initial velocity vector field. For simplicity, we assume that the external force has a scalar potential and is included into the pressure gradient.

(N-S) has been studied since many years and making a great contribution to the research on nonlinear partial differential equations. The mathematical treatment for (N-S) was originally established by Leray [36] who proved the existence of global weak solutions for an arbitrary large data a in $L^2(\mathbb{R}^3)$. Despite enormous efforts of many mathematicians, the problem on uniqueness and regularity of Leray's weak solutions is still an open question. It should be noticed that global existence of the smooth solution for an arbitrary large data was proposed by Clay institute as one of the important 7 mathematical problems in the millennium.

Although the fundamental breakthrough for investigation of (N-S) is difficult to be found, some remarkable partial results have been produced since these

15 years. Masuda [38] extended Serrin's class for uniqueness of weak solutions and made it clear that the class $L^\infty(0,T;L^n(\mathbb{R}^n))$ plays an important role for both uniqueness and regularity of weak solutions. Kozono-Sohr [34] showed that the uniqueness holds in $L^\infty(0,T;L^n)$. Concerning its regularity, Giga [19] and von Wahl [54] proved that every weak solution in $C([0,T);L^n)$ is regular. Then, Kozono-Sohr [35], Beirão da Veiga [4] and Hishida-Izumida [23] showed that if the weak solution u in $L^\infty(0,T;L^n(\mathbb{R}^n))$ has a small gap from the left in time in the norm of L^n, then u is actually regular. Furthermore, Neustupa [42] proved that for $n=3$, every weak solution $L^\infty(0,T;L^3(\mathbb{R}^3))$ possesses only a finitely many singularities in the space. It is still an open question whether every weak solution in $L^\infty(0,T;L^3)$ is regular or not.

On the other hand, the problem on local well-posedness to (N-S) had been considerably established. After the pioneer works of Kieselev-Ladyzenskaya [28], Fujita-Kato [17] and Solonnikov [50], Kato [25] and Giga-Miyakawa [21] obtained the local classical solution to (N-S) provided $a \in L^n(\mathbb{R}^n)$. In their papers, the $L^p - L^q$-estimates for the Stokes semigroup and the theory of fractional powers of the Stokes operator furnished powerful tools which had a great influence upon further developments of other nonlienar partial differential equations. Moreover, the problem on the blow-up of solutions to (N-S) has been a central question in the nonlinear analysis. Leray [36] intended to have constructed the solution which blows up at a finite time by means of the backward self-similar solution. Leray's proposal was negatively solved by Nečas-Ružička-Šverák [40] who proved that there exists no backward self-similar solution to (N-S) within the class of finite energy in $\mathbb{R}^3$. It should be noticed that this blow-up problem is closely related to the open question on regularity of weak solutions in $L^\infty(0,T;L^n(\mathbb{R}^n))$. Among these results, Caffarelli-Kohn-Nirenberg [9] proved the ε-regularity theorem for suitable solutions which satisfy the localized energy inequality. Compared with the semilnear parabolic equations, it is difficult to obtain local regularity of the solution u to (N-S) because another unknown function p requires the information on global behaviour of u in $\mathbb{R}^n$. Kozono [30] overcame this difficulty and obtained the theorem on removable singularities for all weak solutions.

In recent years, it is clarified that the various techniques in the harmonic analysis are very useful to investigation of (N-S). Coifman-Lions-Meyer-Semmes [12] showed that the nonlinear term$u\cdot\nabla u$ belongs to the Hardy space. Inspired by this fundamental fact, Miyakawa [41] gave a sharp decay rate of solutions $u(t)$ in $L^2(\mathbb{R}^n)$ as $t\to\infty$. Kozono-Taniuchi [31] improved the blow-up criterion of Beal-Kato-Majda [2] and Kato-Ponce [26], and showed that the space BMO-functions is suitable to investigate behaviour of vorticity. Some further results were obtained in the Besov space(see [33]). It seems to be possible that the analysis of (N-S) will make a contribution to development of harmonic analysis.

To state our problems, we first introduce some function spaces. We denote by $C_{0,\sigma}^\infty$ the set of all C^∞ vector functions $\phi=(\phi^1,\cdots,\phi^n)$ with compact support in $\mathbb{R}^n$, such that div $\phi=0$. L_σ^r is the closure of $C_{0,\sigma}^\infty$ with respect to the L^r-norm

$\|\cdot\|_r$. $(\cdot,\cdot)$ denotes the duality pairing between L^r and $L^{r'}$, where $1/r + 1/r' = 1$. L^r stands for the usual (vector-valued) L^r-space over $\mathbb{R}^n$, where $1 < r < \infty$. $H^1_{0,\sigma}$ denotes the closure of $C^\infty_{0,\sigma}$ with respect to the norm

$$\|\phi\|_{H^1} = \|\phi\|_2 + \|\nabla\phi\|_2,$$

where $\nabla\phi = (\partial\phi^i/\partial x_j), i, j = 1, \cdots, n$.

For an interval I in $\mathbb{R}^1$ and a Banach space X, $L^p(I;X)$ and $C^m(I;X)$ denote the usual Banach spaces of functions of L^p and C^m-cla ss on I with values in X, respectively, where $1 \le p \le \infty$, $m = 0, 1, \cdots$.

Our definition of a weak solution of (N-S) now reads

Definition 0.1. *Let $a \in L^2_\sigma$. A measurable function u on $\mathbb{R}^n \times (0,T)$ is called a weak solution of (N-S) on $(0,T)$ if*

(i) *$u \in L^\infty(0,T;L^2_\sigma) \cap L^2(0,T;H^1_{0,\sigma})$;*

(ii) *For every $\Phi \in H^1(0,T;H^1_{0,\sigma} \cap L^n)$ with $\Phi(T) = 0$,*

$$\int_0^T \Big\{-(u, \partial_t\Phi) + (\nabla u, \nabla\Phi) + (u\cdot\nabla u, \Phi)\Big\}\, dt = (a, \Phi(0)). \tag{0.1}$$

Concerning existence of the weak solutions, we have by Leray [36] and Hopf [22].

Theorem 0.2. (Leray-Hopf) *For every $a \in L^2_\sigma$, there exists at least one weak solution u of (N-S) on $(0,\infty)$ such that*

$$\|u(t)\|_2^2 + 2\int_0^t \|\nabla u(\tau)\|_2^2\, d\tau \le \|a\|_2^2, \quad 0 \le t < \infty, \tag{0.2}$$

and

$$\|u(t) - a\|_2 \to 0 \quad \text{as } t \to +0.$$

We are interested in the following problems on well-posedness to (N-S);

Problems.

(I) Uniqueness and regularity of weak solutions

(II) Global existence of regular solutions for large data a

(III) Blow-up; dose there exist $T_* < \infty$ such that

$$u(t) \in C^\infty(\mathbb{R}^n) \quad \text{for } 0 < t < T_*, \text{ but } u(T_*) \notin C^\infty(\mathbb{R}^n) \ ?$$

(IV) Stability of solutions; let $u(t)$ be a solution of (N-S) on $(0,\infty)$. For the initial disturbance $b = b(x)$ and the perturbation $f = f(x,t)$ of external force, can we find a solution v of the *perturbed* Navier-Stokes equations

$$\text{(N-S')} \quad \begin{cases} \dfrac{\partial v}{\partial t} - \Delta v + v\cdot\nabla v + \nabla p' = f, \quad \operatorname{div} v = 0 \quad x \in \mathbb{R}^n, t \in (0,\infty), \\ v\,|_{t=0} = a + b, \end{cases}$$

such that

$$\lim_{t\to\infty} \|v(t) - u(t)\|_2 = 0, \quad \lim_{t\to\infty} \|\nabla v(t) - \nabla u(t)\|_2 = 0 \quad ?$$

1. Uniqueness and regularity

Let us introduce the class $L^s(0,T;L^r)$ with the norm $\|\cdot\|_{L^s(0,T;L^r)}$;

$$\|u\|_{L^s(0,T;L^r)} = \left(\int_0^T \|u(t)\|_r^s \, dt\right)^{1/s} = \left(\int_0^T (\int_{\mathbb{R}^n} |u(x,t)|^r dx)^{s/r} \, dt\right)^{1/s}$$

The classical result on uniqueness and regularity of weak solutions in the class $L^s(0,T;L^r)$ was given by Foias, Serrin and Masuda [15], [48], [49], [38]:

Theorem 1.1. (Foias-Serrin-Masuda) *Let $a \in L^2_\sigma$.*

(i) *Let u and v be two weak solutions of (N-S) on $(0,T)$. Suppose that u satisfies*

$$u \in L^s(0,T;L^r) \quad \text{for } 2/s + n/r = 1 \text{ with } n < r \le \infty. \tag{1.1}$$

Assume that v fulfills the energy inequality (0.2) for $0 \le t < T$. Then we have $u \equiv v$ on $[0,T)$.

(ii) *Eery weak solution u of (N-S) in the class (1.1) satisfies*

$$\frac{\partial u}{\partial t}, \quad \frac{\partial^{\alpha_1+\cdots+\alpha_n} u}{\partial x_1^{\alpha_1}\cdots \partial x_n^{\alpha_n}} \in C(\mathbb{R}^n \times (0,T)) \tag{1.2}$$

for all multi-indices $\alpha = (\alpha_1, \cdots, \alpha_n)$ with $|\alpha| = \alpha_1 + \cdots + \alpha_n \le 2$.

Remark 1.2. (i) In Theorem 1.1 (i) , v need not belong to the class (1.1). On the other hand, every weak solution u with (1.1) fulfills the energy identity

$$\|u(t)\|_2^2 + 2\int_0^t \|\nabla u(\tau)\|_2^2 \, d\tau = \|a\|_2^2, \quad 0 \le t \le T. \tag{1.3}$$

It seems to be an interesting question whether every weak solution satisfies the energy inequality (0.2).

(ii) If u is merely in the Leray-Hopf class as in (i) in Definition 0.1, then there exists s_0, r_0 with $2/s_0 + n/r_0 = n/2$ su ch that $u \in L^{s_0}(0,T;L^{r_0})$. For example, we may take $s_0 = 2$ and $r_0 = 2n/(n-2)$. In particular, by Theorem 1.1 with the aid of interpolation inequality

$$\|u\|_{L^{r_0}(\mathbb{R}^2)} \le C\|u\|_{L^2(\mathbb{R}^2)}^{r_0/2} \|\nabla u\|_{L^2(\mathbb{R}^2)}^{1-r_0/2}, \quad 2 < r_0 < \infty \quad \text{for } u \in H^1(\mathbb{R}^2),$$

we see that every weak solution of (N-S) in the 2-dimensional case is unique and regular, so Problems (I), (II) and (III) are completely solved in $\mathbb{R}^2$. Notice that if u is regular, then s and r can be taken arbitrarily large, which makes the quantity $2/s + n/r$ *smaller.*

(iii) The class (1.1) is important from viewpoint of the scaling invariance. It can be easily seen that if $\{u,p\}$ is a pair of the solution to (N-S) on $\mathbb{R}^n \times (0,\infty)$, then so is the family $\{u_\lambda, p_\lambda\}_{\lambda>0}$, where

$$u_\lambda(x,t) \equiv \lambda u(\lambda x, \lambda^2 t), \quad p_\lambda(x,t) \equiv \lambda^2 p(\lambda x, \lambda^2 t).$$

Scaling invariance means that there holds

$$\|u_\lambda\|_{L^s(0,\infty;L^r)} \left(= \lambda^{1-(\frac{2}{s}+\frac{n}{r})}\|u\|_{L^s(0,\infty;L^r)}\right) = \|u\|_{L^s(0,\infty;L^r)} \quad \text{for all } \lambda > 0$$

if and only if

$$2/s + n/r = 1.$$

The solution $\{u,p\}$ with the property that $u_\lambda(x,t) = u(x,t)$, $p_\lambda(x,t) = p(x,t)$ for all $\lambda > 0$ is called a *self-similar* solution. For (N-S), the self-similar solution has the form such as

$$u(x,t) = \frac{1}{\sqrt{t}}U(\frac{x}{\sqrt{t}}), \quad p(x,t) = \frac{1}{t}P(\frac{x}{\sqrt{t}}),$$

where $U = (U^1(y), \cdots, U^n(y))$, $P = P(y)$ is the functions for $y = (y_1, \cdots, y_n) \in \mathbb{R}^n$. More precisely, the above solution is called a *forward self-similar* solution. As for the *backward self-similar* solution, we shall discuss in Section 3.

We shall next deal with the critical case with $s = \infty$ and $r = n$ in (1.1).

Theorem 1.3. (**Masuda** [38], **Kozono-Sohr** [34], [35]) *Let $a \in L^2_\sigma$.*

(i) *(uniqueness) Let u and v be weak solutions of* (N-S). *Suppose that $u \in L^\infty(0,T;L^n)$ and that v satisfies the energy inequality (0.2) for $0 \le t < T$. Then we have $u \equiv v$ on $[0,T)$.*

(ii) *(regularity) There exists a positive constant ε_0 such that if u is a weak solution of* (N-S) *in $L^\infty(0,T;L^n)$ with the property*

$$\limsup_{t\to t_*-0} \|u(t)\|_n^n \le \|u(t_*)\|_n^n + \varepsilon_0 \quad \textit{for } t_* \in (0,T), \tag{1.4}$$

then u satisfies

$$\frac{\partial u}{\partial t}, \quad \frac{\partial^{\alpha_1+\cdots+\alpha_n}u}{\partial x_1^{\alpha_1}\cdots\partial x_n^{\alpha_n}} \in C(\mathbb{R}^n \times (t_*-\rho, t_*+\rho)) \quad \textit{for some } \rho > 0, \tag{1.5}$$

where $\alpha = (\alpha_1, \cdots, \alpha_n)$ is an arbitrary multi-index with $|\alpha| = \alpha_1 + \cdots + \alpha_n \le 2$. In particular, if u has the property (1.4) for every $t_* \in (0,T)$, then u is regular on $\mathbb{R}^n \times (0,T)$ as in (1.2).

Remark 1.4. (i) Masuda [38] proved that if $u \in L^\infty(0,T;L^n)$ is continuous from the right on $[0,T)$ in the norm of L^n, then there holds $u \equiv v$ on $[0,T)$. Later on, Kozono-Sohr [34] showed that every weak solution u in $L^\infty(0,T;L^n)$ of (N-S) on $(0,T)$ becomes necessarily continuous from the right in the norm of L^n.

(ii) By the above theorem, every weak solution in $C([0,T);L^n)$ is unique and regular. This was proved by Giga [19] and von Wahl [54]. In Section 2, we shall give another proof by a different method.

(iii) Recently, Hishida-Izumida [23] improved the condition (1.4). They proved regularity of u under the weaker assumption that

$$\liminf_{t\to t_*-0} \|u(t)\|_n^n \le \|u(t_*)\|_n^n + \varepsilon_0.$$

It seems to be an interesting question whether or not *every* weak solution $u \in L^\infty(0,T;L^n)$ is regular.

Finally in this section, we investigate the size of singular sets of weak solutions in the 3-dimensional case. For a weak solution u in $\mathbb{R}^3 \times (0,T)$ we denote by $S(u)$ the singular set defined by

$$S(u) \equiv \{(x,t) \in \mathbb{R}^3 \times (0,T); u \notin L^\infty(B_\rho(x,t)) \quad \text{for } \forall \rho > 0\},$$

where $B_\rho(x,t) = \{(y,s) \in \mathbb{R}^3 \times (0,T); |y-x| < \rho, |s-t| < \rho\}$. For each $t \in (0,T)$ we set $S_t(u) = \{x \in \mathbb{R}^3; (x,t) \in S(u)\}$.

Theorem 1.5. (**Neustupa** [42]) *Let $n = 3$. There is an absolute constant $\varepsilon_0 > 0$ such that every weak solution u in $L^\infty(0,T;L^3)$ fulfills*

$$\sharp S_t(u) \le \left(\frac{1}{\varepsilon_0} \cdot \sup_{0<\tau<T} \|u(\tau)\|_3\right)^3$$

for all $t \in (0,T)$. Here $\sharp S$ denotes the number of elements of the set S.

Remark 1.6. (i) Caffarelli-Kohn-Nirenberg [9] showed if the weak solution u satisfies the *generalized* energy inequality

$$2\iint_{\mathbb{R}^3\times(0,T)} |\nabla u|^2\phi\, dx\, dt \le \iint_{\mathbb{R}^3\times(0,T)} [|u|^2(\partial_t\phi + \Delta\phi) + (|u|^2 + 2p)u\cdot\nabla\phi]\, dx\, dt \tag{1.6}$$

for all $\phi \in C_0^\infty(\mathbb{R}^3 \times (0,T))$ with $\phi \ge 0$, then $H^1(S) = 0$, where $H^1(S)$ denotes the one-dimensional Hausdorff measure of the set S in the space-time $\mathbb{R}^3 \times (0,\infty)$. Recently, Choe-Lewis [11] obtained a better estimate of Hausdorff dimension of S.

(ii) Taniuchi [53] found a class of weak solutions satisfying (1.6). His class is larger than that of Serrin's (1.1).

Next, we investigate local properties of weak solutions in $\mathbb{R}^3$. Let u be a weak solution of (N-S) on $(0,T)$. We call $(x_0,t_0) \in \mathbb{R}^3 \times (0,T)$ a *regular* point if there are $\delta > 0$ and $\rho > 0$ such that

$$\frac{\partial u}{\partial t}, \frac{\partial^{\alpha_1+\cdots+\alpha_n} u}{\partial x_1^{\alpha_1}\cdots\partial x_n^{\alpha_n}} \in C(B_\delta(x_0) \times (t_0-\rho, t_0+\rho))$$

for all multi-index $\alpha = (\alpha_1,\cdots,\alpha_n)$ with $|\alpha| = \alpha_1 + \cdots + \alpha_n \le 2$. Here $B_\delta(x_0) = \{y \in \mathbb{R}^3; |y-x_0| < \delta\}$. The point (x_0,t_0) is called *singular* unless it is regular. u is called *regular* on a space-time $Q = D \times (a,b)$ if every point of Q is a regular one.

Theorem 1.7. (**Kozono** [29]) *Let $n = 3$. There is an absolute constant $\varepsilon_0 > 0$ with the following property. If u is a weak solution of (N-S) on $(0,T)$ and if u satisfies at $(x_0,t_0) \in \mathbb{R}^3 \times (0,T)$*

$$\sup_{t_0-\rho<t<t_0+\rho} \|u(t)\|_{L^3_W(B_\delta(x_0))} \le \varepsilon_0 \tag{1.7}$$

for some $\delta > 0$ *and* $\rho > 0$, *then* (x_0, t_0) *is a regular point. Here* $\|\cdot\|_{L^3_W(B_\delta(x_0))}$ *denotes the weak* L^3*-norm* $\|u\|_{L^3_W(B_\delta(x_0))} = \sup_{R>0} R\mu\{x \in B_\delta(x_0); |u(x)| > R\}^{\frac{1}{3}}$ (μ; *Lebesgue measure).*

Corollary 1.8. (**Removable Singularities**) *Let* $n = 3$. *There is an absolute constant* ε_0 *with the following property. Suppose that* u *is a weak solution of (N-S) on* $(0, T)$. *If* (x_0, t_0) *is an isolated singular point of* u *satisfying*

$$\limsup_{x\to x_0, t\to t_0} |x - x_0||u(x,t)| < \varepsilon_0, \tag{1.8}$$

then (x_0, t_0) *is a regular point.*

In particular, if u *behaves at* (x_0, t_0) *like*

$$u(x,t) = o(|x - x_0|^{-1}) \quad \text{as } x \to x_0 \tag{1.9}$$

uniformly with respect to t *in some neighbourhood of* t_0, *then* (x_0, t_0) *is a regular point.*

Remark 1.9. (i) Serrin [48] and Takahashi [52] showed that every weak solution u of (N-S) satisfying

$$\int_a^b (\int_D |u(x,t)|^r \, dx)^{\frac{s}{r}} \, dt < \infty \quad \text{on a cylinder } D \times (a,b) \subset \Omega \times (0,T),$$

for $2/s + 3/r \le 1$ with $r > 3$ is of class C^∞ in the space variables. Our theorem deals with the marginal case when $s = \infty$ and $r = 3$. Furthermore, our weak space $L^3_W(D)$ is larger than the usual $L^3(D)$. Under the condition (1.7), we obtain interior regularity of u not only in the space but also in the *space-time* variables, while Serrin [48] imposed the additional assumption that

$$\partial_t u \in L^s(a,b; L^2(D)) \quad \text{for some } s \ge 1.$$

(ii) Caffarelli-Kohn-Nirenberg [9] gave an absolute constant ε_1 with the following property. Let u be a weak solution of (N-S) on $(0, T)$ with the *generalized* energy inequality (1.6). Suppose that u and its the associated pressure p satisfy

$$R^{-2} \iint_{Q_R(x_0,t_0)} (|u|^3 + |u||p|) \, dx \, dt + R^{-13/4} \int_{t_0-R^2}^{t_0} \left(\int_{|x-x_0|<R} |p| \, dx \right)^{5/4} dt$$
$$\le \varepsilon_1, \tag{1.10}$$

where $Q_R(x_0,t_0) = \{(x,t); |x - x_0| < R, t_0 - R^2 < t < t_0\}$ denotes the parabolic cylinder. Then u is regular in $Q_{R/2}(x_0, t_0)$. In Theorem 1.7, we do not need any energy inequality and show that the condition on the pressure p is redundant. Moreover, the advantage of our theorem enables us to handle the singularity (x_0, t_0) of u such as

$$u(x,t) = o(|x - x_0|^{-1}) \quad \text{as } x \to x_0$$

uniformly with respect to t in some neighbourhood of t_0, the case of which is excluded in their paper because for such (x_0, t_0) we have in (1.10)

$$\iint_{Q_R(x_0,t_0)} |u(x,t)|^3\, dx\, dt = \infty.$$

2. Local existence and uniqueness of strong solutions

In this section, we investigate the solution with (1.1). To this end, we define the strong solutions.

Definition 2.1. *Let $a \in L^n_\sigma$. A measurable function defined on $\mathbb{R}^n \times (0,T)$ is called a strong solution of (N-S) on $(0,T)$ if*

(i)

$$u \in C([0,T); L^n_\sigma), \quad \frac{\partial u}{\partial t}, Au \in C((0,T); L^n_\sigma); \tag{2.1}$$

(ii) *u satisfies*

$$\begin{cases} \frac{\partial u}{\partial t} + Au + P(u\cdot\nabla u) = 0, & \text{in } L^n_\sigma \text{ for } 0<t<T, \\ u(0) = a. \end{cases} \tag{2.2}$$

In the above definition, P denotes the Helmholtz-Weyl projection from L^r onto L^r_σ for $1 < r < \infty$. More precisely, $P = \{P_{jk}\}_{j,k=1,\cdots,n}$ can be represented as $P_{jk} = \delta_{jk} + R_j R_k$, where δ_{jk} is the Kronecker symbol and $R_j = F^{-1}(\frac{\sqrt{-1}\xi_j}{|\xi|^2}F)$, $j = 1, \cdots, n$ are the Riesz transforms(F; Fourier transform). $A = -P\Delta$ is the Stokes operator.

Remark 2.2. It is easy to see that every strong solution u of (N-S) on $(0,T)$ is regular as in (1.2).

Concerning the existence and uniqueness of the strong solution, we have

Theorem 2.3. (**Kato** [25], **Giga-Miyakawa** [21], **Brezis** [8]) *For $n < r < \infty$, there is a constant $\gamma = \gamma(n,r) > 0$ with the following property. If the initial data $a \in L^n_\sigma$ and $T_* > 0$ satisfy*

$$\sup_{0<t\le T_*} t^{\frac{n}{2}(\frac{1}{n}-\frac{1}{r})} \|e^{-tA}a\|_r < \gamma \tag{2.3}$$

then there exists a unique strong solution $u(t)$ of (N-S) on $[0, T_*)$. Moreover, such a solution u has the property $t^{\frac{n}{2}(\frac{1}{n}-\frac{1}{r})}u(\cdot) \in C([0,T_*); L^r)$ with

$$\lim_{t\to+0} t^{\frac{n}{2}(\frac{1}{n}-\frac{1}{r})} \|u(t)\|_r = 0. \tag{2.4}$$

If, in addition, $a \in L^n_\sigma \cap L^2_\sigma$ satisfies (2.3), then u is also a weak solution of (N-S) in $(0, T_*)$.

Under the condition (2.3) we can construct a strong solution u on the interval $(0, T_*)$ by the successive approximation. To verify (2.3), we make use of the following $L^p - L^r$-estimates for the Stokes semigroup $\{e^{-tA}\}_{t\geq 0}$;

$$\begin{cases} \|e^{-tA}a\|_r \leq Ct^{-\frac{n}{2}(1/p-1/r)}\|a\|_p, & 1 \leq p \leq r \leq \infty, \\ \|\nabla e^{-tA}a\|_r \leq Ct^{-\frac{n}{2}(1/p-1/r)-1/2}\|a\|_p, & 1 \leq p \leq r < \infty \end{cases} \tag{2.5}$$

hold for all $a \in L^p_\sigma$ and all $t > 0$, where $C = C(n, p, r)$. Hence, if $a \in L^n_\sigma \cap L^r$ for some $n < r < \infty$, then (2.3) can be achieved in such a way that

$$T_* = \left(\frac{\gamma}{C\|a\|_r}\right)^{\frac{2r}{r-n}} \tag{2.6}$$

with the same constant C as in (2.5). If $a \in L^n_\sigma$, by the density argument, for every $\varepsilon > 0$, we can take $\tilde{a} \in C^\infty_{0,\sigma}$ so that $\|a - \tilde{a}\|_n < \varepsilon$. Hence by (2.5) with $p = n$, we have

$$\begin{aligned} t^{\frac{n}{2}(\frac{1}{n}-\frac{1}{r})}\|e^{-tA}a\|_r &\leq t^{\frac{n}{2}(\frac{1}{n}-\frac{1}{r})}\|e^{-tA}(a-\tilde{a})\|_r + t^{\frac{n}{2}(\frac{1}{n}-\frac{1}{r})}\|e^{-tA}\tilde{a}\|_r \\ &\leq C\|a-\tilde{a}\|_n + Ct^{\frac{n}{2}(\frac{1}{n}-\frac{1}{r})}\|\tilde{a}\|_r \\ &\leq C\varepsilon + Ct^{\frac{n}{2}(\frac{1}{n}-\frac{1}{r})}\|\tilde{a}\|_r, \end{aligned} \tag{2.7}$$

which yields $\limsup_{t\to+0} t^{\frac{n}{2}(\frac{1}{n}-\frac{1}{r})}\|e^{-tA}a\|_r \leq C\varepsilon$. Since ε is arbitrary, we obtain

$$\lim_{t\to+0} t^{\frac{n}{2}(\frac{1}{n}-\frac{1}{r})}\|e^{-tA}a\|_r = 0 \tag{2.8}$$

which ensures existence of T_* in (2.3) for $a \in L^n_\sigma$. However, this convergence is *not uniform* for a in any fixed bounded subset of L^n_σ. So, it is not clear whether the interval T_* for existence of strong solution with the initial data $a \in L^n_\sigma$ can be characterized in terms of the L^n-norm of a such as (2.6). To overcome this difficulty, Brezis [8] considered a class of precompact subsets in L^n_σ.

Proposition 2.4. **(Brezis)** *Let $n < r < \infty$. For every precompact set K in L^n_σ there exists a monotone non-decreasing and uniformly bounded function $\delta_r(t; K)$ of $t > 0$ with $\lim_{t\to+0} \delta_r(t; K) = 0$ such that*

$$t^{\frac{n}{2}(\frac{1}{n}-\frac{1}{r})}\|e^{-tA}a\|_r \leq \delta_r(t; K) \tag{2.9}$$

holds for all $a \in K$ and all $t > 0$. In particular, we can take $T_* = T_*(K)$ so that (2.3) holds for all $a \in K$.

Proof. $\delta_r(t; K)$ can be given by the following definition

$$\delta_r(t; K) \equiv \sup_{a\in K}\left(\sup_{0<\tau\leq t} \tau^{\frac{n}{2}(\frac{1}{n}-\frac{1}{r})}\|e^{-\tau A}a\|_r\right).$$

Indeed, since K is precompact in L^n_σ, it is bounded. Hence there is a constant $L > 0$ such that $\|a\|_n \leq L$ for all $a \in K$. By (2.5) we see that the right hand side of the above definition is finite and that $\delta_r(t; K)$ is well-defined with

$$\delta_r(t; K) \leq CL \quad \text{for all } t > 0.$$

This implies uniform boundedness. Obviously by definition, $\delta_r(t;K)$ is a monotone non-decreasing function of $t > 0$. Now, it suffices to show that

$$\lim_{t\to+0} \delta_r(t;K) = 0.$$

Let $U_\varepsilon(a) = \{b \in L^n_\sigma; \|b-a\|_n < \varepsilon\}$. For any $\varepsilon > 0$, there holds $\bar{K} \subset \bigcup_{a\in\bar{K}} U_\varepsilon(a)$. Since $\bar{K}$ is compact, we can select finitely many points $a_1(\varepsilon), a_2(\varepsilon), \cdots, a_m(\varepsilon) \in \bar{K}$ such that $\bar{K} \subset \bigcup_{j=1}^m U_\varepsilon(a_j(\varepsilon))$. Since $C^\infty_{0,\sigma}$ is dense in L^n_σ, we may assume that $a_j(\varepsilon) \in C^\infty_{0,\sigma}$ for all $1 \le j \le m$. Define $M_\varepsilon \equiv \mathrm{Max}\{\|a_1(\varepsilon)\|_r, \cdots, \|a_m(\varepsilon)\|_r\}$. For any $a \in K$ there is some $1 \le j_0 \le m$ such that $a \in U_\varepsilon(a_{j_0}(\varepsilon))$. For such j_0 we have in the same way as in (2.7) with the aid of (2.5)

$$\begin{aligned} \tau^{\frac{n}{2}(\frac{1}{n}-\frac{1}{r})}\|e^{-\tau A}a\|_r &\le C\|a-a_{j_0}(\varepsilon)\|_n + \tau^{\frac{n}{2}(\frac{1}{n}-\frac{1}{r})}\|e^{-\tau A}a_{j_0}(\varepsilon)\|_r \\ &\le C\varepsilon + C\tau^{\frac{n}{2}(\frac{1}{n}-\frac{1}{r})}\|a_{j_0}(\varepsilon)\|_r \\ &\le C\varepsilon + CM_\varepsilon t^{\frac{n}{2}(\frac{1}{n}-\frac{1}{r})} \end{aligned}$$

for all $0 < \tau \le t$. Taking the supremum of the above estimate for $\tau \in (0,t]$ and $a \in K$, we obtain

$$\delta_r(t;K) \le C\varepsilon + CM_\varepsilon t^{\frac{n}{2}(\frac{1}{n}-\frac{1}{r})}.$$

Letting $t \to +0$ in both sides of the above, we have $\limsup_{t\to+0} \delta_r(t;K) \le C\varepsilon$. Since $\varepsilon > 0$ is arbitrary, this implies that

$$\lim_{t\to+0} \delta_r(t;K) = 0.$$

□

Proposition 2.4 has two applications. One is refinement of the classical theorem on uniqueness of strong solutions, and another is simplification of the proof of regularity criterion on weak solutions in $C([0,T);L^n)$. Although both of them are relatively well known for the experts of the Navier-Stokes equations, we give here a sketch of proofs. In particular, we should notice that our investigation is closely related to the question on regularity given by Remark 1.4 (iii).

First, we consider uniqueness of strong solutions in Theorem 2.3. In the classical result of Fujita-Kato [17] and Kato [25], they imposed the restriction (2.4) on the behaviour near $t = 0$ of $\|u(t)\|_r$ for $n < r < \infty$. Later on, Brezis [8] showed that (2.4) is redundant by means of the proposition which states that every strong solution u of (N-S) necessarily fulfills (2.4).

By Duhamel's principle, (2.2) can be reduced to the following integral equation.

$$u(t) = e^{-tA}a - \int_0^t e^{-(t-\tau)A}P(u\cdot\nabla u)(\tau)\,d\tau, \quad 0 < t < T. \tag{2.10}$$

The classical result on existence uniqueness reads as follows.

Theorem 2.5. (**Fujita-Kato** [17], **Kato** [25]) *Let $a \in L^n_\sigma$ and let $n < r < \infty$.*

(i) *If a and T_* satisfy (2.3), then we can construct a solution $u(t)$ of (2.10) on $[0,T_*)$ in the class $C([0,T_*);L^n_\sigma)\cap C((0,T_*);L^r)$ with the property (2.4).*

(ii) *Suppose that u is a solution of (2.10) in $C([0,T);L^n_\sigma)\cap C((0,T);L^r)$. If u satisfies (2.4), then u is the only solution of (2.10).*

To show that (2.4) is redundant for uniqueness, we need

Proposition 2.6. *Let K be a precompact set in L^n_σ and let $n<r<\infty$. Suppose that $\delta_r(t;K)$ is the same function of $t>0$ as in Proposition 2.4. Then there exists $T_*>0$ such that for every $a\in K$ we can construct a solution $u(t)$ of (2.10) on $[0,T_*)$ in the class $C([0,T_*);L^n_\sigma)\cap C((0,T_*);L^r)$. Moreover, such a solution satisfies*

$$t^{\frac{n}{2}(\frac{1}{n}-\frac{1}{r})}\|u(t)\|_r\le 2\delta_r(t;K)\quad \textit{for all } 0<t<T_*. \tag{2.11}$$

In particular, u fulfills (2.4).

Remark 2.7. This proposition asserts that the time-interval T_* of existence of solutions to (2.10) can be taken *uniformly* on each precompact subset K of the initial data in L^n_σ.

Proof of Proposition 2.6. Since $\lim_{t\to+0}\delta_r(t;K)=0$, we can choose $T_*>0$ so that $\delta_r(T_*;K)<\gamma$, where γ is the same constant as in (2.3). Since $\delta_r(t;K)$ is a monotone non-decreasing of t, we have by (2.9) that

$$\sup_{0<t<T_*} t^{\frac{n}{2}(\frac{1}{n}-\frac{1}{r})}\|e^{-tA}a\|_r<\gamma\quad\text{for all } a\in K.$$

Then it follows from Theorem 2.5 (i) that for every $a\in K$ there is a solution $u(t)$ of (2.10) on $[0,T_*)$ in the class $C([0,T_*);L^n_\sigma)\cap C((0,T_*);L^r)$ with the property (2.4). Let us define $M(t)$ by

$$M(t)\equiv\sup_{0<\tau\le t}\tau^{\frac{n}{2}(\frac{1}{n}-\frac{1}{r})}\|u(\tau)\|_r.$$

By (2.4), we see that $M\in C([0,T_*))$. Then by (2.5) and (2.10) there holds

$$\begin{aligned}\|u(t)\|_r &\le \|e^{-tA}a\|_r+\int_0^t\|P\nabla\cdot e^{-(t-\tau)A}(u\otimes u)(\tau)\|_r\,d\tau\\ &\le \|e^{-tA}a\|_r+C\int_0^t(t-\tau)^{-\frac{n}{2r}-\frac{1}{2}}\|u(\tau)\|_r^2\,d\tau\\ &\le \|e^{-tA}a\|_r+CM(t)^2\int_0^t(t-\tau)^{-\frac{n}{2r}-\frac{1}{2}}\tau^{\frac{n}{r}-1}\,d\tau\\ &\le \|e^{-tA}a\|_r+C\beta M(t)^2t^{-\frac{n}{2}(\frac{1}{n}-\frac{1}{r})},\qquad 0<t<T_*,\end{aligned}$$

where $\beta=B(1/2-n/2r,n/r)$, $C=C(n,r)$. Applying Proposition 2.4 to the above estimate, we have

$$t^{\frac{n}{2}(\frac{1}{n}-\frac{1}{r})}\|u(t)\|_r\le\delta_r(t;K)+C\beta M(t)^2,\quad 0<t<T_*.$$

Since both $\delta_r(t;K)$ and $M(t)$ are non-decreasing functions of $t>0$, this implies

$$M(t) \le \delta_r(t;K) + C\beta M(t)^2, \quad 0<t<T_*. \tag{2.12}$$

Since $\lim_{t\to+0} \delta_r(t;K)=0$, we may assume T_* satisfies also

$$\delta_r(T_*;K) < \frac{1}{4C\beta}.$$

Hence by (2.12), there holds

$$M(t) \le \frac{1-\sqrt{1-4C\beta\delta_r(t;K)}}{2C\beta} \, (\le 2\delta_r(t;K)) \tag{2.13}$$

or

$$M(t) \ge \frac{1+\sqrt{1-4C\beta\delta_r(t;K)}}{2C\beta} \left(\ge \frac{1}{2C\beta}\right) \tag{2.14}$$

for all $0<t<T_*$. Since $M(t)$ is continuous on $[0,T_*)$ with $\lim_{t\to+0} M(t)=0$ (see (2.4)), the latter case (2.14) cannot occur. Hence we obtain from (2.13)

$$M(t) \le 2\delta_r(t;K), \quad \text{for all } 0<t<T_*.$$

This proves Proposition 2.6.

Because of Theorem 2.5 (ii), to prove assertion on uniqueness in Theorem 2.3, we may show the following lemma.

Lemma 2.8. (**Brezis** [8]) *Let $a \in L^n_\sigma$ and let $n<r<\infty$. Every solution u of (2.10) in the class $C([0,T);L^n_\sigma)\cap C((0,T);L^r)$ fulfills (2.4).*

Proof. We first define K as

$$K \equiv \{u(t); 0<t<T/2\}.$$

Since $u\in C([0,T);L^n_\sigma)$, K is a precompact subset of L^n_σ. For this K, we take the function $\delta_r(t;K)$ given by Proposition 2.4. Furthermore, by Proposition 2.6 we can take $T_*>0$ and a solution $\tilde{u}(t)$ of (2.10) on $(0,T_*)$ for every initial data $\tilde{a}\in K$. Let us denote this $\tilde{u}(t)$ by

$$\tilde{u}(t) \equiv S(t)\tilde{a}, \quad 0<t<T_*$$

By (2.11), there holds

$$t^{\frac{n}{2}(\frac{1}{n}-\frac{1}{r})}\|S(t)\tilde{a}\|_r \le 2\delta_r(t;K), \quad 0<t<T_* \tag{2.15}$$

for all $\tilde{a}\in K$. Let us take s arbitrarily as $0<s<\text{Min.}\{T/2,T_*\}$. Then we have $u(s)\in K$. Since $u\in C((0,T);L^r)$, we see $\lim_{t\to+0} t^{\frac{n}{2}(\frac{1}{n}-\frac{1}{r})}\|u(t+s)\|_r=0$. Hence it follows from Theorem 2.5 (ii) and definition of the map $S(t)$ that

$$u(t+s)=S(t)u(s), \quad 0<t<T_*.$$

From (2.15) we obtain

$$t^{\frac{n}{2}(\frac{1}{n}-\frac{1}{r})}\|u(t+s)\|_r = t^{\frac{n}{2}(\frac{1}{n}-\frac{1}{r})}\|S(t)u(s)\|_r \le 2\delta_r(t;K), \quad 0<t<T_*.$$

Since $u \in C((0,T);L^r)$, by letting $s \to 0$ in the above estimate we have

$$t^{\frac{n}{2}(\frac{1}{n}-\frac{1}{r})}\|u(t)\|_r \le 2\delta_r(t;K), \quad 0 < t \le T_*.$$

Since $\lim_{t\to+0}\delta_r(t;K) = 0$, this yields

$$\lim_{t\to+0} t^{\frac{n}{2}(\frac{1}{n}-\frac{1}{r})}\|u(t)\|_r = 0.$$

□

Remark 2.9. Furioli - Lemarié Rieusset - Terraneo [18] and Cannone - Planchon [10] obtained some further results on the uniqueness to (2.10). Their method is based on driving a bilinear estimate on the Besov space. Lions - Masmuodi [37] also gave a uniqueness theorem by reducing the problem to that of solvability of the adjoint equation whose coefficient is in $C([0,T);L^n)$.

We shall next apply Proposition 2.4 to the proof of regularity of weak solutions in $C([0,T);L^n)$.

Theorem 2.10. (**Giga** [19], **von Wahl** [54]) *Let* $a \in L^2_\sigma$. *Every weak solution* u *of (N-S) in* $C([0,T);L^n)$ *is regular as in (1.2).*

Proof. Let us define the set K by

$$K = \{u(t); 0 < t < T\}.$$

Since $u \in C([0,T);L^n)$ with $\operatorname{div} u = 0$, K is a precompact subset of L^n_σ. We take some $n < r < \infty$. Then it follows from Proposition 2.4 that there exists $T_* = T_*(K,r)$ such that

$$\sup_{0<t<T_*} t^{\frac{n}{2}(\frac{1}{n}-\frac{1}{r})}\|e^{-tA}a\|_r \le \delta_r(T_*;K) < \gamma, \tag{2.16}$$

for all $a \in K$ where γ is the same constant as in (2.3). Let $\rho \equiv T_*/2$. For every $t_* \in (0,T)$ we have by (2.16) that

$$\sup_{0<t<T_*} t^{\frac{n}{2}(\frac{1}{n}-\frac{1}{r})}\|e^{-tA}u(t_*-\rho)\|_r < \gamma.$$

By Theorem 2.3 and Remark 2.2, there exists a strong solution v of (N-S) with $v|_{t=t_*-\rho} = u(t_*-\rho)$ such that

$$v \in C([t_*-\rho, t_*+\rho);L^n), \quad \frac{\partial v}{\partial t}, \frac{\partial^{\alpha_1+\cdots+\alpha_n} v}{\partial x_1^{\alpha_1}\cdots\partial x_n^{\alpha_n}} \in C(\mathbb{R}^n \times (t_*-\rho, t_*+\rho)) \tag{2.17}$$

where $\alpha = (\alpha_1,\cdots,\alpha_n)$ is an arbitrary multi-index with $|\alpha| = \alpha_1 + \cdots + \alpha_n \le 2$. Notice that v is also a weak solution. Then uniqueness result of Theorem 1.3 (i) yields

$$u(t) \equiv v(t) \quad \text{for } t \in [t_*-\rho, t_*+\rho).$$

Since $t_* \in (0,T)$ can be taken arbitrarily, we can conclude that u is regular as in (1.2). □

To deal with the problem on regularity of weak solutions in $L^\infty(0,T;L^n)$, the above proof proposes us the following question.

Question. For every weak solution u in $L^\infty(0,T;L^n)$ is the set

$$K = \{u(t); 0 < t < T\}$$

precompact in L^n_σ ?

3. Blow-up

3.1. Blow-up self-similar solution

In $\mathbb{R}^3$, Leray [36] intended to have constructed a weak solution $u(t)$ blowing-up at $t = T$ by means of the *backward self-similar* solution:

$$u(x,t) = \begin{cases} \frac{1}{\sqrt{T-t}} U(\frac{x}{\sqrt{T-t}}), & 0 < t < T, \\ 0, & T \le t < \infty, \end{cases}$$

$$p(x,t) = \begin{cases} \frac{1}{T-t} P(\frac{x}{\sqrt{T-t}}), & 0 < t < T, \\ 0, & T \le t < \infty, \end{cases}$$

where $U = (U^1(y), U^2(y), U^3(y))$, $P = P(y)$, $y = (y^1, y^2, y^3) \in \mathbb{R}^3$. It can be easily seen that if $\{u,p\}$ solves (N-S) in $\mathbb{R}^3 \times (0,\infty)$, then $\{U,P\}$ is a solution of

$$\text{(L)} \qquad \begin{cases} -\Delta U + U\cdot\nabla U + \frac{1}{2}y\cdot\nabla U + \frac{1}{2}U + \nabla P = 0, & \text{in } \mathbb{R}^3 \\ \operatorname{div} U = 0 \quad \text{in } \mathbb{R}^3 \end{cases}$$

We can show that u belongs to the Leray-Hopf class(see Definition 0.1 (i)), i.e., $u \in L^\infty(0,\infty;L^2(\mathbb{R}^3))$ with $\nabla u \in L^2(0,\infty;L^2(\mathbb{R}^3))$ if and only if U satisfies

$$U \in H^1(\mathbb{R}^3), \quad \text{i.e.,} \quad U, \nabla U \in L^2(\mathbb{R}^3).$$

Leray proposed the problem whether there exists a nontrivial solution U of (L) in $H^1(\mathbb{R}^3)$?

Remark 3.1. It is important to find a solution U of (L) within the class $H^1(\mathbb{R}^3)$. If we may allow that U grows at infinity, then we can find a nontrivial solution U of (L). Indeed, Majda [39] found a solution U of (L) with the linear growth at infinity. His solution has the form

$$U(y) = Ay, \quad P(y) = < Qy, y >,$$

where $A = \{a_{ij}\}_{1\le i,j\le 3}$ and $Q = \{q_{ij}\}_{1\le i,j\le 3}$ are 3×3 matrices. In this form, $\{U,P\}$ is a solution of (L) if and only if A and Q satisfy

$$A^2 + A + (Q + Q^t) = O, \quad \text{Tr.}A = 0.$$

It is easy to find a pair of solutions $\{A,Q\}$ to the above equation. In this direction, there are some papers which investigate the relation between uniqueness and the growth order at infinity of solutions to (N-S). See, for example, Chae [27], Okamoto [44] and Giga-Inui-Kato-Matsui [20].

Leray's problem is closely related to the regularity criterion of weak solutions in the critical case $L^\infty(0,\infty;L^3(\mathbb{R}^3))$, on which we discussed in Section 1. We obtain from the form of the backward self-similar solution that

$$\|u(t)\|_r = \begin{cases} \dfrac{\|U\|_r}{(T-t)^{\frac{r-3}{2r}}} & \text{for } 0 < t < T, \\ 0 & \text{for } T \le t < \infty, \end{cases}$$

for $3 \le r \le \infty$. Hence if $2/s + 3/r = 1$ with $3 < r \le \infty$, then we have

$$\int_0^T \|u(t)\|_r^s \, dt = \|U\|_r^s \int_0^T \frac{1}{(T-t)^{s\frac{r-3}{2r}}} \, dt = \|U\|_r^s \int_0^T \frac{1}{T-t} \, dt = \infty,$$

which implies that $u \notin L^s(0,T';L^r(\mathbb{R}^3))$ for any $T' \ge T$. This does not contradict Theorem 1.1 (ii). However, since

$$\|u(t)\|_3 = \begin{cases} \|U\|_3 & \text{for } 0 < t < T, \\ 0 & \text{for } T \le t < \infty, \end{cases}$$

there holds $u \in L^\infty(0,\infty;L^3(\mathbb{R}^3))$. Hence if there exists a solution $U(\neq 0) \in H^1(\mathbb{R}^3)$ of (L), then the class $L^\infty(0,T;L^3(\mathbb{R}^3))$ is *not* a regular class of weak solutions to (N-S).

Leray's problem was negatively solved by Nečas-Ružička-Šverák [40]:

Theorem 3.2. (**Nečas-Ružička-Šverák**) *If U is a solution of (L) in $L^3(\mathbb{R}^3)$, then we have*

$$U \equiv 0 \quad \text{in } \mathbb{R}^3.$$

In particular, since $H^1(\mathbb{R}^3) \subset L^2(\mathbb{R}^3) \cap L^6(\mathbb{R}^3) \subset L^3(\mathbb{R}^3)$, (L) possesses only a trivial solution in $H^1(\mathbb{R}^3)$, so there exists no backward self-similar solution blowing up at $t = T$ in the Leray-Hopf class.

Theorem 3.2 suggests us to consider the following question.

Question.

(i) Is every weak solution u in $L^\infty(0,T;L^n)$ *regular* or not ?

(ii) For $2/s + n/r = 1$ with $n < r \le \infty$, is every weak solution u in $L^s_{\mathrm{w}}(0,T;L^r)$ *regular* ? Here L^s_{w} denotes the *weak-L^s* space in time.

3.2. Continuation of $u(t)$ beyond $t = T$

Let us denote by H^s_σ the set of all $u \in H^s \equiv W^{s,2}(\mathbb{R}^n)$ such that $\operatorname{div} u = 0$. Recall the theorem on local existence of strong solutions by Fujita-Kato [17](see also Theorem 2.3). For every $a \in H^s_\sigma$ with $s > n/2 - 1$, there is $T = T(\|a\|_{H^s}) > 0$ and a unique solution $u(t)$ of (N-S) on $[0,T)$ such that

$$u \in C([0,T);H^s_\sigma) \cap C^1((0,T);H^s) \cap C((0,T);H^{s+2}). \tag{3.1}$$

Such a solution u is actually regular as in (1.2). Since $s > n/2 - 1$, there holds an embedding $H^s \subset L^r$ for $r > n$. Hence, the interval $T(\|a\|_{H^s})$ of existence of the strong solution can be concretely characterized by (2.6). We are interested in behaviour of $u(t)$ at $t = T$. More precisely, we want to know whether the solution $u(t)$ really loses its regularity at $t = T$? i.e.,

$$u(T) \notin C^2(\mathbb{R}^n) \quad \text{or} \quad u(t) \to \infty \quad \text{as } t \uparrow T \,?$$

Another possibility is that $u(t)$ can be extended beyond $t = T$. It may be possible that there exists $T' > T$ such that u can be extended to the solution on $[0, T')$ as

$$u \in C([0,T'); H^s_\sigma) \cap C^1((0,T'); H^s) \cap C((0,T'); H^{s+2}). \tag{3.2}$$

Giga [19] and Beale-Kato-Majda [2] gave a criterion on extension of strong solutions. Let $a \in H^s_\sigma$ for $s > n/2 - 1$. Suppose that $u(t)$ is a solution of (N-S) on $[0, T)$ in the class (3.1). If either

$$\int_0^T \|u(t)\|_r^\kappa \, dt < \infty \quad \text{for } 2/\kappa + n/r = 1 \text{ with } n < r \le \infty \tag{3.3}$$

or

$$\int_0^T \|\text{rot } u(t)\|_\infty \, dt < \infty \tag{3.4}$$

is satisfied, then there is $T' > T$ such that u can be continued to the solution on $[0, T')$ in the class (3.2).

We shall obtain more larger class than (3.3) and (3.4) for possibility of extension of the strong solution. To state our results, we need the space of BMO-functions. A locally summable function f belongs to BMO if

$$\|f\|_{BMO} \equiv \sup_B \frac{1}{|B|} \int_B |f(x) - f_B| \, dx < \infty,$$

where B denotes the ball in $\mathbb{R}^n$ and $f_B = \dfrac{1}{|B|} \displaystyle\int_B f(y) \, dy$. In the above definition, supremum is taken for all balls B in $\mathbb{R}^n$.

Remark 3.3. (i) By definition, it is obvious that $L^\infty \subset BMO$. For $f(x) = \log |x|$, we have $f \in BMO$, but $f \notin L^\infty$, so BMO is strictly larger than L^∞.

(ii) There holds the estimate

$$\|\nabla u\|_{BMO} \le C \|\text{rot } u\|_{BMO} \quad \text{for all } u \in H^{s+2}_\sigma \text{ with } s > n/2 - 1. \tag{3.5}$$

First, it should be noted that $H^{\frac{n}{2}} \subset BMO$. Since div $u = 0$, we have by the Biot-Savard Law that

$$\frac{\partial u}{\partial x_j} = R_j(R \times \text{rot } u), \quad j = 1, \cdots, n,$$

where $R_j = F^{-1} \dfrac{i\xi_j}{|\xi|} F (j = 1, \cdots, n)$ denotes the Riesz transforms. Since R_j is bounded in BMO, the above formula yields (3.5). Notice that (3.5) is not true in L^∞ because R_j is not a bounded operator in L^∞.

Theorem 3.4. **(Kozono-Taniuchi** [31]) *Let $s > n/2 - 1$ and let $a \in H^s_\sigma$. Suppose that u is a strong solution of (N-S) in the class (3.1). Assume that u satisfies either of the following conditions:*

$$\int_{\varepsilon_0}^T \|u(t)\|^2_{BMO}\,dt < \infty \quad or \quad \int_{\varepsilon_0}^T \|\mathrm{rot}\ u(t)\|_{BMO}\,dt < \infty$$

for some $0 < \varepsilon_0 < T$. Then u can be continued to the solution on $[0, T')$ in the class (3.2) for some $T' > T$.

Corollary 3.5. *Let u be a solution of (N-S) in the class (3.1) for $s > n/2 - 1$. Suppose that T is maximal, i.e., u cannot be continued in the class (3.2) for any $T' > T$. Then both of the following conditions must be fulfilled*

$$\int_{\varepsilon}^T \|u(t)\|^2_{BMO}\,dt = \int_{\varepsilon}^T \|\mathrm{rot}\ u(t)\|_{BMO}\,dt = \infty$$

for all $0 < \varepsilon < T$. In particular, we have

$$\limsup_{t\uparrow T} \|u(t)\|_{BMO} = \limsup_{t\uparrow T} \|\mathrm{rot}\ u(t)\|_{BMO} = \infty.$$

Furthermore, we can apply Theorem 3.4 to the criterion on uniqueness and regularity of weak solutions.

Corollary 3.6.

(i) (uniqueness) *Let $a \in L^2_\sigma$ and let u, v be two weak solutions of (N-S) on $(0, T)$. Suppose that*

$$u \in L^2(0, T; BMO)$$

and that v satisfies the energy inequality (0.2). Then we have $u \equiv v$ on $[0, T]$.

(ii) (regularity) *Let $a \in L^2_\sigma$. Suppose that u is a weak solution satisfying either of the following conditions*

$$u \in L^2(0, T; BMO) \quad or \quad \mathrm{rot}\ u \in L^1(0, T; BMO).$$

Then u is a solution of (N-S) in the class

$$u \in C([\varepsilon, T); H^s_\sigma) \cap C^1((\varepsilon, T); H^s) \cap C((\varepsilon, T); H^{s+2}), \quad s > n/2 - 1$$

for all $0 < \varepsilon < T$. Actually u is regular in $\mathbb{R}^n \times (0, T)$ as in (1.2).

Remark 3.7. (i) By Theorem 1.1, every weak solution in $L^2(0, T; L^\infty)$ is unique and regular. Our class $L^2(0, T; BMO)$ is *larger* than that one. Beirão da Veiga [4] showed that every weak solution u with

$$\nabla u \in L^\kappa(0, T; L^r) \quad \text{for } 2/\kappa + n/r = 2 \text{ with } 1 < \kappa < \infty,\ n/2 < r < \infty$$

is regular in $\mathbb{R}^n \times (0, T)$ as in (1.2). Corollary 3.6 covers the *borderline case* $\kappa = 1$ and $r = \infty$. Furthermore, it does not need to impose the condition on ∇u but on the *vorticity* rot u. The assumption on $\|\mathrm{rot}\ u\|_{BMO}$ can be replaced by $\|\mathrm{Def}\ u\|_{BMO}$, where Def u denotes the deformation tensor.

(ii) Concerning uniqueness of weak solutions u with rot $u \in L^1(0,T;BMO)$, Ogawa-Taniuch [43] proved that if rot $u \in L\log L(0,T;BMO)$ and if v satisfies the energy inequality (0.2), then there holds $u \equiv v$ on $[0,T)$.

We shall give a sketch of the proof of Theorem 3.4 and Corollary 3.6. Obviously, Corollary 3.5 is an immediate consequence of Theorem 3.4. To this end, we need the following bilinear estimates.

Lemma 3.8. (**Kozono-Taniuch** [31, Lemma 1]) (i) *For $1 < r < \infty$ there is a constant $C = C(n,r)$ such that*

$$\|f\cdot\nabla g\|_r \le C(\|f\|_r\|\nabla g\|_{BMO} + \|\nabla f\|_{BMO}\|g\|_r)$$

holds for all $f, g \in W^{1,r}$ with $\nabla f, \nabla g \in BMO$.

(ii) *Let $\alpha = (\alpha_1, \cdots, \alpha_n)$, $\beta = (\beta_1, \cdots, \beta_n)$ be multi-indices with $|\alpha| = \alpha_1 + \cdots + \alpha_n \ge 1$ and $|\beta| = \beta_1 + \cdots + \beta_n \ge 1$. Then there holds*

$$\|\partial^\alpha f\cdot\partial^\beta g\|_r \le C(\|f\|_{BMO}\|(-\Delta)^{\frac{|\alpha|+|\beta|}{2}}g\|_r + \|(-\Delta)^{\frac{|\alpha|+|\beta|}{2}}f\|_r\|g\|_{BMO})$$

for all $f, g \in BMO\cap W^{|\alpha|+|\beta|,r}$ with $C = C(n,r,\alpha,\beta)$, where $\partial^\alpha = \dfrac{\partial^{|\alpha|}}{\partial x_1^{\alpha_1}\cdots\partial x_n^{\alpha_n}}$.

An immediate consequence of this Lemma is

Lemma 3.9. *Let $u = (u_1, \cdots, u_n) \in W^{1,r}$ for $1 < r < \infty$ with* div $u = 0$ *and* rot $u \in BMO$. *Then we have $u\cdot\nabla u \in L^r$ with*

$$\|u\cdot\nabla u\|_r \le C\|u\|_r\|\mathrm{rot}\ u\|_{BMO}.$$

Indeed, by taking $f = g = u$ in Lemma 3.8 (i), we have

$$\|u\cdot\nabla u\|_r \le C\|u\|_r\|\nabla u\|_{BMO}.$$

From this and (3.5), we obtain Lemma 3.9.

Out line of proof of Theorem 3.4. Since $s > n/2 - 1$, we have by the Sobolev embedding that $H^s \subset L^r$ for $r = 2n/(n-2s) > n$. So by (2.6), the time interval T_* for existence of the strong solution $u(t)$ with the initial data a is characterized as

$$T_* = \frac{C}{\|a\|_{H^s}^{\frac{2}{s-(n/2-1)}}} \quad \text{or} \quad T_* = \frac{C}{\|a\|_r^{2r/(r-n)}},$$

where C is independent of a. Hence, it suffices to prove the following a priori estimate

$$\sup_{\varepsilon_0<t<T}\|u(t)\|_{H^{[s]+1}} \le \|u(\varepsilon_0)\|_{H^{[s]+1}}\exp\left(C\int_{\varepsilon_0}^T\|u\|_{BMO}^2\,dt\right), \tag{3.6}$$

or

$$\sup_{\varepsilon_0<t<T}\|u(t)\|_r \le \|u(\varepsilon_0)\|_r\exp\left(C\int_{\varepsilon_0}^T\|\mathrm{rot}\ u\|_{BMO}\,d\tau\right) \tag{3.7}$$

where $C = C(n,s,r,T)$. We shall first establish (3.7). Let us make use of the integral representation of u

$$u(t) = e^{-(t-\varepsilon_0)A}u(\varepsilon_0) - \int_{\varepsilon_0}^{t} e^{-(t-s)A}P(u\cdot\nabla u)(s)\,ds, \quad \varepsilon_0 < t < T.$$

Since $\|e^{-tA}\|_{B(L^r,L^r)} \le 1$ for all $t > 0$ and since $\|P\|_{B(L^r,L^r)} = C(n,r) < \infty$, it follows from Lemma 3.9 that

$$\begin{aligned}\|u(t)\|_r &\le \|u(\varepsilon_0)\|_r + C\int_{\varepsilon_0}^{t} \|u\cdot\nabla u\|_r\,d\tau \\ &\le \|u(\varepsilon_0)\|_r + C\int_{\varepsilon_0}^{t} \|\text{rot } u\|_{BMO}\|u\|_r\,d\tau, \quad \varepsilon_0 < t < T.\end{aligned}$$

From this and the Gronwall inequality, we obtain

$$\|u(t)\|_r \le \|u(\varepsilon_0)\|_r \exp\left(C\int_{\varepsilon_0}^{t} \|\text{rot } u\|_{BMO}\,d\tau\right), \quad \varepsilon_0 < t < T,$$

which implies (3.7).

We next prove (3.6). Set

$$v = \partial^\alpha u = \frac{\partial^{|\alpha|}u}{\partial x_1^{\alpha_1}\cdots\partial x_n^{\alpha_n}}.$$

Applying ∂^α to (N-S), we have for v the equation

$$\frac{\partial v}{\partial t} - \Delta v + u\cdot\nabla v + \nabla q = G, \quad \varepsilon_0 < t < T, \tag{3.8}$$

where $q = \partial^\alpha p$ and

$$G = -\sum_{|\beta|\le|\alpha|-1} {}_\alpha C_\beta \partial^{\alpha-\beta}u\cdot\nabla(\partial^\beta u). \tag{3.9}$$

Taking the inner product in L^2 between (3.8) and v, and then integrating the result identity on the time interval (ε_0, t), we obtain

$$\|v(t)\|_2^2 + 2\int_{\varepsilon_0}^{t} \|\nabla v\|_2^2\,d\tau \le \|v(\varepsilon_0)\|_2^2 + 2\int_{\varepsilon_0}^{t} \|G\|_2\|v\|_2\,d\tau. \tag{3.10}$$

Here we have used

$$(u\cdot\nabla v, v) = 0, \qquad (\nabla q, v) = 0.$$

By Lemma 3.8 (ii), we have

$$\|G\|_2 \le C\|u\|_{BMO}\|(-\Delta)^{\frac{|\alpha|+1}{2}}u\|_2,$$

from which and (3.10) it follows that

$$\begin{aligned}
&\|\partial^\alpha u(t)\|_2^2 + 2\int_{\varepsilon_0}^t \|\nabla(\partial^\alpha u)\|_2^2\, d\tau \\
\le\ & \|\partial^\alpha u(\varepsilon_0)\|_2^2 + C\int_{\varepsilon_0}^t \|u\|_{BMO}\|(-\Delta)^{\frac{|\alpha|+1}{2}} u\|_2\|\partial^\alpha u\|_2\, d\tau \\
\le\ & \|\partial^\alpha u(\varepsilon_0)\|_2^2 + \int_{\varepsilon_0}^t \|(-\Delta)^{\frac{|\alpha|+1}{2}} u\|_2^2\, d\tau + C\int_{\varepsilon_0}^t \|u\|_{BMO}^2\|\partial^\alpha u\|_2^2\, d\tau \quad (3.11)
\end{aligned}$$

with C independent of t.

Summing over α with $0 \le |\alpha| \le [s]+1$, we have

$$\|u(t)\|_{H^{[s]+1}}^2 \le \|u(\varepsilon_0)\|_{H^{[s]+1}}^2 + C\int_{\varepsilon_0}^t \|u\|_{BMO}^2\|u\|_{H^{[s]+1}}^2\, d\tau \tag{3.12}$$

for all $\varepsilon_0 \le t < T$. Now the Gronwall inequality yields

$$\|u(t)\|_{H^{[s]+1}} \le \|u(\varepsilon_0)\|_{H^{[s]+1}} \exp\left(C\int_{\varepsilon_0}^t \|u\|_{BMO}^2\, dt\right), \quad \varepsilon_0 < t < T \tag{3.13}$$

and we have the desired a priori estimate (3.6). This proves Theorem 3.4.

Outline of the proof of Corollary 3.6. (i) First, we need to show that u satisfies the energy identity (1.3). For this purpose, we make use of Coifman-Lions-Meyer-Semmes [12]; for $u \in H^1_\sigma$ there holds

$$u\cdot\nabla u \in \boldsymbol{H}^1 \quad \text{with} \quad \|u\cdot\nabla u\|_{\boldsymbol{H}^1} \le C\|u\|_2\|\nabla u\|_2, \tag{3.14}$$

where $\boldsymbol{H}^1$ denotes the Hardy space. Since $(\boldsymbol{H}^1)^* = BMO$, we have

$$\left|\int_0^T (u\cdot\nabla u, \Phi)\, dt\right| \le C\|u\|_{L^\infty(0,T;L^2)}\|\nabla u\|_{L^2(0,T;L^2)}\|\Phi\|_{L^2(0,T;BMO)}$$

for all $\Phi \in L^2(0,T;BMO)$. This estimate enables us to choose the test functions Φ of (0.1) from $L^2(0,T;BMO)$ and we can show the energy identity (1.3). In the same manner, we obtain

$$\int_0^t \{2(\nabla u, \nabla v) + (v\cdot\nabla v, u) - (u\cdot\nabla v, u)\}\, d\tau = -(u(t), v(t)) + \|a\|_2^2. \tag{3.15}$$

Addition of (3.15) (multiplied by -2), (1.3) for u and (0.2) for v yields

$$\begin{aligned}
\|w(t)\|_2^2 + 2\int_0^t \|\nabla w\|_2^2\, d\tau &\le 2\int_0^t (w\cdot\nabla v, u)\, d\tau \\
&= 2\int_0^t (w\cdot\nabla w, u)\, d\tau, \qquad (3.16)
\end{aligned}$$

where $w = v - u$. By (3.14), we have

$$\begin{aligned}\left|\int_0^t (w\cdot\nabla w, u)\, d\tau\right| &\le C\int_0^t \|w\cdot\nabla w\|_{\boldsymbol{H}^1}\|u\|_{BMO}\, d\tau\\ &\le C\int_0^t \|w\|_2\|\nabla w\|_2\|u\|_{BMO}\, d\tau\\ &\le \int_0^t \|\nabla w\|_2^2\, d\tau + C\int_0^t \|w\|_2^2\|u\|_{BMO}^2\, d\tau.\end{aligned}$$

Hence it follows from (3.16) that

$$\|w(t)\|_2^2 \le C\int_0^t \|w\|_2^2\|u\|_{BMO}^2\, d\tau, \quad 0 \le t < T.$$

Since $u \in L^2(0,T;BMO)$ and since $w(0) = 0$, the Gronwall inequality yields

$$\|w(t)\|_2^2 = 0, \quad 0 \le t < T$$

from which we obtain the desired uniqueness.

(ii) Next, we shall prove regularity for u. For every $0 < \varepsilon < T$, there exists $t_0 \in (0, \varepsilon)$ such that $u(t_0) \in L^r$ for $n < r < \infty$. Notice that $L^2 \cap BMO \subset L^q$ for all $2 \le q < \infty$. Then by Theorem 2.3 with (2.6), there exist T_* and a unique smooth solution $\tilde{u}$ on $[t_0, T_*)$ with $\tilde{u}|_{t=t_0} = u(t_0)$. Since $\tilde{u}$ is the smooth solution and since u is a weak solution with the energy identity (1.3), it follows from Theorem 1.1 that

$$u(t) \equiv \tilde{u}(t) \quad \text{for } t \in [t_0, T_*).$$

Actually, there holds $T_* = T$. Suppose that $T_* < T$. Then $\tilde{u}(t)$ cannot be extended to the smooth solution beyond $t = T_*$. On the other hand, since $\tilde{u} = u$ on $[0, T_*)$, we have by the assumption that

$$\int_0^{T_*} \|\tilde{u}(t)\|_{BMO}^2\, dt < \infty \quad \text{or} \quad \int_0^{T_*} \|\mathrm{rot}\ \tilde{u}(t)\|_{BMO}\, dt < \infty.$$

Hence it follows from Theorem 3.4 that there exists $T' > T_*$ such that $\tilde{u}$ is the smooth solution on $[t_0, T')$, which contradicts the hypothesis that T_* is the maximum existence time. This proves Corollary 3.6.

3.3. Further results on the Besov space

Theorem 3.4 can be extended if we introduce the Besov space. This approach is also useful to the Euler equations for the ideal fluids. For definition of the Besov space $B^s_{p,q}$ and the homogeneous Besov space $\dot{B}^s_{p,q}$ for $s \in \mathbb{R}$, $1 \le p, q \le \infty$, we first introduce the Littlewood-Paley decomposition $\{\varphi_k\}_{k=-\infty}^{\infty}$. Let us take $\phi = \phi(\xi) \in C_0^\infty(\mathbb{R}^n_\xi)$ satisfying

$$\mathrm{supp}\ \phi \subset \{\xi \in \mathbb{R}^n; 1/2 \le |\xi| \le 2\}, \qquad \sum_{k=-\infty}^{\infty} \phi(2^{-k}\xi) \equiv 1 \quad \text{for } \xi \ne 0. \tag{3.17}$$

We define $\varphi_k = \varphi_k(x), \psi = \psi(x)$ by

$$\hat{\varphi}_k(\xi) = \phi(2^{-k}\xi), \quad k = 0, \pm 1, \pm 2, \cdots, \quad \hat{\psi}(\xi) = 1 - \sum_{k=1}^{\infty} \phi(2^{-k}\xi).$$

Obviously, φ, $k = 0, \pm 1, \pm 2, \cdots$ and ψ are rapidly decreasing C^∞-functions and there holds supp $\psi \in \{|\xi| \le 2\}$. By (3.17), we have $f = \sum_{k=-\infty}^{\infty} \varphi_k * f$ for all rapidly decreasing C^∞-functions f.

Let $s \in \mathbb{R}$, $1 \le p, q \le \infty$. The homogeneous Besov space $\dot{B}^s_{p,q}$ and the Besov space $B^s_{p,q}$ are defined by

$$\dot{B}^s_{p,q} \equiv \{f : \mathbb{R}^n \to \mathbb{C}; \|f\|_{\dot{B}^s_{p,q}} = \left(\sum_{k=-\infty}^{\infty} (2^{ks}\|\varphi_k * f\|_p)^q\right)^{\frac{1}{q}} < \infty\},$$

$$B^s_{p,q} \equiv \{f : \mathbb{R}^n \to \mathbb{C}; \|f\|_{B^s_{p,q}} = \|\psi * f\|_p + \left(\sum_{k=1}^{\infty}(2^{ks}\|\varphi_k * f\|_p)^q\right)^{\frac{1}{q}} < \infty\}.$$

Notice that for noninteger $s > 0$, we have $W^{s,p} = B^s_{p,p}$, where $W^{s,p} \equiv \{f : \mathbb{R}^n \to \mathbb{C}; \|f\|_{W^{s,p}} < \infty\}$ is the fractional ordered Sobolev space with the norm

$$\|f\|_{W^{s,p}} = \|f\|_{W^{[s],p}} + \sum_{|\alpha|=[s]} \left(\iint_{\mathbb{R}^n \times \mathbb{R}^n} \frac{|D^\alpha f(x) - D^\alpha f(y)|^p}{|x-y|^{(s-[s])p+n}}\, dx\, dy\right)^{\frac{1}{p}}.$$

It can be shown that for $s > 0$, $1 \le p, q \le \infty$ there holds $B^s_{p,q} = L^p \cap \dot{B}^s_{p,q}$ with the equivalence of norms

$$\|f\|_{B^s_{p,q}} \cong \|f\|_p + \|f\|_{\dot{B}^s_{p,q}}.$$

Out result on continuation of the strong solution now reads

Theorem 3.10. (**Kozono-Ogawa-Taniuchi** [33]) *Let $s > n/2 - 1$ and let $a \in H^s_\sigma$. Suppose that u is a strong solution of (N-S) in the class (3.1). Assume that u satisfies either of the following conditions:*

$$\int_{\varepsilon_0}^{T} \|u(t)\|^2_{\dot{B}^0_{\infty,\infty}}\, dt < \infty \quad or \quad \int_{\varepsilon_0}^{T} \|\mathrm{rot}\; u(t)\|_{\dot{B}^0_{\infty,\infty}}\, dt < \infty$$

for some $0 < \varepsilon_0 < T$. Then u can be continued to the solution on $[0, T')$ in the class (3.2) for some $T' > T$.

In the same way as in Corollary 3.6, we can apply Theorem 3.10 to the criterion on uniqueness and regularity of weak solutions.

Corollary 3.11. *Let $a \in L^2_\sigma$. Suppose that u is a weak solution satisfying the energy inequality of the strong form*

$$\|u(t)\|_2 + 2\int_s^t \|\nabla u(\tau)\|_2^2\, d\tau \le \|u(s)\|_2^2 \tag{3.18}$$

for almost all $s > 0$ and all t such that $t > s$. If either of the following conditions is satisfied

$$u \in L^2(0,T;\dot{B}^0_{\infty,\infty}) \quad or \quad \text{rot } u \in L^1(0,T;\dot{B}^0_{\infty,\infty}),$$

then u is a solution of (N-S) in the class

$$u \in C([\varepsilon,T);H^s_\sigma) \cap C^1((\varepsilon,T);H^s) \cap C((\varepsilon,T);H^{s+2}), \quad s > n/2-1$$

for all $0 < \varepsilon < T$. Actually u is regular in $\mathbb{R}^n \times (0,T)$ as in (1.2).

Remark 3.12. (i) Since $BMO \subset \dot{B}^0_{\infty,\infty}$, we may regard Theorem 3.10 as an extension of Theorem 3.4. On the other hand, in Corollary 3.11 we need to impose the energy inequality of strong form on the weak solution u. If $n \leq 4$, for every $a \in L^2_\sigma$ there exists at least one weak solution of (N-S) with the energy inequality of strong form(see Leray [36] and Kato [25]).

(ii) For the Euler equations, Kozono - Taniuchi [32] and Kozono - Ogawa - Taniuchi [33] proved the same extension criterion on the solution as in Theorem 3.10 provided rot $u \in L^1(0,T;BMO)$ or rot $u \in L^1(0,T;\dot{B}^0_{\infty,\infty})$.

(iii) The weak and strong solutions to (N-S) in the Besov space and other function spaces were systematically investigated by Amann [1].

To prove Theorem 3.10, we need the following the logarithmic Sobolev inequality.

Lemma 3.13. *For $1 \leq p < \infty$, $n/p < s < \infty$ and $1 \leq \rho \leq q \leq \infty$, there is a constant $C = C(n,s,p,q,\rho)$ such that*

$$\|f\|_{\dot{B}^s_{\infty,\rho}} \leq C\left(1 + \|f\|_{\dot{B}^0_{\infty,r}} \log^{\frac{1}{\rho}-\frac{1}{r}}(1 + \|f\|_{B^s_{p,q}})\right) \tag{3.19}$$

for all $f \in B^s_{p,q} \cap \dot{B}^0_{\infty,r}$ with $\rho \leq r \leq \infty$. Here the constant C is independent of r.

Remark 3.14. (i) Brezis - Gallouet [6], Brezis - Wainger [7] and Ozawa [45] showed that for $1 < p,r < \infty$, $k = n/r$, $s > n/p$ there holds

$$\|f\|_\infty \leq C\left\{1 + \|f\|_{W^{\frac{n}{r},r}}(1+\log(e+\|f\|_{W^{s,p}}))^{1-1/r}\right\}, \tag{3.20}$$

for all $f \in W^{n/r,r} \cap W^{s,p}$. Since the embeddings

$$\dot{B}^0_{\infty,1} \subset L^\infty, \quad W^{s,p} \subset B^s_{p,\infty}, \quad W^{n/r,r} \subset \dot{B}^0_{\infty,r}.$$

hold, by taking $\rho = 1$ and $q = \infty$ in (3.19), we obtain (3.20).

(ii) Beale-Kato-Majda [2] and Kozono-Taniuchi [31] showed that for $1 < p < \infty$, $s > n/p$, there holds

$$\|f\|_\infty \leq C\{1 + \|f\|_{BMO}(\log(e+\|f\|_{W^{s,p}}))\} \quad \text{for all } f \in W^{s,p}, \tag{3.21}$$

$$\|\nabla u\|_\infty \leq C\{1 + \|\text{rot } u\|_{BMO}\log(e+\|u\|_{W^{s+1,p}})\}, \quad \text{for for all } u \in W^{s+1,p} \text{ with div } u = 0. \tag{3.22}$$

Since the embeddings

$$W^{s,p} \subset B^s_{p,\infty}, \quad \dot{B}^0_{\infty,1} \subset L^\infty \subset BMO \subset \dot{B}^0_{\infty,\infty}$$

hold, by taking $\rho = 1$, $q = r = \infty$, we have (3.21). Letting $f = \nabla u$ in (3.21), with the aid of the estimate $\|\nabla u\|_{BMO} \le C\|\mathrm{rot}\ u\|_{BMO}$ (see (3.5)), we have (3.22).

We shall give an outline of then proof of Theorem 3.10. The proof of Corollary 3.11 is parallel to that as we have derived Corollary 3.6 from Theorem 3.4.

Outline of the proof of Theorem 3.10. Let us first assume $u \in L^2(\varepsilon_0, T; \dot{B}^0_{\infty,\infty})$. In the same way as in (3.10), (3.12) and (3.13), we have for $m > n/2$

$$\|u(t)\|_{H^m} \le \|u(\varepsilon_0)\|_{H^m} \exp\left(C\int_{\varepsilon_0}^t \|u\|^2_{BMO}\, dt\right), \quad \varepsilon_0 < t < T. \tag{3.23}$$

It is known by Frazier - Jawerth - Weiss [16, p. 42] that

$$\|u\|_{BMO} \cong \|u\|_{\dot{F}^0_{2,\infty}} \equiv \sup_{x\in\mathbb{R}^n}\left(\sum_{j=-\infty}^{\infty} |\varphi_j * u(x)|^2\right)^{1/2} \le \|u\|_{\dot{B}^0_{\infty,2}}$$

Then it follows from (3.23) and Lemma 3.13 with $p = 2$, $s = m$, $\rho = q = 2$ and $r = \infty$ that

$$\begin{aligned} &\|u(t)\|_{H^m} \\ \le\ & \|u(\varepsilon_0)\|_{H^m} \exp\left(C\int_{\varepsilon_0}^t \{1 + \|u(\tau)\|^2_{\dot{B}^0_{\infty,\infty}} \log(1 + \|u(\tau)\|_{H^m})\}\, d\tau\right) \end{aligned}$$

for all $\varepsilon_0 < t < T$. By taking $y(t) \equiv \log(1 + \|u(t)\|_{H^m})$, we obtain from the above estimate

$$y(t) \le y(\varepsilon_0) + C\int_{\varepsilon_0}^t \left(1 + \|u(\tau)\|^2_{\dot{B}^0_{\infty,\infty}} y(\tau)\right) d\tau, \quad \varepsilon_0 < t < T.$$

Now the Gronwall inequality yields

$$y(t) \le y(\varepsilon_0)\exp\left\{C\int_{\varepsilon_0}^t \left(1 + \|u(\tau)\|^2_{\dot{B}^0_{\infty,\infty}}\right) d\tau\right\}, \quad \varepsilon_0 < t < T,$$

which implies an apriori estimate of $\|u(t)\|_{H^m}$ as

$$\sup_{\varepsilon_0<t<T} \|u(t)\|_{H^m} \le (\|u(\varepsilon_0)\|_{H^m} + 1)\exp\left(\exp\left\{C\int_0^T \left(1 + \|u(\tau)\|^2_{\dot{B}^0_{\infty,\infty}}\right) d\tau\right\}\right).$$

Since the time interval for existence of local solutions is concretely characterized as in (2.6), this estimates causes a desired continuation of u on $[0, T')$.

We next prove under the assumption rot $u \in L^1(\varepsilon_0, T; \dot{B}^0_{\infty,\infty})$. We return to (3.8) and (3.9). Notice that the function G is rewritten as

$$G = \partial^\alpha(u\cdot\nabla u) - u\cdot\nabla^\alpha u.$$

Let $m' \geq [n/2]+2$, there holds $H^{m'-1} \subset L^\infty$, and hence we have by the commutator estimate

$$\|\partial^\alpha(fg) - f\partial^\alpha g\|_2 \leq C\|f\|_{H^{m'}}\|g\|_\infty + \|\nabla f\|_\infty \|g\|_{H^{m'-1}}, \quad |\alpha| \leq m'$$

for for all $f \in H^{m'}s$ and for all $g \in H^{m'-1}$. Choice of $f = u$, $g = \nabla u$ in the above estimate yields

$$\|G\|_2 \leq C\|u\|_{H^{m'}}\|\nabla u\|_\infty, \quad |\alpha| \leq m'. \tag{3.24}$$

Now, taking the inner product in L^2 between (3.8) and $\partial^\alpha u$, we have by (3.24) and the Schwarz inequality that

$$\frac{1}{2}\frac{d}{dt}\|\partial^\alpha u\|_2^2 \leq |(G, \partial^\alpha u)| \leq \|G\|_2 \|\partial^\alpha u\|_2 \leq C\|\nabla u\|_\infty \|u\|_{H^{m'}}\|\partial^\alpha u\|_2. \tag{3.25}$$

Here we have used again the identities

$$(u \cdot \nabla v, v) = -(u \cdot \nabla v, v) = 0, \quad (\nabla q, v) = 0,$$

which hold for for all $u, v \in H^{m'}$ with div $u =$ div $v = 0$.

Summing both sides of (3.25) for $|\alpha| \leq m'$, we have

$$\frac{d}{dt}\|u(t)\|_{H^{m'}} \leq C\|\nabla u(t)\|_\infty \|u(t)\|_{H^s}, \quad \varepsilon_0 < t < T.$$

By the Gronwall inequality, there holds

$$\|u(t)\|_{H^{m'}} \leq \|u(\varepsilon_0)\|_{H^{m'}} \exp\left(C\int_{\varepsilon_0}^t \|\nabla u(\tau)\|_\infty \, d\tau\right), \quad \varepsilon_0 < t < T.$$

Since $m'-1 > n/2$, we apply Lemma 3.13 with $\rho = 1$, $p = q = 2$, $s = m'-1$ and $r = \infty$ to the estimate of $\|\nabla u\|_\infty$, and obtain from the above

$$\begin{aligned} &\|u(t)\|_{H^{m'}} \\ \leq\ &\|u(\varepsilon_0)\|_{H^{m'}} \exp\left\{C\int_{\varepsilon_0}^t \left(1 + \|\nabla u(\tau)\|_{\dot{B}^0_{\infty,\infty}} \log(1 + \|u(\tau)\|_{H^{m'}})\right) d\tau\right\}. \end{aligned}$$

Define $\tilde{y}(t) \equiv \log(1 + \|u(t)\|_{H^{m'}})$ and we have

$$\tilde{y}(t) \leq \tilde{y}(0) + C\int_{\varepsilon_0}^t \left(1 + \|\nabla u(\tau)\|_{\dot{B}^0_{\infty,\infty}} y(\tau)\right) d\tau, \quad \varepsilon_0 < t < T.$$

Again by the Gronwall inequality,

$$\tilde{y}(t) \leq \tilde{y}(\varepsilon_0) \exp\left\{C\int_{\varepsilon_0}^T \left(1 + \|\nabla u(\tau)\|_{\dot{B}^0_{\infty,\infty}}\right) d\tau\right\} \tag{3.26}$$

for for all $t \in [\varepsilon_0, T)$.

On the other hand, since the Riesz transforms are $R = (R_1, \cdots, R_n)$ are bounded in $\dot{B}^0_{\infty,\infty}$ and since $\nabla u = R(R \times \text{rot } u)$, implied by the Biot-Savard law(see (3.5)), we have

$$\|\nabla u\|_{\dot{B}^0_{\infty,\infty}} \leq C\|\text{rot } u\|_{\dot{B}^0_{\infty,\infty}}. \tag{3.27}$$

Then it follows from (3.26) and (3.27) that

$$\tilde{y}(t) \le \tilde{y}(\varepsilon_0) \exp\left\{ C \int_{\varepsilon_0}^{T} \left(1 + \|\mathrm{rot}\, u(\tau)\|_{\dot{B}^0_{\infty,\infty}}\right) d\tau \right\} \tag{3.28}$$

for all $t \in [\varepsilon_0, T)$, which implies

$$\sup_{\varepsilon_0 < t < T} \|u(t)\|_{H^{m'}}$$
$$\le (\|u(\varepsilon_0)\|_{H^{m'}} + 1) \exp\left(\exp\left\{ C \int_{\varepsilon_0}^{T} \left(1 + \|\mathrm{rot}\, u(\tau)\|_{\dot{B}^0_{\infty,\infty}}\right) d\tau \right\}\right).$$

This apriori estimate yields the desired continuation.

4. Stability

In this section we shall consider the problem (IV) on stability of solutions in $\mathbb{R}^3$. Stability problem was originally proposed by Leray [36]. He gave a question whether every weak solution u of (N-S) on $(0, \infty)$ satisfies $\|u(t)\|_2 \to 0$ as $t \to \infty$. After 50 years of Leray' question, Masuda [38] solved his problem affirmatively provided u fulfills the energy inequality (3.18) of *strong form*. We may regard Masuda's results as stability around the rest motion. Here we shall discuss on the solution in the class (1.1).

Theorem 4.1. **(Kozono** [30]**)** *Let $n = 3$ and let $a \in L^2_\sigma$. Assume that u is a weak solution of (N-S) in the class*

$$u \in L^\alpha(0, \infty; L^r) \quad \text{for } \frac{2}{\alpha} + \frac{3}{r} = 1 \text{ with } r > 3. \tag{4.1}$$

Suppose that $b \in L^2_\sigma$, $f \in L^1(0, \infty; L^2)$. Then for every weak solution v of (N-S') satisfying the energy inequality

$$\|v(t)\|_2^2 + 2\int_0^t \|\nabla v\|_2^2\, d\tau \le \|a + b\|_2^2 + 2\int_0^t (f, v)\, d\tau, \quad 0 \le t < \infty,$$

there holds

$$\int_t^{t+1} \|v(\tau) - u(\tau)\|_2^2\, d\tau \to 0 \quad \text{as } t \to \infty.$$

In addition, if the weak solution v satisfies the energy inequality of the stronger form:

$$\|v(t)\|_2^2 + 2\int_s^t \|\nabla v\|_2^2\, d\tau \le \|v(s)\|_2^2 + 2\int_s^t (f, v)\, d\tau \tag{4.2}$$

for almost every $s \ge 0$, including $s = 0$, and all t such that $s < t < \infty$, then there holds

$$\|v(t) - w(t)\|_2 \to 0 \quad \text{as } t \to \infty.$$

If we impose additional assumptions on f, then we obtain an explicit rate of convergence.

Theorem 4.2. *In addition to the hypothesis of Theorem 4.1, assume that*

$$\|f(t)\|_2 = O(t^{-1}) \quad \text{as } t \to \infty.$$

Then every weak solution v of (N-S') with (4.2) converges to u as

$$\|v(t) - u(t)\|_r = o(t^{-\frac{3}{2}(\frac{1}{2}-\frac{1}{r})}) \quad \text{for } 2 \le r < \infty,$$
$$\|v(t) - u(t)\|_\infty = O(t^{-\frac{3}{4}}),$$
$$\|\nabla v(t) - \nabla u(t)\|_r = o(t^{-\frac{3}{2}(\frac{1}{2}-\frac{1}{r})-\frac{1}{2}}) \quad \text{for } 2 \le r < \infty$$

as $t \to \infty$.

Remark 4.3. (i) In $\mathbb{R}^3$, Beirão da Veiga-Secchi [3] investigated the solution u of (N-S) in the class

$$u \in L^\infty(0,\infty; L^{r+2}) \quad \text{for } r > 3.$$

They showed that if $f \equiv 0$ and if $b \in L^1 \cap L^r$ with $\|b\|_r$ small, then there is a unique solution v of (N-S') satisfying

$$\|v(t) - u(t)\|_r \le C(1+t)^{-3/4} \quad \text{for all } t > 0.$$

(ii) Ponce-Racke-Sideris-Titi [47] treated the solution u of (N-S) in the class (4.1) and obtained a constant $\delta > 0$ depending on u such that if

$$\|b\|_{H^1} + \int_0^\infty (\|f(t)\|_2 + \|f(t)\|_2^2)\, dt \le \delta,$$

then there is a unique solution v of (N-S') with

$$\sup_{t>0} \|v(t) - u(t)\|_{H^1} \le M(\delta),$$

where $M(\delta)$ is a constant with the property $\lim_{\delta\to 0} M(\delta) = 0$.

(iii) Although their estimates hold globally in time, smallness assumption on the perturbations b and f is required. On the other hand, our theorems hold for large t, but we do not need any restriction on b and f. It follows from Theorem 1.1 that every weak solution u in the class (4.1) is unique and regular. Theorem 4.1 guarantees also its stability; it is clarified that well-posedness of the Navier-Stokes equations can be unified in the class (4.1).

References

[1] Amann, H., *On the strong solvability of the Navier-Stokes equations.* J. Math. Fluid Mech. **2** (2000), 16–98 .

[2] Beale, J.T., Kato, T., Majda, A., *Remarks on the breakdown of smooth solutions for the 3-D Euler equations.* Commu. Math. Phys. **94** (1984), 61–66 .

[3] H. Beirão da Veiga and P. Secchi, *L^p-stability for strong solutions of the Navier-Stokes equations in the whole space.* Arch. Rational Mech. Anal. **98** (1987), 65–70.

[4] Beirão da Veiga, H., *A new regularity class for the Navier-Stokes equations in $\mathbb{R}^n$.* Chin. Ann. of Math. **16 B** (1995), 407–412 .

[5] J. Bergh and J. Löfström, *Interpolation Spaces.* Berlin-Heidelberg-New York: Springer-Verlag 1976.

[6] Brezis, H., Gallouet, T., *Nonlinear Schrödinger evolution equations.* Nonlinear Anal. TMA **4** (1980), 677–681.

[7] Brezis, H., Wainger, S., *A note on limiting cases of Sobolev embeddings and convolution inequalities.* Comm. Partial Differential Equations **5** (1980), 773–789.

[8] Brezis, H., *Remarks on the preceding paper by M. Ben-Artzi, "Global solutions of two-dimensional Navier-Stokes and Euler equations".* Arch. Rational Mech. Anal. **128** (1994), 359–360.

[9] Caffarelli, L., Kohn, R., Nirenberg, L., *Partial regularity of suitable weak solutions of the Navier-Stokes equations.* Comm. Pure Appl. Math. **35** (1982), 771–831.

[10] Cannone, M., Planchon, F., *On the regularity of the bilinear term for solutions to the incompressible Navier-Stokes equations.* Rev. Mat. Iberoamericana **16** (2000), 1–16.

[11] Choe, H.J., Lewis, J.L., *On the singular set in the Navier-Stokes equations.* J. Funct. Anal. **175** (2000), 348–369.

[12] Coifman, R., Lions., P.L., Meyer, Y., Semmes, S., *Compensated compactness and Hardy spaces.* J. Math. Pures. Appl. **72** (1993), 247–286.

[13] Coifman, R., Meyer, Y., *Au delá des opérateurs pseudodifférentieles.* Astérisque **57**, Société Mathématique de France, 1978.

[14] Fefferman, C., Stein, E. M., *H^p spaces of several variables.* Acta Math. **129** (1972), 137–193.

[15] C. Foias, *Une remarque sur l'unicité des solutions des equations de Navier-Stokes en dimension n,* Bull. Soc. Math. France **89** (1961), 1–8.

[16] Frazier, Mi., Jawerth, B., Weiss, G., *Littlewood-Paley theory and the study of function spaces.* CBMS Regional Conference Series in Mathematics, **79** American Mathematical Society, Providence, RI, 1991

[17] Fujita, H., Kato, T., *On the Navier-Stokes initial value problem 1.* Arch. Rational Mech. Anal. **16** (1964), 269–315.

[18] Furioli, G, Lemarié-Rieusset, P-G, Terraneo, E., *Sur l'unicite dans $L^3(R^3)$ des solutions "mild" des equations de Navier-Stokes.* C. R. Acad. Sci. Paris Ser. I Math. **325** (1997), 1253–1256.

[19] Giga, Y., *Solutions for semilinear parabolic equations in L^p and regularity of weak solutions of the Navier-Stokes system.* J. Differential Eq. **62** (1986), 182–212.

[20] Giga, Y., Inui, K., Kato, J., Matsui, S., *Remarks on the uniqueness of bounded solutions of the Navier-Stokes equations.* Preprint ser. Hokkaido Univ. Math. no. 497.

[21] Giga, Y., Miyakawa, T., *Solution in L_r of the Navier-Stokes initial value problem.* Arch. Rational Mech. Anal.**89** (1985), 267–281.

[22] Hopf, E., *Über die Anfangswertaufgabe für die hydrodyanamischen Grundgleichungen.* Math. Nach. **4** (1950/51), 213–231.

[23] Hishida, T., Izumida, K., *Remarks on a regularity criterion for weak solutions to the Navier-Stokes equations in $\mathbb{R}^n$* Analysis **20** (2000), 191–200.

[24] John, F., Nirenberg, L., *On functions of bounded mean oscillation.* Comm. Pure Appl. Math. **14** (1961), 415–426.

[25] Kato, T., *Strong L^p-solutions of the Navier-Stokes equation in $\mathbb{R}^m$, with applications to weak solutions.* Math. Z. **187** (1984), 471–480.

[26] Kato, T., Ponce, G., *Commutator estimates and the Euler and Navier-Stokes equations.* Comm. Pure Appl. Math. **41** (1988), 891–907.

[27] Kim, N, Chae, D., *On the uniqueness of the unbounded classical solutions of the Navier-Stokes and associated equations.* J. Math. Anal. Appl. **186** (1994), 91–96.

[28] Kiselev, A.A., Ladyzenskaya, O.A, *On the existence and uniqueness of the solution of the nonstationary problem for a viscous incompressible fluid.* Izv. Akad. Nauk SSSR, Ser. Mat. **21** (1957), 655–680.

[29] Kozono, H., *Removable singularities of weak solutions to the Navier-Stokes equations.* Communications in Partial Differential Equations **23** (1998), 949–966.

[30] Kozono, H., *Asymptotic stability of large solutions with large perturbation to the Navier-Stokes equations.* J. Func. Anal. **176** (2000), 153–197.

[31] Kozono, H., Taniuchi, Y., *Bilinear estimates in BMO and the Navier-Stokes equations.* Math. Z. **235** (2000), 173–194.

[32] Kozono, H., Taniuchi, Y., *Limiting case of the Sobolev inequality in BMO, with application to the Euler equations.* Commun. Math. Phy. **214**(2000), 191–200.

[33] Kozono, H., Ogawa, T., Taniuchi, Y., *The critical Sobolev inequalities in Besov spaces and regularity criterion to some semi-linear evolution equations.* to appear in Math. Z.

[34] Kozono, H., Sohr, H., *Remark on uniqueness of weak solutions to the Navier-Stokes equations.* Analysis **16** (1996), 255–271.

[35] Kozono, H., Sohr, H., *Regularity criterion on weak solutions to the Navier-Stokes equations.* Advances in Diff. Eq. **2** (1997), 535–554.

[36] Leray, J., *Sur le mouvement d'un liquide visqeux emplissant l'espace.* Acta. Math. **63** (1934), 193–248.

[37] Lions, P.L., Masmoudi, N., *Unicité des solutions faibles de Navier-Stokes dans* $L^N(\Omega)$. C. R. Acad. Sci. Paris Ser. I Math. **327** (1998), 491–496.

[38] Masuda, K., *Weak solutions of Navier-Stokes equations.* Tohoku Math. J. **36** (1984), 623–646.

[39] Majda, A., *Vorticities and the mathematical theory of incompressible fluid flow.* Comm. Pure Appl. Math. **34** (1986), S187–S220.

[40] Nečas, J., Ružička, M., Šverák, V., *On Leray's self-similar solutions of the Navier-Stokes equations.* Acta Math. **176** (1996), 283–294.

[41] Miyakawa, T., *Hardy spaces of solenoidal vector fields, with applications to the Navier-Stokes equations.* Kyushu J. Math. **50** (1996), 1–64.

[42] Neustupa, J., *Partial regularity of weak solutions to the Navier-Stokes equations in the class $L^\infty(0,T;L^3(\Omega))$.* J. Math. Fluid Mech. **1** (1999), 309–325.

[43] Ogawa, T.,Taniuchi, Y., *Remarks on uniqueness and blow-up criterion to the Euler equations in the generalized Besov spaces.* J. Korean Math. Soc.**37** (2000), 1007–1019.

[44] Okamoto, H., *Exact solutions of the Navier-Stokes equations via Leray's scheme.* Japan J. Indust. Appl. Math. **14** (1997), 169–197.

[45] Ozawa, T., *On critical cases of Sobolev's inequalities.* J. Func. Anal. **127** (1995), 259–269.

[46] Ponce, G., *Remarks on a paper by J. T. Beale, T. Kato and A. Majda.* Commun. Math. Phys. **98** (1985), 349–353.

[47] G. Ponce, R. Racke, T. C. Sideris and E. S. Titi, *Global stability of large solutions to the 3D Navier-Stokes equations.* Commun. Math. Phys. **159** (1994), 329–341.

[48] Serrin, J., *On the interior regularity of weak solutions of the Navier-Stokes equations.* Arch. Rational Mech. Anal. **9** (1962),187–195.

[49] Serrin, J., *The initial value problem for the Navier-Stokes equations.* Nonlinear Problems, R. E. Langer ed., Madison: University of Wisconsin Press, (1963) 69–98.

[50] Solonnikov, V.A., *Estimates for the solutions of a nonstationary Navier-Stokes equations.* J. Soviet Math. **8** (1977), 1–116.

[51] Stein, E. M., *Harmonic Analysis.* Princeton University Press 1993.

[52] Takahashi, S., *On interior regularity criteria for weak solutions of the Navier-Stokes equations.* Manuscripta Math.**69** (1990), 237–254.

[53] Taniuchi, Y., *On generalized energy equality of the Navier-Stokes equations.* Manuscripta Math.**94** (1997), 365–384.

[54] Wahl, W. von, *Regularity of weak solutions of the Navier-Stokes equations.* Proceedings of the 1983 Summer Institute on Nonlinear Functional Analysis and Applications, Proc. Symposia in Pure Mathematics **45**, F. E. Brower ed., Providence Rhode Island : Amer. Math. Soc. (1986), 497–503.

Mathematical Institute,
Tohoku University,
Sendai, 980-8578, Japan
E-mail address: `kozono@math.tohoku.ac.jp`

Anisotropic and Geometric Criteria for Interior Regularity of Weak Solutions to the 3D Navier–Stokes Equations

Jiří Neustupa and Patrick Penel

Abstract. We formulate sufficient conditions for regularity of a suitable weak solution $(v; p_*)$ in a sub–domain D of the time–space cylinder Q_T in Section 3. The conditions are anisotropic in the sense that the assumptions about v_1, v_2 (the first two components of velocity) differ from the assumptions about the third component of velocity v_3. The question what types of deformations of infinitely small volumes of the fluid support regularity and what types contribute to a blow–up is studied in Section 4. Finally, we mention some open problems in Section 5.

1. Introduction

Suppose that Ω is a domain in $\mathbb{R}^3$ and T is a positive number. We denote $Q_T = \Omega\times\,]0,T[$. The Navier–Stokes initial–boundary value problem for a viscous incompressible fluid with the homogeneous Dirichlet–type boundary condition is defined by the equations

$$\frac{\partial v}{\partial t} + (v\cdot\nabla)v = f - \nabla p_* + \nu\,\Delta v \qquad \text{in } Q_T, \tag{1}$$

$$\operatorname{div} v = 0 \qquad \text{in } Q_T, \tag{2}$$

and by the conditions

$$v = 0 \qquad \text{on } \partial\Omega\times\,]0,T[\,, \tag{3}$$

$$v\big|_{t=0} = v_0 \tag{4}$$

where $v = (v_1, v_2, v_3)$ and p_* denote the unknown velocity and pressure, f is an external body force and $\nu > 0$ is the viscosity coefficient. ∇ and Δ denote operator nabla, respectively the Laplace operator, with respect to the spatial variables.

The equation of continuity (2) expresses the conservation of mass in an incompressible fluid. The Navier–Stokes equation (1) expresses the conservation of momentum. It is a vector equation and it can also be written in components, as a system of three equations. (This is why we usually speak about the Navier–Stokes equations.) Many of important and so far known qualitative properties of the problem (1)–(4) are described in the book [36] by R. Temam or in the recent work G. P. Galdi [11]. We briefly remind some of them.

We will use these function spaces: H will be the closure of the set $\{u \in C_0^\infty(\Omega)^3;\ \operatorname{div} u = 0\}$ in $L^2(\Omega)^3$ and H_0^1 will denote the closure of the same set in $W^{1,2}(\Omega)^3$. Note that if Ω has a bounded and locally lipschitzian boundary $\partial\Omega$ or if Ω is a half–space then

$H = \{u \in L^2(\Omega)^3;\ \operatorname{div} u = 0$ in Ω (in the sense of distributions) and $(u \cdot n)|_{\partial\Omega} = 0$ (in the sense of traces)$\}$ and

$H_0^1 = \{u \in W^{1,2}(\Omega)^3;\ \operatorname{div} u = 0$ a.e. in Ω and $u|_{\partial\Omega} = 0$ (in the sense of traces)$\}$.

(See [10], Chapter III.) The norm and the scalar product in H will be denoted in the same way as the norm and the scalar product in $L^2(\Omega)$, i.e. $\|\,.\,\|_2$ and $(.\,,\,.)_2$. The norm in $L^k(\Omega)$ will be denoted by $\|\,.\,\|_k$. The norms of vector–valued or tensor–valued functions will be denoted in the same way as the norms of scalar–valued functions.

An anisotropic Lebesgue space $L^{r,s}(Q_T)$ is defined as $L^r(0,T;\ L^s(\Omega))$. A function g (defined a.e. in Q_T) is said to be in $L^{r,s}(D)$ (where D is a sub–domain of Q_T) if $\chi_D\, g \in L^{r,s}(Q_T)$. ($\chi_D$ is the characteristic function of D.)

If we multiply formally equation (1) by function $\varphi \in C_0^\infty(\Omega \times [0,T[)^3$, such that $\operatorname{div}\varphi = 0$ in Q_T, and we integrate on Q_T, we obtain

$$\int_0^T \int_\Omega \left[v \cdot \frac{\partial\varphi}{\partial t} - (v \cdot \nabla)v \cdot \varphi - \nu\, \nabla v \cdot \nabla\varphi \right] dx\, dt$$
$$= \; - \int_0^T \int_\Omega f \cdot \varphi\, dx\, dt \; - \int_\Omega v_0 \cdot \varphi(x,0)\, dx\,. \tag{5}$$

If $v_0 \in H$ and $f \in L^2(Q_T)^3$ then a measurable function $v\ :\ \Omega \times [0,T[\longrightarrow \mathbb{R}^3$ is called *a weak solution* of the problem (1)–(4) if

a) $v \in L^2(0,T;\ H_0^1) \cap L^\infty(0,T;\ H)$ and

b) v satisfies the integral identity (5) for all functions $\varphi \in C_0^\infty(\Omega \times [0,T[)^3$ such that $\operatorname{div}\varphi = 0$ in Q_T.

An *associated pressure* to the weak solution v is a scalar function p_* such that v and p_* satisfy equation (1) in Q_T in the sense of distributions. Each weak solution v can be redefined on a set of zero measure so that it becomes a weakly continuous mapping of $[0,T[$ to H. (See [11].)

The existence of a weak solution of the problem (1)–(4) is known already for a long time (see J. Leray [21], E. Hopf [14] and others):

The problem (1)–(4) has at least one weak solution v which satisfies a so called energy inequality

$$\|v(.\,,t)\|_2^2 + 2\nu \int_0^t \|\nabla v(.\,,\sigma)\|_2^2\, d\sigma \le 2 \int_0^t \big(v(.\,,\sigma),\, f(.\,,\sigma)\big)_2\, d\sigma + \|v_0\|_2^2 \tag{6}$$

for all $t \in [0,T[$.

If Ω is a bounded domain or an exterior domain, a half–space or if it coincides with the whole $\mathbb{R}^3$ then the weak solution v can be constructed so that is satisfies a so called *strong energy inequality*

$$\|v(.\,,t)\|_2^2 + 2\nu \int_s^t \|\nabla v(.\,,\sigma)\|_2^2\, d\sigma \le 2 \int_s^t \big(v(.\,,\sigma),\, f(.\,,\sigma)\big)_2\, d\sigma + \|v_0\|_2^2 \qquad (7)$$

for a.a. $s \in [0,T\,]$ and all $t \in [s,T[$. (See J. Leray [21], G. P. Galdi and P. Maremonti [9], H. Sohr, W. von Wahl and M. Wiegner [33], T. Miyakawa and H. Sohr [23] and W. Borchers and T. Miyakawa [2].)

The weak solution is so far known to be unique only if it belongs to a certain class of functions which, however, does not cover the whole space $L^2(0,T;\, H_0^1) \cap L^\infty(0,T;\, H)$. The results of G. Prodi [29], J. L. Lions and G. Prodi [22], H. Kozono and H. Sohr [16] and others lead to the proposition:

Let u and v be two weak solutions of the problem (1)–(4). If u satisfies the energy inequality (6) and

a) $v \in L^{p,q}(Q_T)^3$ *where* $2 \le p \le +\infty$, $3 < q \le +\infty$ *and* $2/p + 3/q \le 1$ *or*

b) $v \in L^{\infty,3}(Q_T)^3$ *and v is right L^3–continuous as a function of t on* $[0,T[$

then $u = v$ a.e. in Q_T.

Moreover, if Ω is a bounded or an exterior domain with a "smooth" boundary or if it is a half–space then assumption b) can be replaced by

b′) $v \in L^{\infty,3}(Q_T)^3$.

The theory of regularity of weak solutions represents an extensive field which involves many results about the regularity up to the boundary for various types of domain Ω, about the interior regularity and about a so called partial regularity. Of many authors who contributed to the development of this theory, we can mention e.g. J. Leray [21], J. Serrin [30], E. B. Fabes, B. F. Jones and N. M. Rivière [7], H. Sohr [31], W. von Wahl [37] and Y. Giga [12]. We formulate the main results in the case when $f = 0$. However, all statements remain also valid if $f \neq 0$ and f is "smooth enough".

Let v be a weak solution of the problem (1)–(4) with $f = 0$ and let

a) $v \in L^{p,q}(Q_T)^3$ *where* $2 \le p \le +\infty$, $3 < q \le +\infty$ *and* $2/p + 3/q = 1$ *or*

b) $v \in C^0([0,T];\, L^3(\Omega)^3)$.

Then, if $\partial\Omega$ is uniformly of the class C^m (for $m \ge 2$) and $0 < s < T$, we have $\dfrac{\partial^k v}{\partial t^k} \in L^2(s,T;\, W^{l,2}(\Omega)^3)$ *for all $k \ge 0$ and $l = 2, \dots, m$.*

If $\partial\Omega$ is uniformly of the class C^∞ then $v \in C^\infty(\overline{\Omega} \times\,]0,T])^3$.

The interior regularity results say:

If v is a weak solution of the problem (1)–(4) with $f = 0$, $\Omega' \times]t_1, t_2[$ is a bounded domain which is a strict sub–domain of Q_T and

$$v \in L^{p,q}(\Omega' \times]t_1, t_2[)^3 \text{ where } 2 \leq p \leq +\infty, 3 < q \leq +\infty \text{ and } 2/p + 3/q = 1$$

then v has all spatial derivatives bounded in compact subsets of $\Omega' \times]t_1, t_2[$.

A point $(x, t) \in Q_T$ is called a *regular point* of a weak solution v of (1)–(4) if there exists a neighborhood U of (x, t) in Q_T such that $\chi_U\, v \in L^\infty(Q_T)^3$. Points of Q_T which are not regular are called *singular.* We shall denote by $S(v)$ the set of all singular points of v. It is obvious that $S(v)$ is closed in Q_T.

The existence of a unique classical solution of the problem (1)–(4) if all input data are "smooth enough" was proved in the case of two–dimensional flows by J. Leray [21], O. A. Ladyzhenskaya [18] and others. The same result is also known for axially symmetric flows in the cases when $\Omega = \mathbb{R}^3$, the external force f as well as the initial velocity v_0 are axially symmetric and the tangential components of v_0 and f are equal to zero (see e.g. S. Leonardi, J. Málek, J. Nečas and M. Pokorný [20]). The existence of a unique classical solution can also be proved for general axially symmetric flows in the case that the axis of symmetry is outside Ω with a positive distance from Ω (see the note in O. A. Ladyzhenskaya [18]).

The existence of a unique regular solution for a general three–dimensional flow is so far known only locally in time (see K. K. Kiselev and O. A. Ladyzhenskaya [15], V. A. Solonnikov [34], J. Heywood [13], G. P. Galdi [11]) or under the assumption that v_0 and f are in appropriate norms "small enough" (see O. A. Ladyzhenskaya [19], J. G. Heywood [13] and others). The main results say:

Suppose that $\partial\Omega$ is uniformly of the class C^2 and $v_0 \in H_0^1$. Then there exists $T > 0$ and a weak solution v of the problem (1)-(4) (with $f = 0$) such that $v \in L^\infty(0, T;\, H_0^1) \cap L^2(0, T;\, W^{2,2}(\Omega)^3)$. Number T is bounded from below by a constant which depends only on $\|\nabla v_0\|_2$, ν and the C^2–regularity of Ω. If Ω is bounded or $\Omega = \mathbb{R}^3$ then $T \geq c_1\, \nu^3 / \|\nabla v_0\|_2^4$. Moreover, there exists a decreasing function d such that if

$$\|v_0\|_2 \;\leq\; d(\|\nabla v\|_2)$$

then T can be taken arbitrarily large. If Ω is bounded or $\Omega = \mathbb{R}^3$ then

$$d(\|\nabla v\|_2) \;=\; c_2\, \nu^2 / \|\nabla v_0\|_2$$

where c_2 depends only on Ω.

The next proposition, which is mainly due to J. Leray [21], C. Foias and R. Temam [8] and J. Heywood [13], says that a weak solution (in a domain with a smooth boundary) is in fact smooth up to a "small" set of instants of time:

Suppose that $\partial\Omega$ is uniformly of the class C^∞ and v is a weak solution of the problem (1)–(4) (with $f = 0$), satisfying the strong energy inequality (7). Then

$$]0, T[= \bigcup_{\gamma \in \Gamma}]a_\gamma, b_\gamma[\cup G \tag{8}$$

where set Γ is at most countable, the intervals $]a_\gamma, b_\gamma[$ are mutually disjoint, v is of the class C^∞ on $\Omega\times\,]a_\gamma, b_\gamma[$ for each $\gamma \in \Gamma$ and the $\frac{1}{2}$–Hausdorff measure of set G is finite.

The existence of a regular solution of the problem (1)–(4) on the time interval $]0, T[$ of an arbitrary length T, in the case of v_0 and f of an arbitrary size, remains an open problem. However, on the other hand, the appearance of a singular point in a weak solution (in the case when all input data are "smooth") has also not been confirmed yet! J. Leray proposed in 1934 in the work [21] to construct a singularity in the solution v, p_* of the problem (1)–(4) in the case $\Omega = \mathbb{R}^3$. He suggested to look for v and p_* in the form

$$v(x,t) = \frac{1}{\sqrt{2a(t_0-t)}}\, V\biggl(\frac{x-x_0}{\sqrt{2a(t_0-t)}}\biggr), \quad p_*(x,t) = \frac{1}{2a(t_0-t)}\, P_*\biggl(\frac{x-x_0}{\sqrt{2a(t_0-t)}}\biggr)$$

where $a > 0$ and V, P_* are new unknown functions which do not depend explicitly on t. They satisfy the system

$$(V\cdot\nabla)V \;=\; -\nabla P_* + \nu\,\Delta V \,-\, a\,V \,-\, a\,(x\cdot\nabla)V \qquad \text{in } \mathbb{R}^3, \tag{9}$$

$$\operatorname{div} V \;=\; 0 \qquad \text{in } \mathbb{R}^3. \tag{10}$$

The existence of a non–trivial solution $V \neq 0$ of the problem (9), (10) would easily lead to the singularity (or, in other words, the blow–up) in the solution v, p_* at the point (x_0, t_0). However, J. Nečas, M. Růžička and V. Šverák [24] proved that the only solution of (9), (10) such that $V \in L^3(\mathbb{R}^3)^3$ is the zero solution $V = 0$.

If we formally multiply the Navier–Stokes equation (1) (where f is considered to be zero) by a non–negative function ϕ, which is infinitely differentiable in Q_T and has a compact support in Q_T, and integrate by parts, we can obtain a so called *generalized energy inequality*

$$\int_0^T\!\!\int_\Omega \nu|\nabla v|^2\,\phi\,dx\,dt \le \int_0^T\!\!\int_\Omega \Bigl[\,|v|^2\Bigl(\frac{\partial\phi}{\partial t} + \nu\Delta\phi\Bigr) \,+\, (|v|^2 + 2p_*)\,v\cdot\nabla\phi\Bigr]\,dx\,dt.$$

The generalized energy inequality plays a fundamental role in treatments of local properties of weak solutions. However, it is not known whether all weak solutions (or even those of them which satisfy the strong energy inequality (7)) satisfy the generalized energy inequality.

L. Caffarelli, R. Kohn and L. Nirenberg [4] introduced the notion of a so called suitable weak solution: A pair $(v; p_*)$ of measurable functions on Q_T is said to be a *suitable weak solution* of the problem (1)–(4) (with $f = 0$) if

1. v is a weak solution of the problem (1)–(4),
2. p_* is an associated pressure, $p_* \in L^{5/4}(Q_T)$, and
3. $(v; p_*)$ satisfies the generalized energy inequality for every infinitely differentiable non-negative function ϕ on Q_T with a compact support in Q_T.

The existence of a suitable weak solution was proved in [4] under the assumption that the initial velocity v_0 is in H and if Ω is bounded, it is also in $W^{2/5,5/4}(\Omega)^3$.

This additional condition $v_0 \in W^{2/5,5/4}(\Omega)^3$ was later removed by Y. Taniuchi (see [35]) who proved the existence of a suitable weak solution for $v_0 \in H$. Thus, the suitable weak solution is a natural type of a weak solution of the problem (1)–(4).

Except for the generalized energy inequality, the other reason why the suitable weak solutions represent an interesting and important type of weak solutions of (1)–(4) is that the singular set $S(v)$ of each suitable weak solution $(v; p_*)$ has a 1–dimensional parabolic measure equal to zero. (See L. Caffarelli, R. Kohn and L. Nirenberg [4].) Since the parabolic measure dominates the Hausdorff measure, this result implies that the 1–dimensional Hausdorff measure of $S(v)$ equals zero.

H. Beirão da Veiga [1] proved regularity of a weak solution to the problem (1)–(4) by imposing conditions on the vorticity $\omega = \operatorname{curl} v$. D .Chae and H. J. Choe [5] obtained a similar result, however they used assumptions only about two components of vorticity:

If $\Omega = \mathbb{R}^3$, $v_0 \in H$, $\omega_0 = \operatorname{curl} v_0 \in H$, $\omega = (\omega_1, \omega_2, \omega_3)$, ω_1 and $\omega_2 \in L^{p,q}(Q_T)$ with $1 < p < +\infty$, $\frac{3}{2} < q < +\infty$, $2/p + 3/q \leq 2$ or if the norms of ω_1 and ω_2 in $L^{\infty,3/2}(Q_T)$ are sufficiently small then v is a classical solution on $]0, T[$.

D. Chae and H. J. Choe [5] also proved regularity by imposing conditions on gradients of two components of velocity:

If $\Omega = \mathbb{R}^3$, $v_0 \in H^1_0$, all components of ∇v_1 and ∇v_2 belong to $L^{p,q}(Q_T)$ with $2 \leq p \leq +\infty$, $3 \leq q \leq +\infty$, $2/p + 3/q \leq 1$ then v is a classical solution on $]0, T[$.

P. Constantin and C. Fefferman [6] proved regularity of a weak solution v under certain assumptions about the behaviour of the quantity $\omega/|\omega|$. More details on this result can be found in this book in the chapter written by C.Bardos.

The question whether the components of velocity v are coupled in such a way that a hypothetical singularity can develop either in all of them or in none of them was studied by J. Neustupa and P. Penel in [26] and by J. Neustupa, P. Penel and A. Novotný in [27]. The result of [27] says:

If D is a sub–domain of Q_T and $(v; p_*)$ is a suitable weak solution of the problem (1)–(4) with $f = 0$ such that v_3 (the third component of v) belongs to $L^{r,s}(D)$ where $4 \leq r \leq +\infty$, $6 < s \leq +\infty$ and $2/r + 3/s \leq \frac{1}{2}$ then v has no singular points in D.

We study a similar problem in Section 3. We show that the class of functions where v is regular need not be defined by means of the same requirements on all components of v: We can assume something more on one component and something less on the other two components. We call the conditions of this type *anisotropic conditions*. Theorem 1 proved in Section 3 represents the generalization of results from [26] and [27].

Section 4 deals with the question what types of deformations of infinitely small volumes in a moving fluid "support" the regularity and what types "contribute" to a hypothetical blow–up in a suitable weak solution $(v; p_*)$ of the problem (1)–(4).

Finally, we formulate some open problems in Section 5.

2. Auxiliary results

We suppose that $(v; p_*)$ is a suitable weak solution of the problem (1)–(4). We will use some additional assumptions about v in a sub–domain D of Q_T in Sections 3 and 4. Since $S(v)$ is closed in Q_T, $S(v) \cap \overline{D'}$ is a closed set for each domain D' in D such that $D' \subset \overline{D}' \subset D$. Suppose further that D' is such a domain. Denote by $\mathcal{T}(D')$, respectively G', the projection of D', respectively the projection of $S(v) \cap \overline{D'}$, onto the time axis. Then the 1–dimensional Hausdorff measure of G' is zero and

$$\mathcal{T}(D') = \bigcup_{\gamma \in \Gamma'}]a'_\gamma, b'_\gamma[\cup G' \tag{11}$$

where the intervals $]a'_\gamma, b'_\gamma[$ are mutually disjoint and set Γ' is at most countable. In accordance with the terminology from in [11], we will call time instants b'_γ *D'–epochs of irregularity*. Existence of a singular point of solution $(v; p_*)$ in D' implies the existence of at least one D'–epoch of irregularity and vice versa. Thus, let us further assume that t_0 is a D'–epoch of irregularity of solution $(v; p_*)$ and (x_0, t_0) is a singular point of $(v; p_*)$ in D'. Then, naturally, t_0 is equal to b'_γ for some $\gamma \in \Gamma'$. We will later show that these assumptions are in a contradiction with the conditions of Theorem 1 in Section 3 or Theorem 2 in Section 4. However, we need several auxiliary lemmas at first.

Lemma 1. *There exist positive numbers τ, ϵ_1, ϵ_2 such that $\epsilon_1 < \epsilon_2$ and*

1. *τ is so small that $a'_\gamma < b'_\gamma - \tau = t_0 - \tau$,*
2. *$\overline{B_{\epsilon_2}(x_0)} \times [t_0 - \tau, t_0] \subset D'$,*
3. *$\{(\overline{B_{\epsilon_2}(x_0)} - B_{\epsilon_1}(x_0)) \times [t_0 - \tau, t_0]\} \cap S(v) = \emptyset$,*
4. *v and all its spatial derivatives are bounded on $(\overline{B_{\epsilon_2}(x_0)} - B_{\epsilon_1}(x_0)) \times [t_0 - \tau, t_0]$.*

($B_{\epsilon_1}(x_0)$ and $B_{\epsilon_2}(x_0)$ denote the open balls in $\mathbb{R}^3$ with the center x_0 and the radii ϵ_1, ϵ_2.) Lemma 1 was proved in [26]. The main ideas of the proof are:

The existence of $\epsilon_0 > 0$ and $\tau > 0$ such that $\overline{B_{\epsilon_0}(x_0)} \times [t_0 - \tau, t_0] \subset D'$ and $a'_\gamma < b'_\gamma - \tau = t_0 - \tau$ follows from the fact that D' is open. It is obvious that ϵ_0 can be chosen to be as small as we need.

The fact that the 1–dimensional Hausdorff measure of $S(v)$ is zero implies the existence of $\epsilon_1 \in]0, \epsilon_0[$ such that $A(\epsilon_1) \cap S(v) = \emptyset$ where

$$A(\epsilon_1) = \{(x,t) \in D';\ |x - x_0| = \epsilon_1,\ t \in [t_0 - \tau, t_0]\}.$$

$A(\epsilon_1)$ is compact and $D' \cap S(v)$ is closed in D'. Hence there exists $\epsilon_2 \in]\epsilon_1, \epsilon_0[$ such that $A(\epsilon) \cap S(v) = \emptyset$ for all $\epsilon \in [\epsilon_1, \epsilon_2]$. This implies statement 3 of the lemma.

Each point of the set $(\overline{B_{\epsilon_2}(x_0) - B_{\epsilon_1}(x_0))} \times [t_0 - \tau, t_0]$ is a regular point of solution v. Thus, there exists an open set $D'' \subset D'$ such that $(\overline{B_{\epsilon_2}(x_0) - B_{\epsilon_1}(x_0))} \times [t_0 - \tau, t_0] \subset D''$ and v is essentially bounded on D''. Using the interior regularity results of J. Serrin [30] (see also G. P. Galdi [11]), we can obtain statement 4 of the lemma. □

Let us further denote for simplicity $B_1 = B_{\epsilon_1}(x_0)$ and $B_2 = B_{\epsilon_2}(x_0)$. Put $\epsilon_3 = (2\epsilon_1 + \epsilon_2)/3$, $\epsilon_4 = (\epsilon_1 + 2\epsilon_2)/3$ and $B_3 = B_{\epsilon_3}(x_0)$, $B_4 = B_{\epsilon_4}(x_0)$.

The local regularity results (J. Serrin [30] and G. P. Galdi [11]) do not directly imply that $\partial v / \partial t$ or p_* are also bounded on $(\overline{B_2} - B_1) \times [t_0 - \tau, t_0]$. Therefore we prove the next lemma.

Lemma 2. *Suppose that Ω' is a bounded domain which is strictly contained in Ω (i.e. $\Omega' \subset \overline{\Omega'} \subset \Omega$) and the solution $(v;p)$ has no singular points in $\overline{\Omega'} \times [t_1, t_2]$. Then p_*, respectively $\partial v/\partial t$, have all spatial derivatives in $L^\alpha(t_1, t_2;\ L^\infty(\Omega'))$, respectively in $L^\alpha(t_1, t_2;\ L^\infty(\Omega')^3)$, for each α such that $1 < \alpha < 2$.*

Proof. There exists $\zeta > 0$ such that v and all its spatial derivatives are bounded in $U_\zeta(\overline{\Omega'}) \times [t_1, t_2]$. ($U_\zeta(\overline{\Omega'})$ is the ζ–neighborhood of $\overline{\Omega'}$.) Denote by ψ an infinitely differentiable cut–off function defined in Ω, which is equal to one on $U_{\zeta/4}(\overline{\Omega'})$, equal to zero on $\Omega - U_{3\zeta/4}(\overline{\Omega'})$ and $0 \le \psi(x) \le 1$ for $x \in U_{3\zeta/4}(\overline{\Omega'}) - U_{\zeta/4}(\overline{\Omega'})$. The product $\psi\, p_*$ satisfies for a.a. $(x,t) \in \Omega \times [t_1, t_2]$ the identity

$$\psi(x)\, p_*(x,t) = -\frac{3}{4\pi} \int_{\mathbb{R}^3} \frac{1}{|x-y|} \left[\Delta(\psi p_*) \right](y,t)\, dy\,.$$

If we use the integration by parts and the equation

$$\Delta p_*(y,t) = -\frac{\partial^2}{\partial y_i\, \partial y_j}\, v_i(y,t)\, v_j(y,t)$$

(which we can obtained by applying the operator of divergence to equation (1)), we can derive that

$$\psi(x)\, p_*(x,t) = p_*^I(x,t) + p_*^{II}(x,t)$$

where

$$p_*^I(x,t) = -\frac{3}{4\pi}\int_{\mathbb{R}^3} \frac{1}{|x-y|}\,\frac{\partial^2}{\partial y_i\,\partial y_j}\Big[\psi\, v_i\, v_j\Big](y,t)\,dy\,,$$

$$p_*^{II}(x,t) = -\frac{3}{2\pi}\int_{\mathbb{R}^3} \frac{x_i-y_i}{|x-y|^3}\left[\frac{\partial\psi}{\partial y_j}\,v_i\,v_j\right](y,t)\,dy$$

$$-\frac{3}{4\pi}\int_{\mathbb{R}^3}\frac{1}{|x-y|}\left[\frac{\partial^2\psi}{\partial y_i\,\partial y_j}\,v_i\,v_j\right](y,t)\,dy + \frac{3}{4\pi}\int_{\mathbb{R}^3}\frac{x_i-y_i}{|x-y|^3}\left[\frac{\partial\psi}{\partial y_i}\,p_*\right](y,t)\,dy$$

$$+\frac{3}{4\pi}\int_{\mathbb{R}^3}\frac{1}{|x-y|}\,[\,\Delta\psi\, p_*\,](y,t)\,dy.$$

Using the boundedness of v and its spatial derivatives on $\operatorname{supp}\psi\times[t_1,t_2]$, we obtain

$$\big|\nabla^k p_*^I(x,t)\big| \le \left|\frac{3}{4\pi}\int_{\mathbb{R}^3}\frac{1}{|x-y|}\,\nabla_y^k\,\frac{\partial^2}{\partial y_i\,\partial y_j}\Big[\psi\,v_i\,v_j\Big](y,t)\,dy\right| \le c_3(k)$$

for $k=0,1,\dots$ and for a.a. $(x,t)\in\Omega\times[t_1,t_2]$. The constants $c_3(k)$ are independent of t.

Furthermore, the integrals in the equation for p_*^{II} can be considered only on the domain $\{y\in\operatorname{supp}\nabla\psi\}\subset U_{3\zeta/4}(\overline{\Omega'})-U_{\zeta/4}(\overline{\Omega'})$ where v and its spatial derivatives are bounded and $|x-y|\ge\zeta/4$ for $x\in\overline{\Omega'}$. Thus, there exist constants $c_4(k)$ and $c_5(k)$ (independent of t) such that

$$|\nabla^k p_*^{II}(x,t)| \le c_4(k)\int_{\operatorname{supp}\nabla\psi}|p_*(y,t)|\,dy + c_5(k)$$

for $k=0,1,\dots$ and $x\in\overline{\Omega'}$. This implies:

$$\int_{t_1}^{t_2}\Big[\max_{x\in\overline{\Omega'}}\big|\nabla^k p_*^{II}(x,t)\big|\Big]^\alpha dt \le c_6(k)\int_{t_1}^{t_2}\bigg(\int_{\operatorname{supp}\nabla\psi}|p_*(y,t)|\,dy\bigg)^\alpha dt + c_7(k)$$

$$\le c_8(k)\int_{t_1}^{t_2}\bigg(\int_{U_\zeta(\overline{\Omega'})}|p_*(y,t)|^\beta\,dy\bigg)^{\alpha/\beta} dt + c_7(k) \qquad (12)$$

where β is chosen so that $2/\alpha+3/\beta=3$. Due to the results of Y. Taniuchi [35](Lemma 5.1) and H. Kozono [17] (Lemma 7.2), $p_*\in L^\alpha(t_1,t_2;\,L^\beta(U_\zeta(\overline{\Omega'})))$ for $1<\alpha<2$, $\frac{3}{2}<\beta<3$ such that $2/\alpha+3/\beta=3$. Hence the last integral in (12) is finite. Thus, the statement of the lemma about p_* is proved. The statement about $\partial v/\partial t$ now easily follows from the Navier–Stokes equation (1). □

We will often use the spaces $L^{a,b}(B_2\times\,]t_0-\tau,t_0[\,)\equiv L^a(t_0-\tau,t_0;\,L^b(B_2))$. In order to avoid complicated expressions, we shall abbreviate their denotation to $L^{a,b}$. We shall also denote by $|||\,.\,|||_{a,b}$ the norm in $L^{a,b}$. $|||\,.\,|||_{(\infty,2)\cap(2,6)}$ will mean the sum $|||\,.\,|||_{\infty,2}+|||\,.\,|||_{2,6}$. Analogously, $\|\,.\,\|_k$ will denote the norm in $L^k(B_2)$.

The restrictions of functions defined a.e. in Q_T to subsets of Q_T will be denoted by the same letters. Thus, for example, $v \in L^\infty(t_0-\tau, t_0; W^{1,2}(B_2)^3)$ is the statement about the restriction of v to $B_2\times\,]t_0-\tau, t_0[$.

Lemma 3. *If* $g \in L^{\infty,2}\cap L^{2,6}$, $2\le a\le +\infty$, $0\le b\le 6$ *and* $\frac{3}{2}\le 2/a+3/b\le\frac{5}{2}$ *then*

$$\|\!|g|\!\|_{a,b} \;\le\; \|\!|g|\!\|_{2,2}^{2/a+3/b-3/2}\;\|\!|g|\!\|_{(\infty,2)\cap(2,6)}^{5/2-(2/a+3/b)} \qquad (\text{if } b\ge 2), \tag{13}$$

$$\|\!|g|\!\|_{a,b} \;\le\; \|\!|g|\!\|_{2,2}^{2/a}\;\|\!|g|\!\|_{(\infty,2)\cap(2,6)}^{1-(2/a)} \qquad (\text{if } b< 2). \tag{14}$$

Proof. We have

$$\begin{aligned}
\|\!|g|\!\|_{a,b} &= \left[\int_{t_0-\tau}^{t_0}\left(\int_{B_2}|g|^\alpha\,|g|^\beta\,|g|^{b-\alpha-\beta}\,dx\right)^{a/b} dt\right]^{1/a}\\
&\le \left[\int_{t_0-\tau}^{t_0}\left(\int_{B_2}|g|^2\,dx\right)^{\alpha a/2b}\left(\int_{B_2}|g|^2\,dx\right)^{\beta a/2b}\left(\int_{B_2}|g|^6\,dx\right)^{\frac{b-\alpha-\beta}{6}\,\frac{a}{b}} dt\right]^{1/a}\\
&\le \|\!|g|\!\|_{\infty,2}^{\alpha/b}\left[\int_{t_0-\tau}^{t_0}\left(\int_{B_2}|g|^2\,dx\right)^{a\beta/2b}\left(\int_{B_2}|g|^6\,dx\right)^{\frac{b-\alpha-\beta}{6}\,\frac{a}{b}} dt\right]^{1/a}\\
&\le \|\!|g|\!\|_{\infty,2}^{\alpha/b}\left[\int_{t_0-\tau}^{t_0}\int_{B_2}|g|^2\,dx\,dt\right]^{\beta/2b}\left[\int_{t_0-\tau}^{T}\left(\int_{B_2}|g|^6\,dx\right)^{1/3} dt\right]^{\frac{1}{2}\,(1-\alpha/b-\beta/b)}\\
&\le \|\!|g|\!\|_{\infty,2}^{\alpha/b}\,\|\!|g|\!\|_{2,2}^{\beta/b}\,\|\!|g|\!\|_{2,6}^{1-\alpha/b-\beta/b} \le \|\!|g|\!\|_{2,2}^{\beta/b}\,\|\!|g|\!\|_{(\infty,2)\cap(2,6)}^{1-\beta/b}
\end{aligned}\tag{15}$$

where $0\le\alpha\le 2,\ 0\le\beta\le 2,\ 0\le b-\alpha-\beta\le 6,$

$$\frac{\alpha}{2}+\frac{\beta}{2}+\frac{b-\alpha-\beta}{6}\le 1 \qquad\text{and}\qquad \frac{a\beta}{2b}+\frac{a(b-\alpha-\beta)}{2b}=1.$$

The last equation gives $\alpha = b(a-2)/a$. Substituting this value to the previous inequalities, we obtain the two conditions

$$\frac{2}{a}-\frac{6}{b}\le\frac{\beta}{b}\le\frac{2}{a} \qquad\text{and}\qquad \frac{\beta}{b}\le\frac{2}{a}+\frac{3}{b}-\frac{3}{2}.$$

These conditions are fulfilled if we choose

$$\frac{\beta}{b}=\frac{2}{a}+\frac{3}{b}-\frac{3}{2}\quad(\text{if } b\ge 2) \qquad\text{and}\qquad \frac{\beta}{b}=\frac{2}{a}\quad(\text{if } b<2).$$

If we substitute this into (15), we obtain (13) and (14). □

We will now localize the Navier–Stokes initial–boundary value problem to B_2 in the spatial variables. We will therefore use an infinitely differentiable cut–off function η such that $\eta=0$ on $\mathbb{R}^3-B_4$, $\eta=1$ on $\overline{B_3}$ and $0\le\eta\le 1$ on $\overline{B_4}-B_3$. Since the product ηv does not satisfy the equation of continuity, we put $u=\eta v - V$ where V is an appropriate function such that $\operatorname{div} V = \operatorname{div}(\eta v) = \nabla\eta\cdot v$. The existence of function V is given by this lemma:

Lemma 4. *If $\mathcal{D}$ is a bounded Lipschitz domain in $\mathbb{R}^3$, $1 < a < +\infty$ and $m \in (\mathbb{N} \cup \{0\})$ then there exists a linear operator R from $W_0^{m,a}(\mathcal{D})$ into $W_0^{m+1,a}(\mathcal{D})^3$ with the properties:*

1. $\operatorname{div} Rf = f$ *for all* $f \in W_0^{m,a}(\mathcal{D})$ *such that* $\int_{\mathcal{D}} f\, dx = 0$,
2. $\exists\, c_9 > 0 : \|\nabla^{m+1} Rf\|_{L^a(\mathcal{D})^{3m+2}} \leq c_9 \, \|\nabla^m f\|_{L^a(\mathcal{D})^{3m}}$ *for all* $f \in W_0^{m,a}(\mathcal{D})$.

Lemma 4 follows from G. P. Galdi [10] (Theorem 3.2, Chap. III.3) and from W. Borchers and H. Sohr [3] (Theorem 2.4).

We set $V(.\,,t) = R(\nabla\eta \cdot v(.\,,t))$ where R is the operator from Lemma 4 (with $\mathcal{D} = B_2$). It follows from the proof of Lemma 4 (see W. Borchers and H. Sohr [3], pp. 73–76) that since $\nabla\eta$ has a compact support in $B_2 - \overline{B_1}$, $V(.\,,t)$ also has a compact support in $B_2 - \overline{B_1}$. Since

$$\int_{B_2} \nabla\eta \cdot v \, dx \;=\; \int_{B_2} \operatorname{div}(\eta v)\, dx \;=\; \int_{\partial B_2} \eta v \cdot n \, dS \;=\; 0$$

(where n is the outer normal vector to ∂B_2), $\operatorname{div} V = \nabla\eta \cdot v$ in $B_2 \times\,]t_0 - \tau, t_0 + \tau[$. Moreover, it follows from item 2 of Lemma 4 and the smoothness of v on $\operatorname{supp}(\nabla\eta) \times [t_0 - \tau, t_0]$ (see item 4 of Lemma 1) that all spatial derivatives of V are bounded on $\overline{B_2} \times [t_0 - \tau, t_0]$.

It can be verified that u $(= \eta v - V)$ satisfies in a strong sense the equations

$$\frac{\partial u}{\partial t} + (u \cdot \nabla) u \;=\; h \;-\; \nabla(\eta\, p_*) \;+\; \nu\, \Delta u \tag{16}$$

$$\operatorname{div} u \;=\; 0 \tag{17}$$

where

$$\begin{aligned} h \;=\; & -\frac{\partial V}{\partial t} - V \cdot \nabla(\eta\, v) - (\eta\, v) \cdot \nabla V + V \cdot \nabla V + (\eta\, v \cdot \nabla\eta)\, v \\ & - \eta\,(1 - \eta)\, v \cdot \nabla v - 2\nu\, \nabla\eta \cdot \nabla v - \nu\, v\, \Delta\eta + \nu\, \Delta V + p_*\, \nabla\eta. \end{aligned}$$

u satisfies the initial condition

$$u(.\,, t_0 - \tau) \;=\; \eta\, v(.\,, t_0 - \tau) \;-\; V(.\,, t_0 - \tau) \tag{18}$$

and the boundary condition

$$u = 0 \tag{19}$$

on $\partial B_2 \times\,]t_0 - \tau, t_0[$. Moreover, since $\eta\, v(.\,,t)$ and $V(.\,,t)$ have a compact support in B_2 for all $t \in\,]t_0 - \tau, t_0[$, u has all derivatives equal to zero on $\partial B_2 \times\,]t_0 - \tau, t_0[$.

If we use Lemma 2 with $\Omega' = B_2 - \overline{B_1}$ and $[t_1, t_2] = [t_0 - \tau, t_0]$ and take into account that $\operatorname{supp} \nabla\eta \subset B_2 - \overline{B_1}$, we can observe that $\nabla\eta \cdot \partial v / \partial t$ has all spatial derivatives in $L^\alpha(t_0 - \tau, t_0;\, L^\infty(B_2))$ for each α between 1 and 2. Thus, $\partial V / \partial t$ has all spatial derivatives in $L^\alpha(t_0 - \tau, t_0;\, L^\infty(B_2)^3)$, too. The same also holds about $p_*\, \nabla\eta$. All other terms in function h have all spatial derivatives bounded in $B_2 \times$

$[t_0 - \tau, t_0]$. We can therefore conclude that function h has all its spatial derivatives in $L^\alpha(t_0 - \tau, t_0;\ L^\infty(B_2)^3)$ for every $\alpha \in]1, 2[$. Moreover, h has a compact support in $\big(B_2 - \overline{B_1}\big) \times [t_0 - \tau, t_0]$.

The components of u will be denoted by u_1, u_2 and u_3. Partial derivatives of u_i with respect to x_j will be denoted by $u_{i,j}$. All these partial derivatives (for $i, j = 1, 2, 3$) belong to $L^{2,2}$. $\omega = (\omega_1, \omega_2, \omega_3)$ will denote $\operatorname{curl} u$.

Lemma 5. *Suppose that* $u_3 \in L^{r,s}$ *for* $2 \le r \le +\infty,\ 3 < s \le +\infty,\ 2/r + 3/s \le 1$. *Then there exist positive constants* c_{10} *and* c_{11} *such that*

$$|||\omega_3|||_{(\infty,2)\cap(2,6)} \le c_{10}\, |||u_3|||_{r,s}\, |||\nabla u|||_{(\infty,2)\cap(2,6)}^{2/r+3/s} + c_{11}\,. \tag{20}$$

Proof. Suppose that $t \in]t_0 - \tau, t_0[$ is a fixed instant of time at first. Applying operator curl to equation (16), we obtain a vector equation for ω. Its third component is

$$\frac{\partial \omega_3}{\partial t} + u \cdot \nabla \omega_3 = h_{2,1} - h_{1,2} + \omega \cdot \nabla u_3 + \nu\, \Delta \omega_3\,. \tag{21}$$

Multiplying equation (21) by ω_3, integrating on $B_2 \times]t_0 - \tau, t_0[$ and applying the integration by parts, we can obtain

$$\frac{1}{2}\frac{d}{dt}\int_{B_2} \omega_3^2\, dx + \nu \int_{B_2} |\nabla \omega_3|^2\, dx = \int_{B_2} (h_{2,1} - h_{1,2})\, \omega_3\, dx - \int_{B_2} \omega\, u_3 \cdot \nabla \omega_3\, dx,$$

$$\frac{1}{2}\frac{d}{dt}\int_{B_2} \omega_3^2\, dx + \nu \int_{B_2} |\nabla \omega_3|^2\, dx \le c_{12} + c_{13} \int_{B_2} |\omega|^2\, u_3^2\, dx,$$

$$\begin{aligned} |||\omega_3|||^2_{(\infty,2)\cap(2,6)} &\le \int_{B_2} \omega_3^2(\,.\,, t_0 - \tau)\, dx + c_{14} + c_{15} \int_{t_0-\tau}^{t_0} \int_{B_2} |\omega|^2\, u_3^2\, dx \\ &\le \|\omega_3(\,.\,, t_0 - \tau)\|_2^2 + c_{14} + c_{15}\, |||u_3|||^2_{r,s}\, |||\omega|||^2_{a,b} \\ &\le \|\omega_3(\,.\,, t_0 - \tau)\|_2^2 + c_{14} + c_{16}\, |||u_3|||^2_{r,s}\, |||\nabla u|||^2_{a,b} \end{aligned}$$

where

$$\frac{2}{r} + \frac{2}{a} = 1, \qquad \frac{2}{s} + \frac{2}{b} = 1 \qquad \text{and} \qquad \frac{2}{a} + \frac{3}{b} = \frac{5}{2} - \left(\frac{2}{r} + \frac{3}{s}\right).$$

We have used the fact that h is supported in $\big(B_2 - \overline{B_1}\big) \times [t_0 - \tau, t_0]$ where ω_3 is bounded. Applying now Lemma 3 to ∇u, we obtain estimate (20). □

Lemma 6. *If* $0 < a \le +\infty,\ 0 < b < +\infty,\ \frac{1}{2} \le 2/a + 3/b \le \frac{3}{2}$ *then there exists* $c_{17} > 0$ *such that*

$$|||u|||_{a,b} \le c_{17}\, |||\nabla u|||_{(\infty,2)\cap(2,6)}^{3/2-(2/a+3/b)}\,. \tag{22}$$

Proof. Suppose that $b \geq 6$ at first. Then we can apply Lemma 3 and we obtain

$$|||u|||_{a,b} \leq c_{17}\, |||\nabla u|||_{a,c} \leq c_{17}\, |||\nabla u|||^{5/2-(2/a+3/c)}_{(\infty,2)\cap(2,6)} = c_{17}\, |||\nabla u|||^{3/2-(2/a+3/b)}_{(\infty,2)\cap(2,6)}$$

where $3/c = 1 + 3/b$. If $b < 6$ then

$$|||u|||_{a,b} \leq |||u|||^{A}_{a,b'}\, |||u|||^{B}_{a,6}$$

where

$$\frac{2}{a} + \frac{3}{b'} = \frac{3}{2}, \qquad A = \frac{b'(6-b)}{b(6-b')}, \qquad B = \frac{6(b-b')}{b(6-b')}.$$

Then

$$|||u|||_{a,b} \leq c_{17}\, |||u|||^{B}_{a,6} \leq c_{17}\, |||\nabla u|||^{B}_{a,2} \leq c_{17}\, |||\nabla u|||^{B(1-2/a)}_{(\infty,2)\cap(2,6)}$$

(by Lemma 3). Easy calculations show that the exponent $B(1-2/a)$ is equal to $\frac{3}{2} - (2/a + 3/b)$. □

3. Anisotropic conditions for regularity

The main result of this section is the next theorem:

Theorem 1. *Suppose that $(v; p_*)$ is a suitable weak solution of the problem (1)–(4), D is an open sub–domain of Q_T and $v = (v_1, v_2, v_3)$. Suppose further that $v_1, v_2 \in L^{p,q}_{loc}(D)$ and $v_3 \in L^{r,s}_{loc}(D)$ where*

(i) $2 \leq p \leq +\infty, \quad 2 \leq q \leq +\infty,$

(ii) $2 \leq r < \infty, \quad 3 \leq s \leq +\infty, \quad \dfrac{2}{r} + \dfrac{3}{s} \leq 1,$

(iii) $\left(\dfrac{2}{p} + \dfrac{3}{q}\right) + \left(\dfrac{2}{r} + \dfrac{3}{s}\right) \leq 2,$

(iv) $\dfrac{2}{p} + \dfrac{2}{r} \leq 1, \quad \dfrac{2}{q} + \dfrac{2}{s} < 1.$

Then $(v; p_*)$ has no singular points in D.

Remark 1. v, as a weak solution, belongs to $L^{\infty,2}(D)^3 \cap L^{2,6}(D)^3$. This implies, by interpolation, that $v \in L^{p,q}(D)^3$ for all $2 \leq p \leq +\infty$, $2 \leq q \leq 6$ such that

$$\frac{2}{p} + \frac{3}{q} \leq \frac{3}{2}. \tag{23}$$

Thus, we can assume without loss of generality that numbers p and q in Theorem 1 satisfy inequality (23). If r and s satisfy the condition

$$\frac{2}{r} + \frac{3}{s} \leq \frac{1}{2}.$$

(which is identical with the corresponding condition from [27]) then we can always find p, q satisfying (23) with "=" instead of "≤" and also satisfying all assumptions of Theorem 1. Thus, Theorem 1 represents a generalization of the theorem from [27].

Remark 2. On the other hand, if

$$\frac{2}{r} + \frac{3}{s} = 1$$

then (iii) expresses the requirement that p and q satisfy

$$\frac{2}{p} + \frac{3}{q} \leq 1\,, \tag{24}$$

i.e. we obtain the usual Prodi–Serrin conditions on all components of velocity. Condition (iv) is in this particular situation rather restrictive and it can be shown that it is not necessary, however we need it in a general case – see Step 3 of the proof of Theorem 1.

Remark 3. Assume that a suitable weak solution $(v; p_*)$ of the problem (1)–(4) develops a singularity at time t_0. Then Theorem 1 says that all the components of v must blow up at time t_0: one in the $L^{r,s}$–norm and two in the $L^{p,q}$–norm (with appropriate r, s, p, q).

Proof of Theorem 1. We can assume without loss of generality that condition (iii) holds with "=" instead of "≤", i.e.

$$\left(\frac{2}{p} + \frac{3}{q}\right) + \left(\frac{2}{r} + \frac{3}{s}\right) = 2. \tag{25}$$

Then (ii) and (23) imply that

$$\frac{1}{2} \leq \frac{2}{r} + \frac{3}{s} \leq 1, \qquad 1 \leq \frac{2}{p} + \frac{3}{q} \leq \frac{3}{2}\,. \tag{26}$$

Assume that D', t_0, x_0, τ, B_1, B_2, u, ω, $L^{a,b}$, $|||\,.\,|||_{a,b}$, etc., have the same meaning as in Section 2. We will denote

$$\mu_1 = |||u_1|||_{p,q}, \qquad \mu_2 = |||u_2|||_{p,q}, \qquad \mu_3 = |||u_3|||_{r,s}\,.$$

Equation (16) can be written, as in [27], in the form

$$\frac{\partial u}{\partial t} + F_1(u) + \ldots + F_6(u) = h - \nabla(\eta\, p_* + \tfrac{1}{2}\, u_3^2) + \nu\, \Delta u \tag{27}$$

where the terms $F_1(u)$, …, $F_6(u)$ have the following meaning:

$$\begin{aligned}
F_1(u) &= (\omega_2\, u_3\,,\, -\omega_1\, u_3\,,\, 0),\\
F_2(u) &= (-\omega_3\, u_2\,,\, \omega_3\, u_1\,,\, 0),\\
F_3(u) &= (u_1\, u_{1,1}\,,\, u_1\, u_{1,2}\,,\, 0),\\
F_4(u) &= (u_2\, u_{2,1}\,,\, u_2\, u_{2,2}\,,\, 0),\\
F_5(u) &= (0\,,\, 0\,,\, u_1\, u_{3,1}),\\
F_6(u) &= (0\,,\, 0\,,\, u_2\, u_{3,2}).
\end{aligned}$$

We will now estimate the $L^{2,2}$–norms of $F_1(u), \ldots, F_6(u)$. c_{18} and c_{19} will be generic constants – i.e. constants whose values may change from line to line. They can depend on μ_1, μ_2 and μ_3. δ will be a positive number which will be needed to be sufficiently small later. Constant c_{19} will depend on δ and we shall stress this by writing $c_{19}(\delta)$.

Step 1. $\quad |||F_1(u)|||_{2,2}^2 = \displaystyle\int_{t_0-\tau}^{t_0}\int_{B_2}(\omega_1^2+\omega_2^2)\,u_3^2\,dx\,dt \le |||u_3|||_{r,s}^2\ |||\nabla u|||_{a,b}^2$

where

$$\frac{2}{r}+\frac{2}{a}=1\,, \qquad \frac{2}{s}+\frac{2}{b}=1 \qquad \text{and} \qquad \frac{2}{a}+\frac{3}{b}=\frac{5}{2}-\left(\frac{2}{r}+\frac{3}{s}\right)\ge\frac{3}{2}\,.$$

Applying Lemma 3 to the norm of ∇u, we obtain

$$|||F_1(u)|||_{2,2}^2 \le \mu_3^2\ c_{18}\ |||\nabla u|||_{(\infty,2)\cap(2,6)}^{2(2/r+3/s)}\,.$$

Step 2. Assume for simplicity that $p<+\infty$. Then

$$|||F_2(u)|||_{2,2}^2$$

$$= \int_{t_0-\tau}^{t_0}\int_{B_2}\omega_3^2\,(u_1^2+u_2^2)\,dx\,dt \le |||\omega_3|||_{a,b}^2\ \left(|||u_1|||_{\alpha,\beta}^2+|||u_2|||_{\alpha,\beta}^2\right) \tag{28}$$

where

$$\frac{1}{a}+\frac{1}{\alpha}=\frac{1}{2}\,, \qquad \frac{1}{b}+\frac{1}{\beta}=\frac{1}{2}\,.$$

We can choose $\alpha=p$ and β such that $2/a+3/b=3/2$. Then

$$\frac{1}{a}=\frac{1}{2}-\frac{1}{p}\,, \qquad \frac{3}{\beta}=1-\frac{2}{p}\,.$$

Applying Lemma 3 and Lemma 5, we obtain:

$$|||\omega_3|||_{a,b}^2 \le c_{18}\ |||\omega_3|||_{(\infty,2)\cap(2,6)}^2 \le c_{18}\,\mu_3^2\ |||\nabla u|||_{(\infty,2)\cap(2,6)}^{2(2/r+3/s)} + c_{18}\,.$$

The norm of u_1 in (28) can be estimated:

$$|||u_1|||_{p,\beta}^2 = \left[\int_{t_0-\tau}^{t_0}\left(\int_{B_2}|u_1|^{\beta\kappa}\,|u_1|^{\beta(1-\kappa)}\,dx\right)^{p/\beta}dt\right]^{2/p}$$

$$\le\ |||u_1|||_{p,q}^{2\kappa}\ |||u_1|||_{p,l}^{2(1-\kappa)} = \mu_1^{2\kappa}\ |||\nabla u_1|||_{p,l}^{2(1-\kappa)} \tag{29}$$

where $0<\kappa\le q/\beta$ (≤ 1),

$$\frac{\kappa}{q}+\frac{1-\kappa}{l}=\frac{1}{\beta}\,, \qquad \text{and} \qquad l<+\infty.$$

Suppose that $2/p+3/q>1$ at first. Then $\beta>q$ and $\kappa<1$. We will apply Lemma 6 to the norm of u_1 in $L^{p,l}$. The condition $\frac{1}{2}\le 2/p+3/l$ leads to

$$2\left(\frac{2}{p}+\frac{3}{q}-\frac{1}{2}\right)\kappa\le 1. \tag{30}$$

The condition $2/p+3/l \leq \frac{3}{2}$ is easily satisfied due to (26). Lemma 3 now gives

$$|||u_1|||_{p,l}^{2(1-\kappa)} \leq c_{18}\, |||\nabla u_1|||_{(\infty,2)\cap(2,6)}^{2(1-\kappa)\,(3/2-2/p-3/l)}.$$

Hence

$$|||\omega_3|||_{a,b}^2\, |||u_1|||_{p,\beta}^2 \leq c_{18}\, \mu_1^{2\kappa}\, \mu_3^2\, |||\nabla u|||_{(\infty,2)\cap(2,6)}^{d_1} + c_{18}\, \mu_1^{2\kappa}\, |||\nabla u|||_{(\infty,2)\cap(2,6)}^{d_2} \tag{31}$$

where

$$\begin{aligned} d_1 &= 2\left(\frac{2}{r}+\frac{3}{s}\right)+2\,(1-\kappa)\left(\frac{3}{2}-\frac{2}{p}-\frac{3}{l}\right), \\ d_2 &= 2\,(1-\kappa)\left(\frac{3}{2}-\frac{2}{p}-\frac{3}{l}\right). \end{aligned}$$

It can be verified that $0 \leq d_1 \leq 2$ if and only if

$$\kappa\left[\frac{3}{2}-\left(\frac{2}{p}+\frac{3}{q}\right)\right] \geq \left(\frac{2}{r}+\frac{3}{q}\right)-\frac{1}{2}. \tag{32}$$

Standard calculations show that due to (25), there exists $\kappa \in]0, q/\beta]$ which satisfies the conditions (30) and (32).

If $2/p+3/q=1$ then $\beta=q$, we can choose $\kappa=1$ and we can obtain (31) with $d_1=2$.

The analogous estimate to (31) can also be derived with u_2 instead of u_1. Substituting these estimates into (28) and interpolating the power with the exponent d_1 between 2 and d_2, we obtain

$$|||F_2(u)|||_{2,2}^2 \leq c_{18}\,(\mu_1^{2\kappa}+\mu_2^{2\kappa})\,\left[\mu_3^2\, |||\nabla u|||_{(\infty,2)\cap(2,6)}^2 + |||\nabla u|||_{(\infty,2)\cap(2,6)}^{d_2}\right]$$

where $0 \leq d_2 < 2$. The same inequality can be derived if $p=+\infty$, too.

Step 3.

$$\begin{aligned} |||F_3(u)|||_{2,2}^2 &= \int_{t_0-\tau}^{t_0}\int_{B_2}(u_1^2\, u_{1,1}^2+u_1^2\, u_{1,2}^2)\, dx\, dt \\ &= \tfrac{1}{3}\int_{t_0-\tau}^{t_0}\int_{B_2}[(u_1^3)_{,1}\, u_{1,1}+(u_1^3)_{,2}\, u_{1,2}]\, dx\, dt \\ &= -\tfrac{1}{3}\int_{t_0-\tau}^{t_0}\int_{B_2}[u_1^3\, u_{1,11}+u_1^3\, u_{1,22}]\, dx\, dt. \end{aligned}$$

Using the equation $u_{1,11}+u_{1,22}=-u_{3,31}-\omega_{3,2}$ which follows immediately from the equation of continuity (13) and the definition of ω_3, we get

$$\begin{aligned}|||F_3(u)|||_{2,2}^2 &= \tfrac{1}{3}\int_{t_0-\tau}^{t_0}\int_{B_2}[u_1^3\,u_{3,31}+u_1^3\,\omega_{3,2}]\,dx\,dt \\ &= -\int_{t_0-\tau}^{t_0}\int_{B_2}[u_1^2\,u_{1,1}\,u_{3,3}+u_1^2\,u_{1,2}\,\omega_3]\,dx\,dt = \int_{t_0-\tau}^{t_0}\int_{B_2}2\,u_1\,u_{1,3}\,u_{1,1}\,u_3\,dx\,dt \\ &+ \int_{t_0-\tau}^{t_0}\int_{B_2}u_1^2\,u_{1,13}\,u_3\,dx\,dt - \int_{t_0-\tau}^{t_0}\int_{B_2}u_1^2\,u_{1,2}\,\omega_3\,dx\,dt.\end{aligned}$$

Let us denote

$$\begin{aligned}I_1 &= \int_{t_0-\tau}^{t_0}\int_{B_2}2\,u_1\,u_{1,3}\,u_{1,1}\,u_3\,dx\,dt,\\ I_2 &= \int_{t_0-\tau}^{t_0}\int_{B_2}u_1^2\,u_{1,13}\,u_3\,dx\,dt,\\ I_3 &= \int_{t_0-\tau}^{t_0}\int_{B_2}u_1^2\,u_{1,2}\,\omega_3\,dx\,dt.\end{aligned}$$

The absolute values of the integrals I_1, I_2 and I_3 can be successively estimated in this way:

$$|I_1| \leq c_{18}\,|||u_3|||_{r,s}\,|||\nabla u_1|||_{a,b}^2\,|||u_1|||_{p,q} \tag{33}$$

where

$$\frac{2}{a}+\frac{1}{p}+\frac{1}{r}=1 \qquad \text{and} \qquad \frac{2}{b}+\frac{1}{q}+\frac{1}{s}=1. \tag{34}$$

(34) and (25) imply that

$$\frac{2}{a}+\frac{3}{b}=\frac{3}{2}. \tag{35}$$

Estimating the norm of ∇u_1 by means of Lemma 3, we obtain

$$|||\nabla u_1|||_{a,b}^2 \leq c_{18}\,|||\nabla u_1|||_{(\infty,2)\cap(2,6)}^2\,. \tag{36}$$

Substituting (36) into (33), we get

$$|I_1| \leq c_{18}\,\mu_1\,\mu_3\,|||\nabla u_1|||_{(\infty,2)\cap(2,6)}^2\,. \tag{37}$$

The estimation of the integral I_2 is the point where we need assumption (iv) of Theorem 1. Applying the Hölder inequality, we obtain

$$\begin{aligned}|I_2| &\leq \delta\int_{t_0-\tau}^{t_0}\int_{B_2}|u_{1,13}|^2\,dx\,dt + c_{19}(\delta)\int_{t_0-\tau}^{t_0}\int_{B_2}u_1^4\,u_3^2\,dx\,dt\\ &\leq \delta\,|||\nabla^2 u|||_{2,2}^2 + c_{19}(\delta)\,|||u_1|||_{p,q}^2\,|||u_3|||_{r,s}^2\,|||u_1|||_{a,b}^2\end{aligned} \tag{38}$$

where

$$\frac{2}{p} + \frac{2}{r} + \frac{2}{a} = 1, \qquad \text{and} \qquad \frac{2}{q} + \frac{2}{s} + \frac{2}{b} = 1. \tag{39}$$

(Numbers a and b are not identical with a and b from (33)–(36).) These conditions and assumption (iv) imply that $a \geq 2$ and $2 \leq b < +\infty$. An elementary calculation gives

$$2\left(\frac{2}{a} + \frac{3}{b}\right) = 5 - 2\left(\frac{2}{p} + \frac{3}{q}\right) - 2\left(\frac{2}{r} + \frac{3}{s}\right) = 1. \tag{40}$$

We can now apply Lemma 6 to the norm of u_1 in $L^{a,b}$ and we obtain

$$|||u_1|||_{a,b}^2 \leq c_{18}\, |||\nabla u_1|||_{(\infty,2)\cap(2,6)}^{2(3/2-2/a-3/b)} = c_{18}\, |||\nabla u_1|||_{(\infty,2)\cap(2,6)}^2 \,. \tag{41}$$

Inequalities (38) and (41) imply

$$|I_2| \leq \delta\, |||\nabla^2 u|||_{2,2}^2 + c_{19}(\delta)\, \mu_1^2\, \mu_3^2\, |||\nabla u|||_{(\infty,2)\cap(2,6)}^2 \,. \tag{42}$$

Finally, I_3 can be estimated as follows:

$$\begin{aligned} |I_3| \ &\leq\ \frac{1}{2} \int_{t_0-\tau}^{t_0} \int_{B_2} (u_1^2\, u_{1,2}^2 + u_1^2\, \omega_3^2)\, dx\, dt \\ &\leq\ \tfrac{1}{2}\, |||F_3(u)|||_{2,2}^2 + \frac{1}{2} \int_{t_0-\tau}^{t_0} \int_{B_2} u_1^2\, \omega_3^2\, dx\, dt\,. \end{aligned}$$

The last integral on the right–hand side is of the same type as the integral in (28). Thus, proceeding in the same way as in the estimation of $|||F_2(u)|||_{2,2}^2$, we obtain:

$$\begin{aligned} |I_3| &\leq \tfrac{1}{2}\, |||F_3(u)|||_{2,2}^2 \\ &\quad + c_{18}\, (\mu_1^{2\kappa} + \mu_2^{2\kappa})\, \big[\, \mu_3^2\, |||\nabla u|||_{(\infty,2)\cap(2,6)}^2 + |||\nabla u|||_{(\infty,2)\cap(2,6)}^{d_2} \big]. \end{aligned} \tag{43}$$

Estimates (37), (42) and (43) yield

$$\begin{aligned} |||F_3(u)|||_{2,2}^2 \ &\leq\ \delta\, |||\nabla^2 u|||_{2,2}^2 + c_{18}\, \mu_1\, \mu_2\, |||\nabla u|||_{(\infty,2)\cap(2,6)}^2 \\ &\quad + c_{19}(\delta)\, \mu_1^2\, \mu_2^2\, |||\nabla u|||_{(\infty,2)\cap(2,6)}^2 \\ &\quad + c_{18}\, (\mu_1^{2\kappa} + \mu_2^{2\kappa})\, \big[\, \mu_3^2\, |||\nabla u|||_{(\infty,2)\cap(2,6)}^2 + |||\nabla u|||_{(\infty,2)\cap(2,6)}^{d_2} \big]. \end{aligned}$$

The norm $|||F_4(u)|||_{2,2}$ can be estimated in the same way as the norm $|||F_3(u)|||_{2,2}$.

Step 4.

$$
\begin{aligned}
|||F_5(u)|||_{2,2}^2 &= \int_{t_0-\tau}^{t_0}\int_{B_2} u_1^2\, u_{3,1}^2\, dx \;=\; -2\int_{t_0-\tau}^{t_0}\int_{B_2} u_1\, u_{1,1}\, u_3\, u_{3,1}\, dx\, dt \\
&\quad - \int_{t_0-\tau}^{t_0}\int_{B_2} u_1^2\, u_{3,11}\, u_3\, dx\, dt \;=\; \int_{t_0-\tau}^{t_0}\int_{B_2} u_{1,1}^2\, u_3^2\, dx\, dt \\
&\quad + \int_{t_0-\tau}^{t_0}\int_{B_2} u_1\, u_{1,11}\, u_3^2\, dx\, dt \;-\; \int_{t_0-\tau}^{t_0}\int_{B_2} u_1^2\, u_{3,11}\, u_3\, dx\, dt.
\end{aligned}
$$

Let us denote the last three integrals by I_4, I_5 and I_6. They can be successively estimated in this way:

$$|I_4| \leq |||u_3|||_{r,s}^2\; |||\nabla u|||_{a,b}^2$$

where

$$\frac{2}{r}+\frac{2}{a}=1\,, \qquad \frac{2}{s}+\frac{2}{b}=1 \qquad \text{and} \qquad \frac{2}{a}+\frac{3}{b}=\frac{5}{2}-\left(\frac{2}{r}+\frac{3}{s}\right)\geq\frac{3}{2}\,.$$

Hence we can apply Lemma 3 and we get

$$|I_4| \leq c_{18}\,\mu_3^2\; |||\nabla u|||_{L^{\infty,2}\cap L^{2,6}}^{2(2/r+3/s)}\,.$$

Further, we have

$$
\begin{aligned}
|I_5| &\leq \delta\, |||u_{1,11}|||_{2,2}^2 + c_{19}(\delta)\int_{t_0-\tau}^{t_0}\int_{B_2} u_1^2\, u_3^4\, dx\, dt \\
&\leq \delta\, |||\nabla^2 u|||_{2,2}^2 + c_{19}(\delta)\, |||u_3|||_{r,s}^4\; |||u_1|||_{a,b}^2
\end{aligned}
$$

where

$$\frac{4}{r}+\frac{2}{a}=1\,, \qquad \frac{4}{s}+\frac{2}{b}=1 \qquad \text{and} \qquad \frac{2}{a}+\frac{3}{b}=\frac{5}{2}-2\left(\frac{2}{r}+\frac{3}{s}\right)\leq\frac{3}{2}\,.$$

Hence

$$|I_5| \leq \delta\, |||\nabla^2 u|||_{2,2}^2 + c_{19}(\delta)\,\mu_3^4\,.$$

I_6 can be estimated in the same way as I_3. Thus, we have

$$
\begin{aligned}
|||F_5(u)|||_{2,2}^2 &\leq c_{18}\,\mu_3^2\; |||\nabla u|||_{(\infty,2)\cap(2,6)}^{2(2/r+3/s)} + \delta\, |||\nabla^2 u|||_{2,2}^2 + c_{19}(\delta)\,\mu_3^4 \\
&\quad + \tfrac{1}{2}\, |||F_3(u)|||_{2,2}^2 + c_{18}\left(\mu_1^{2\kappa}+\mu_2^{2\kappa}\right)\left[\mu_3^2\; |||\nabla u|||_{(\infty,2)\cap(2,6)}^2 + |||\nabla u|||_{(\infty,2)\cap(2,6)}^{d_2}\right].
\end{aligned}
$$

The norm $|||F_6(u)|||_{2,2}$ can be estimated in the same way as $|||F_5(u)|||_{2,2}$.

Step 5. The derived estimates of $F_1(u),\ \dots,\ F_6(u)$ give:

$$
\begin{aligned}
|||F_1(u)+\dots+F_6(u)|||_{2,2}^2 &\leq c_{18}\, \|h\|_{2,2}^2 + \delta\, |||\nabla^2 u|||_{2,2}^2 + c_{18}\,\mu_3^2\; |||\nabla u|||_{(\infty,2)\cap(2,6)}^{2(2/r+3/s)} \\
&\quad + \left[c_{18}\,(\mu_1+\mu_2)\,\mu_3 + c_{19}(\delta)\,(\mu_1^2+\mu_2^2)\,\mu_3^2 + (\mu_1^{2\kappa}+\mu_2^{2\kappa})\,\mu_3^2\right] |||\nabla u|||_{(\infty,2)\cap(2,6)}^2 \\
&\quad + c_{18}\left(\mu_1^{2\kappa}+\mu_2^{2\kappa}\right) |||\nabla u|||_{(\infty,2)\cap(2,6)}^{d_2} + c_{19}(\delta)\,\mu_3^4\,. \qquad (44)
\end{aligned}
$$

The problem (16)–(19) can be considered as a Stokes problem with the external force $h - F_1(u) - \cdots - F_6(u)$. Its solution leads to the estimate

$$|||\nabla u|||_{\infty,2}^2 + |||\nabla^2 u|||_{2,2}^2 \leq c_{20}\, |||F_1(u) + \ldots + F_6(u)|||_{2,2}^2$$
$$+ c_{21} \int_{t_0-\tau}^{t_0} \int_{B_2} |h \cdot \Delta u| \, dx\, dt + c_{22}\, \|\nabla u(\,.\,, t_0 - \tau)\|_2^2$$

with some positive constants c_{20}–c_{22}. The integral on the right–hand side is finite because h is supported in $\left(\overline{B_2} - B_1\right) \times [t_0 - \tau, t_0]$ where, due to Lemma 2, Δv is bounded and, consequently, Δu is also bounded. If we estimate the norm of $F_1(u) + \cdots + F_6(u)$ by means of (44), we obtain

$$|||\nabla u|||_{\infty,2}^2 + |||\nabla^2 u|||_{2,2}^2 \leq c_{18}\, \|h\|_{2,2}^2 + \delta\, |||\nabla^2 u|||_{2,2}^2 + c_{18}\, \mu_3^2\, |||\nabla u|||_{(\infty,2)\cap(2,6)}^{2(2/r+3/s)}$$
$$+ \left[c_{18}\, (\mu_1 + \mu_2)\, \mu_3 + c_{19}(\delta)\, (\mu_1^2 + \mu_2^2)\, \mu_3^2 + (\mu_1^{2\kappa} + \mu_2^{2\kappa})\, \mu_3^2 \right] |||\nabla u|||_{(\infty,2)\cap(2,6)}^2$$
$$+ c_{18}\, \left(\mu_1^{2\kappa} + \mu_2^{2\kappa}\right) |||\nabla u|||_{(\infty,2)\cap(2,6)}^{d_2} + c_{19}(\delta)\, \mu_3^4 + c_{23} + c_{22}\, \|\nabla u(\,.\,, t_0 - \tau)\|_2^2 .$$

We can choose δ arbitrarily small. Moreover, by choosing τ small enough, we can also make μ_3 to be arbitrarily small. Then the terms with powers $2(2/r + 3/s)$ or 2 on the right hand side can be absorbed by the left hand side and we obtain

$$|||\nabla u|||_{\infty,2}^2 + |||\nabla^2 u|||_{2,2}^2 \leq c_{18}\, \|h\|_{2,2}^2 + c_{18}\, (\mu_1^{2\kappa} + \mu_2^{2\kappa})\, |||\nabla u|||_{(\infty,2)\cap(2,6)}^{d_2}$$
$$+ c_{19}\, \mu_3^4 + c_{23} + c_{22}\, \|\nabla u(\,.\,, t_0 - \tau)\|_2^2 .$$

Thus, since $u = v - V$ on $B_3 \times\,]t_0 - \tau, t_0[$, we have proved that

$$v \in L^\infty(t_0 - \tau, t_0;\, W^{1,2}(B_3)^3). \tag{45}$$

Now we can use e.g. the regularity criterion from [4] (p. 776, Proposition 2) which says that if

$$\lim_{\delta \to 0+} \sup\, \frac{1}{\delta} \int_{t_0-\delta^2}^{t_0} \int_{|x-x_0|<\delta} |\nabla v|^2 \, dx\, dt = 0 \tag{46}$$

then (x_0, t_0) is a regular point of the solution $(v; p_*)$. Since (46) easily follows from (45), we can see that the chosen point (x_0, t_0) is a regular point, indeed. This is the contradiction with the assumption that t_0 is a D'–epoch of irregularity of solution $(v; p_*)$ and (x_0, t_0) is a singular point of $(v; p_*)$ in D'. Thus, solution $(v; p_*)$ has no D'–epoch of irregularity and so it is regular in D'. D' was an arbitrary domain, strictly contained in D. Thus, $(v; p_*)$ has no singular points in D. This completes the proof of Theorem 1. □

4. A geometric regularity criterion

Assume that D is a sub–domain of Q_T, t_0 is a D–epoch of irregularity of a suitable weak solution $(v; p_*)$ of the problem (1)–(4) and x_0 is a point in Ω such that $(x_0, t_0) \in D$. We can localize the problem to the backward neighborhood of point

(x_0, t_0) in the same way as in Section 2. Thus, we put $u = \eta v - V$ where u is a strong solution of the problem (16)–(19) on $B_2 \times]t_0 - \tau, t_0[$. We will use the notation from Section 2, too.

Suppose that $t \in]t_0 - \tau, t_0[$. Multiplying equation (16) by Δu and integrating on B_2, we obtain

$$\frac{d}{dt}\frac{1}{2}\int_{B_2} |\nabla u|^2\, dx + \nu \int_{B_2} |\Delta u|^2\, dx = -\int_{B_2} h \cdot \Delta u\, dx + \int_{B_2} (u \cdot \nabla u) u \cdot \Delta u\, dx$$

$$= -\int_{B_2} h \cdot \Delta u\, dx - \int_{B_2} u_{i,j}\, u_{i,k}\, u_{j,k}\, dx\,. \tag{47}$$

Let us denote $e_{ij} = \frac{1}{2}(u_{i,j} + u_{j,i})$ and $\epsilon_{ij} = \frac{1}{2}(u_{i,j} - u_{j,i})$. (e_{ij}) is the symmetric and (ϵ_{ij}) is the skew–symmetric part of the tensor $(u_{i,j})$. The last integral on the right–hand side of (47) can be written in the form

$$-\int_{B_2} u_{i,j}\, u_{i,k}\, u_{j,k}\, dx = -\int_{B_2} e_{ij}\, u_{i,k}\, u_{j,k}\, dx = -\int_{B_2} e_{ij}\, (e_{ik} + \epsilon_{ik})\, (e_{jk} + \epsilon_{jk})\, dx$$

$$= -\int_{B_2} e_{ij}\, e_{ik}\, e_{jk}\, dx - \int_{B_2} e_{ij}\, e_{ik}\, \epsilon_{jk}\, dx - \int_{B_2} e_{ij}\, \epsilon_{ik}\, e_{jk}\, dx - \int_{B_2} e_{ij}\, \epsilon_{ik}\, \epsilon_{jk}\, dx\,.$$

Using the symmetry of (e_{ij}), the skew–symmetry of (ϵ_{ij}) and interchanging j with k, we obtain:

$$e_{ij}\, e_{ik}\, \epsilon_{jk} = -e_{ij}\, e_{ik}\, \epsilon_{kj} = -e_{ik}\, e_{ij}\, \epsilon_{jk} = -e_{ij}\, e_{ik}\, \epsilon_{jk}\,.$$

Hence $e_{ij}\, e_{ik}\, \epsilon_{jk} = 0$. Similarly, $e_{ij}\, \epsilon_{ik}\, e_{jk} = 0$. Thus,

$$-\int_{B_2} u_{i,j}\, u_{i,k}\, u_{j,k}\, dx = -\int_{B_2} e_{ij}\, e_{ik}\, e_{jk}\, dx - \int_{B_2} e_{ij}\, \epsilon_{ik}\, \epsilon_{jk}\, dx\,. \tag{48}$$

The last integral can be expressed by means of ω_1, ω_2, ω_3 (the components of $\omega = \operatorname{curl} u$):

$$-\int_{B_2} e_{ij}\, \epsilon_{ik}\, \epsilon_{jk}\, dx = \frac{1}{4}\int_{B_2} e_{ij}\, \omega_i\, \omega_j\, dx\,. \tag{49}$$

Hence

$$\frac{d}{dt}\frac{1}{2}\int_{B_2} |\nabla u|^2\, dx + \nu \int_{B_2} |\Delta u|^2\, dx$$

$$= -\int_{B_2} h \cdot \Delta u\, dx - \int_{B_2} e_{ij}\, e_{ik}\, e_{jk}\, dx + \frac{1}{4}\int_{B_2} e_{ij}\, \omega_i\, \omega_j\, dx\,. \tag{50}$$

Applying operator curl to equation (16), we obtain

$$\frac{\partial \omega}{\partial t} + (u \cdot \nabla)\omega = (\omega \cdot \nabla) u + \nu \Delta \omega + \operatorname{curl} h\,.$$

Multiplying this equation by ω and integrating on B_2, we obtain

$$\frac{d}{dt}\frac{1}{2}\int_{B_2} |\omega|^2\, dx + \nu \int_{B_2} |\nabla \omega|^2\, dx = \int_{B_2} e_{ij}\, \omega_i\, \omega_j\, dx + \int_{B_2} \operatorname{curl} h \cdot \omega\, dx\,. \tag{51}$$

We can easily verify that

$$\int_{B_2} |\omega|^2\, dx = \int_{B_2} |\nabla u|^2\, dx \qquad \text{and} \qquad \int_{B_2} |\nabla\omega|^2\, dx = \int_{B_2} |\Delta u|^2\, dx\,.$$

Thus, (51) implies

$$\frac{d}{dt}\frac{1}{2}\int_{B_2} |\nabla u|^2\, dx + \nu \int_{B_2} |\Delta u|^2\, dx = \int_{B_2} e_{ij}\, \omega_i\, \omega_j\, dx + \int_{B_2} \operatorname{curl} h \cdot \omega\, dx\,. \quad (52)$$

Multiplying equation (52) by $-\frac{1}{4}$ and summing with equation (50), we obtain

$$\begin{aligned} &\frac{d}{dt}\frac{3}{8}\int_{B_2} |\nabla u|^2\, dx + \frac{3\nu}{4}\int_{B_2} |\Delta u|^2\, dx \\ &= -\int_{B_2} e_{ij}\, e_{ik}\, e_{jk}\, dx - \int_{B_2} h\cdot\Delta u\, dx - \frac{1}{4}\int_{B_2} \operatorname{curl} h\cdot\omega\, dx \\ &= -\int_{B_2} e_{ij}\, e_{ik}\, e_{jk}\, dx - \frac{5}{4}\int_{B_2} h\cdot\Delta u\, dx\,. \end{aligned} \quad (53)$$

Assume that x is a given point in B_2 for a while. Then the system of coordinates can be chosen so that the tensor (e_{ij}) has a diagonal representation at point x:

$$(e_{ij})(x,t) = \begin{pmatrix} \lambda_1 & 0 & 0 \\ 0 & \lambda_2 & 0 \\ 0 & 0 & \lambda_3 \end{pmatrix}$$

where λ_1, λ_2, λ_3 are the eigenvalues of $(e_{ij})(x,t)$. We can suppose without the loss of generality that

$$\lambda_1 \le \lambda_2 \le \lambda_3\,. \quad (54)$$

Let us denote by e^1, e^2, e^3 the corresponding eigenvectors. The eigenvectors are (or can be chosen to be) orthogonal because (e_{ij}) is symmetric. Due to the equation of continuity (17), we have

$$\lambda_1 + \lambda_2 + \lambda_3 = 0. \quad (55)$$

Then

$$(e_{ij}\, e_{ik}\, e_{jk})(x,t) = \lambda_1^3 + \lambda_2^3 + \lambda_3^3 = 3\,\lambda_1\,\lambda_2\,\lambda_3\,. \quad (56)$$

The product $\lambda_1\,\lambda_2\,\lambda_3$ is an invariant of the tensor $(e_{ij})(x,t)$ and so it is independent of the choice of the system of coordinates. Hence equation (56) holds in all points $x \in B_2$ and we obtain

$$\begin{aligned} \frac{d}{dt}\frac{3}{8}\int_{B_2} |\nabla u|^2\, dx + \frac{3\nu}{4}\int_{B_2} |\Delta u|^2\, dx &= -3\int_{B_2} \lambda_1\,\lambda_2\,\lambda_3\, dx - \frac{5}{4}\int_{B_2} h\cdot\Delta u\, dx\,, \\ &\le -3\int_{B_2} \lambda_1\,\lambda_2\,\lambda_3\, dx + c_{24}\,\|h(.\,,t)\|_1\,. \end{aligned} \quad (57)$$

(The reason why the integral of $h{\cdot}\Delta u$ can be estimated by $c_{24}\,\|h(.\,,t)\|_1$ is explained in the previous section in the part Step 5.) Note that λ_1, λ_2, λ_3 are functions of x and t. Function u can blow up at the point (x_0, t_0) only if the right–hand side of (57) tends to $+\infty$ in a sequence $\{t_n\}$ of instants of time such that $t_n \to t_0-$ as $n \to +\infty$. It is therefore obvious that

a) points (x,t) where $\lambda_1\,\lambda_2\,\lambda_3 \geq 0$ do not contribute to the blow–up

and on the other hand

b) points (x,t) where $\lambda_1\,\lambda_2\,\lambda_3 < 0$ contribute to the blow–up.

This has an easy geometric interpretation:

a) $\lambda_1\,\lambda_2\,\lambda_3 \geq 0$, together with (54), implies that λ_1, $\lambda_2 \leq 0$ and $\lambda_3 \geq 0$. This means that infinitely small volumes of the fluid are compressed in a two–dimensional sub–space of $\mathbb{R}^3$ which is generated by eigenvectors e^1, e^2 and they are stretched in a one–dimensional sub–space, generated by eigenvector e^3.

b) The inequality $\lambda_1\,\lambda_2\,\lambda_3 < 0$ and (54) imply that $\lambda_1 < 0$ and λ_2, $\lambda_3 > 0$. This means that the infinitely small volumes of the fluid are compressed in a one–dimensional sub–space, generated by eigenvector e^1, and they are stretched in a two–dimensional sub–space, generated by eigenvectors e^2 and e^3.

We are now in a position that we can formulate the theorem. Since we wish to formulate it in terms of solution v and not u $(= \eta v - V)$, we will speak about eigenvalues ζ_1, ζ_2 and ζ_3 of the symmetric tensor $\left(\frac{1}{2}\,(v_{i,j} + v_{j,i})\right)$ rather than about eigenvalues λ_1, λ_2, λ_3 of the tensor $(e_{ij}) = \left(\frac{1}{2}\,(u_{i,j} + u_{j,i})\right)$. One can observe that in order to obtain regularity, it is sufficient to control the middle eigenvalue, respectively only its positive part.

Theorem 2. *Suppose that D is an open sub–domain of Q_T, $(v; p_*)$ is a suitable weak solution of the problem (1)–(4), $\zeta_1 \leq \zeta_2 \leq \zeta_3$ are the eigenvalues of the tensor $\left(\frac{1}{2}\,(v_{i,j} + v_{j,i})\right)$ and $\zeta_2 = \zeta_2^I + \zeta_2^{II}$ where*

(v) $\zeta_2^I \in L^{\infty,a}_{loc}(D)$ *for some $a > \frac{3}{2}$ and*

(vi) $(\zeta_2^{II})_+ \in L^{1,+\infty}_{loc}(D)$.

Then the solution $(v; p_*)$ is regular in D.

$(\zeta_2^{II})_+$ denotes the positive part of ζ_2^{II}.

Proof. Suppose that (x_0, t_0) and $u = \eta v - V$ have the same meaning as in Section 2 and as in the previous part of this Section 4. Remind that $B_1 = B_{\epsilon_1}(x_0)$ and $B_2 = B_{\epsilon_2}(x_0)$ where numbers ϵ_1 and ϵ_2 are given by Lemma 1. Then we can derive (57). Eigenvalues ζ_1, ζ_2 and ζ_3 satisfy, due to the equation of continuity and analogously to (55), the equation

$$\zeta_1 + \zeta_2 + \zeta_3 = 0.$$

Thus, since $\zeta_1 \le \zeta_2 \le \zeta_3$, the product $\zeta_1 \zeta_3$ is non–positive. The eigenvalues λ_1, λ_2, λ_3 of (e_{ij}) coincide with the eigenvalues ζ_1, ζ_2, ζ_3 of $\left(\frac{1}{2}(v_{i,j} + v_{j,i})\right)$ on B_1 and λ_1, λ_2, λ_3 are bounded (due to the smoothness of v) on $(B_2 - B_1)\times\,]t_0 - \tau, t_0[$. Hence

$$-3\int_{B_2} \lambda_1\,\lambda_2\,\lambda_3\,dx \le -3\int_{B_1} \zeta_1\,\zeta_2\,\zeta_3\,dx + c_{25} = -3\int_{B_1} \zeta_1\,(\zeta_2^I + \zeta_2^{II})\,\zeta_3\,dx + c_{25}$$

$$\le -3\left(\int_{B_1} |\zeta_2^I|^a\,dx\right)^{1/a}\left(\int_{B_1} dx\right)^{1/b}\left(\int_{B_1} \zeta_1^3\,\zeta_3^3\,dx\right)^{1/3}$$

$$-\ 3\,\|(\zeta_2^{II})_+(t)\|_\infty \int_{B_1} \zeta_1\,\zeta_3\,dx + c_{25}$$

where $1/a + 1/b + 1/3 = 1$. $b < +\infty$ because $a > \frac{3}{2}$. Then we have

$$-3\int_{B_2} \lambda_1\,\lambda_2\,\lambda_3\,dx \le c_{26}\,\epsilon_1^{3/b}\left(\int_{B_1} |\zeta_2^I|^a\,dx\right)^{1/a}\left(\int_{B_1} |\nabla v|^6\,dx\right)^{1/3}$$

$$+\,c_{27}\,\|(\zeta_2^{II})_+(t)\|_\infty \int_{B_1} |\nabla v|^2\,dx + c_{25}$$

$$\le c_{26}\,\epsilon_1^{3/b}\left(\int_{B_1} |\zeta_2^I|^a\,dx\right)^{1/a}\left(\int_{B_2} |\nabla u|^6\,dx\right)^{1/3}$$

$$+\,c_{27}\,\|(\zeta_2^{II})_+(t)\|_\infty \int_{B_2} |\nabla u|^2\,dx + c_{25}$$

$$\le c_{28}\,\epsilon_1^{3/b}\int_{B_2} |\Delta u|^2\,dx + c_{27}\,\|(\zeta_2^{II})_+(t)\|_\infty \int_{B_2} |\nabla u|^2\,dx + c_{25}$$

Since number ϵ_1 can be achieved to be so small that $c_{28}\,\epsilon_1^{3/b} < \frac{1}{2}\nu$, we have

$$\frac{d}{dt}\frac{3}{8}\int_{B_2} |\nabla u|^2\,dx + \left(\frac{\nu}{2} - c_{28}\,\epsilon_1^{3/b}\right)\int_{B_2} |\Delta u|^2\,dx$$

$$\le c_{27}\,\|(\zeta_2^{II})_+(t)\|_\infty \int_{B_2} |\nabla u|^2\,dx + c_{25} + c_{24}\,\|h(.\,,t)\|_1\,. \tag{58}$$

We can now use the fact that $\|(\zeta_2^{II})_+\|_\infty \in L^1(]t_0 - \tau, t_0[)$ as a function of t and we can integrate inequality (58) with respect to t. Thus, we prove that the norm $\|\nabla u(t)\|_2$ is, as a function of t, bounded on the interval $]t_0 - \tau, t_0[$. This implies (45). Proceeding in the same way as at the end of the proof of Theorem 1, we can show that (x_0, t_0) is a regular point of the solution $(v; p_*)$. Finally, this also means that $(v; p_*)$ can have no singular point in D. □

It follows from the proof of Theorem 2 that condition (v) can be replaced by condition

(v)′ *the norm of* ζ_2^I *in* $L^{\infty,3/2}(D'')$ *tends to zero as* $m_4(D'') \to 0$ *for each subdomain* D'' *of* D

or in other words

(v)″ *the measure of* D'' *defined as the norm of* ζ_2^I *in* $L^{\infty,3/2}(D'')$ *is continuous with respect to the Lebesgue measure* m_4 *in* D.

5. Some open problems

The Navier–Stokes equation (1) is known already since the 19th century. It was derived under the assumptions that the fluid is a continuous medium, the stress tensor depends linearly on the tensor of velocity of deformations (Newton's law) and moreover, under an a priori assumption that velocity v and pressure p_* have a certain smoothness. However, the existence of solutions with this smoothness , in a general three–dimensional case, still remains an open problem! This problem is closely connected with the discussion whether the Navier–Stokes equations represent a "good" mathematical model for motion of a viscous incompressible fluid: There exist mechanisms in real fluids which do not enable the speed of motion to increase above all limits. It is highly desirable to know whether the mathematical model, based on the Navier–Stokes equations and used worldwide in thousands of calculations, also involves such mechanisms or whether, on the other hand, it admits solutions with singularities. This question is considered to be one of the most challenging open problems of applied mathematics for the new century. Let us clearly formulate the question:

Problem 1. *"Suppose that all the input data of the problem (1)–(4) are "smooth". (The input data are domain* Ω*, the initial velocity* v_0 *and the external force* f*.* f *can also be considered to be zero.) Does there exist a classical solution* v *of (1)–(4)?"* Or at least: *"Does there exist a weak solution which has no singular points in* Q_T*?"*

Problem 2. Another fundamental question is the question of uniqueness: *"Is a weak solution of the problem (1)–(4) unique?"* Or *"Is it unique at least in the class of solutions that satisfy the energy inequality (6) or the strong energy inequality (7)?"*

Problem 3. $L^{\infty,3}(Q_T)^3$ is a class where the uniqueness of a weak solution v is known to hold (due to H. Kozono and H. Sohr [16] – see also Section 1). As to regularity of v, it is only known that the weak solution v which belongs to this class can have singular points at instants of time whose number is at most countable (G. P. Galdi [11]) and the number of singular points on each time level is at most finite (J. Neustupa [25]). The questions *"Is the weak solution* v *which finds itself in* $L^{\infty,3}(Q_T)^3$ *fully regular?"* or at least *"Is the number of instants of time where* *it can have singular points at most finite?"* are open. The problem *"Is the theorem from [27] (where the component v_3 of velocity is assumed to be in $L^{r,s}(D)$ with $4 \le r \le +\infty$, $6 < s \le +\infty$ and $2/r + 3/s \le \frac{1}{2}$) also valid with $r = +\infty$ and $s = 6$?"* is of the same nature.

Problem 4. Our definition of a regular point or a singular point relates only to points inside Q_T. However, it is natural to extend this definition to points on $\overline{\Omega} \times]0,T[$. Suppose that $(v; p_*)$ is a suitable weak solution of (1)–(4). The question *"Does the Caffarelli–Kohn–Nirenberg theorem, saying that the 1–dimensional Hausdorff measure of $S(v)$ is zero, remain valid if $S(v)$ also involves possible singular points of v on $\partial\Omega \times]0,T[$?"* represents an important problem which is still unsolved. Analogously, *"If component v_3 of v satisfies the assumptions from paper [27] in the whole Q_T (i.e. $v_3 \in L^{r,s}(Q_T)$ for some $4 \le r \le +\infty$, $6 < s \le +\infty$ such that $2/r + 3/s \le \frac{1}{2}$), does this imply that v has no singular points in $\overline{\Omega} \times]0,T[$?"*

Problem 5. If domain Ω, initial velocity v_0 and external force f are "smooth" and axially symmetric then the existence of a unique regular solution $(v; p_*)$ is known, as it was mentioned in Section 1, in a special case when $\Omega = \mathbb{R}^3$ and the tangential components of v_0 and f are zero. J. Neustupa and M. Pokorný [28] proved for a general axially symmetric suitable weak solution $(v; p_*)$:

If $f = 0$, D is a sub–domain of Q_T and the radial component of v has its negative part in $L^{r,s}(D)$ for some $2 \le r \le +\infty$, $3 < s \le +\infty$ such that $2/r + 3/s \le 1$ or the tangential component of v belongs to $L^{r,s}(D)$ where

- *either* $6 \le s \le +\infty$, $20/7 \le r \le +\infty$ *and* $2/r + 3/s \le 7/10$
- *or* $24/5 < s < 6$, $10 < r \le +\infty$ *and* $2/r + 3/s \le 1 - 9/(5s)$

then $(v; p_*)$ has no singular points in D.

If the solution $(v; p_*)$ has a singular point (x_0, t_0) then x_0 must lay on the axis of symmetry. (Otherwise we can easily derive a contradiction with the Caffarelli–Kohn–Nirenberg theorem saying that the 1–dimensional Hausdorff measure of the set of singular points of $(v; p_*)$ is zero.) However, the question *"Does a singular point in a general axially symmetric flow exist?"* is still unsolved.

Problem 7. As it was mentioned in Section 1, D. Chae and H. J. Choe [5] proved regularity of a weak solution under assumptions about two components of vorticity ω and P. Constantin and C. Fefferman [6] proved regularity by imposing conditions on $\omega/|\omega|$. A natural question is: *"Is it possible to prove regularity by imposing conditions only on one component of ω?"*

Problem 8. Results from [27] and also Theorem 1 (Section 3) imply regularity of the suitable weak solution $(v; p_*)$ in domain $D \Subset Q_T$, if $v_3 \in L^{r,s}(D)$ if $4 \le r \le +\infty$, $6 < s \le +\infty$ and $2/r + 3/s \le \frac{1}{2}$. A natural question, which follows from the comparison of these conditions with the usual Prodi–Serrin conditions, is: *"Is it possible to improve this result so that only $2/r + 3/s \le 1$ is assumed?"*

Problem 9. The question which can be only of a technical character and concerns Theorem 1 is *"Is it possible to prove Theorem 1 without condition (iv)?"* Another question which arises in connection with Theorem 1 and is motivated by the fact that regularity can be proved by imposing the Prodi–Serrin conditions only on two components of velocity (the authors were pointed out to this fact by H. J. Choe) is: *"Is it possible to improve Theorem 1 so that we only assume that $v_1 \in L^{p,q}_{loc}(D)$ instead of v_1, $v_2 \in L^{p,q}_{loc}(D)$?"*

Problem 10. We will use the denotation from Theorem 2. The two situations

a) $\zeta_1 \zeta_2 \zeta_3 \geq 0$ and

b) $\zeta_1 \zeta_2 \zeta_3 < 0$

have the same geometric interpretations as the cases a) $\lambda_1 \lambda_2 \lambda_3 \geq 0$ and b) $\lambda_1 \lambda_2 \lambda_3 < 0$ which are explained before Theorem 1. The following question now naturally arises: *"Is it possible that the deformations of the type a) and of the type b) of infinitely small volumes in the fluid are in the large in a balance so that the integral of $\zeta_1 \zeta_2 \zeta_3$ on B_1 remains of the same order as the integral of $|\nabla v|^2$?"* (See also the proof of Theorem 2.) The positive answer would obviously lead to the regularity of solution v. It seems to us that the answer depends on deeper geometric properties of the three–dimensional space.

We are convinced that answers to the formulated questions would represent important and interesting contributions to the theory of regularity of weak solutions to the Navier–Stokes equations.

Acknowledgement. The research was supported by the University of Toulon–Var and in the case of the first author, it was also supported by the Grant Agency of the Czech Republic (grant No. 201/99/0267) and by the research plan of the Ministry of Education of the Czech Republic No. MSM 98/21000010.

References

[1] H. Beirão da Veiga, *Concerning the regularity problem for the solutions of the Navier–Stokes equations,* C.R. Acad. Sci. Paris, t. 321, Série I (1995), 405–408.

[2] W. Borchers and T. Miyakawa, *Decay for the Navier–Stokes flow in halfspaces,* Math. Ann. **282** (1988), 139–155.

[3] W. Borchers and H. Sohr, *On the equations* $\operatorname{rot} v = g$ *and* $\operatorname{div} u = f$ *with zero boundary conditions,* Hokkaido Math. J. **19** (1990), 67–87.

[4] L. Caffarelli, R. Kohn and L. Nirenberg, *Partial regularity of suitable weak solutions of the Navier–Stokes equations,* Comm. on Pure and Apl. Math. **35**, (1982), 771–831.

[5] D. Chae and H. J. Choe, *Regularity of solutions to the Navier–Stokes equation,* Electronic J. of Diff. Equations **5**, (1999), 1–7.

[6] P. Constantin and C. Fefferman, *Direction of vorticity and the problem of global regularity for the Navier-Stokes equations,* Indiana Univ. Math. Journal **42**, No. 3 (1993), 775–789.

[7] E. B. Fabes, B. F. Jones and N. M. Rivière, *The initial value problem for the Navier–Stokes equations with data in L^p*, Arch. Rat. Mech. Anal. **45** (1972), 222–240.

[8] C. Foias and R. Temam, *Some analytic and geometric properties of the solutions of the evolution Navier–Stokes equations,* J. Math. Pures et Appl. **58** (1979), 339–368.

[9] G. P. Galdi and P. Maremonti, *Monotonic decreasing and asymptotic behaviour of the kinetic energy for weak solutions of the Navier–Stokes equations in exterior domains,* Arch. Rat. Mech. Anal. **94** (1986), 253–266.

[10] G. P. Galdi, *An Introduction to the Mathematical Theory of the Navier–Stokes Equations, Vol. I: Linearized Steady Problems,* Springer–Verlag, New York–Berlin–Heidelberg (1994).

[11] G. P. Galdi, *An Introduction to the Navier–Stokes Initial–Boundary Value Problem,* in *Fundamental Directions in Mathematical Fluid Mechanics,* editors G.P.Galdi, J.Heywood, R.Rannacher, series "Advances in Mathematical Fluid Mechanics", Vol. 1, Birkhauser–Verlag, Basel (2000), 1–98.

[12] Y. Giga, *Solutions for semilinear parabolic equations in L^p and regularity of weak solutions of the Navier–Stokes equations,* J. of Diff. Equations **61** (1986), 186–212.

[13] J. G. Heywood, *The Navier–Stokes equations: On the existence, uniqueness and decay of solutions,* Indiana Univ. Math. J. **29** (1980), 639–681.

[14] E. Hopf, *Über die Anfangswertaufgabe für die Hydrodynamischen Grundgleichungen,* Math. Nachr. **4** (1950), 213–231.

[15] K. K. Kiselev and O. A. Ladyzhenskaya, *On existence and uniqueness of the solutions of the non–stationary problem for a viscous incompressible fluid,* Izv. Akad. Nauk SSSR bf 21 (1957), 655–680 (in Russian).

[16] H. Kozono and H. Sohr, *Remark on uniqueness of weak solutions to the Navier–Stokes equations,* Analysis **16** (1996), 255–271.

[17] H. Kozono, *Uniqueness and regularity of weak solutions to the Navier–Stokes equations,* Lecture Notes in Num. and Appl. Anal. **16** (1998), 161–208.

[18] O. A. Ladyzhenskaya, *The global solution of the boundary value problem for the Navier–Stokes equations in the case of two spatial variables,* Comm. on Pure and Apl. Math. **12** (1959), 427–433.

[19] O. A. Ladyzhenskaya, *Uniqueness and smoothness of generalized solutions of the Navier–Stokes equations,* Zap. Nauch. Sem. LOMI **5** (1967), 169–185 (in Russian).

[20] S. Leonardi, J. Málek, J. Nečas and M. Pokorný, *On the results of Uchovskii and Yudovich on axially symmetric flows of a viscous fluid in* $\mathbb{R}^3$, Zeitschrift für Angew. Anal. **18**, (1999), 639–649.

[21] J. Leray, *Sur le mouvements d'un liquide visqueux emplissant l'espace,* Acta Mathematica **63** (1934), 193–248.

[22] J. L. Lions and G. Prodi, *Un théorème d'existence et unicité dans les équations de Navier–Stokes en dimension 2,* Compte Rend. Acad. Sci. Paris **248** (1959), 3519–3521.

[23] T. Miyakawa and H. Sohr, *On energy inequality, smoothness and large time behaviour in L^2 for weak solutions of the Navier–Stokes equations in exterior domains,* Math. Z. **199** (1988), 455–478.

[24] J. Nečas, M. Růžička and V. Šverák, *On Leray's self–similar solutions of the Navier–Stokes equations,* Acta Mathematica **176** (1996), 283–294.

[25] J. Neustupa, *Partial regularity of weak solutions to the Navier–Stokes equations in the class* $L^\infty(0,T;\, L^3(\Omega)^3)$, J. of Mathematical Fluid Mechanics **1** (1999), 309–325.
[26] J. Neustupa and P. Penel, *Regularity of a suitable weak solution to the Navier–Stokes equations as a consequence of regularity of one velocity component,* in *Nonlinear Applied Analysis,* editors A.Sequeira et al., Plenum Press, New York (1999), 391–402.
[27] J. Neustupa, A. Novotný and P. Penel, *An interior regularity of a weak solution to the Navier–Stokes equations in dependence on one component of velocity,* to appear in *Topics in Mathematical Fluid Mechanics,* a special issue of Quaderni di Matematica (2001).
[28] J. Neustupa and M. Pokorný, *An interior regularity criterion for an axially symmetric suitable weak solution to the Navier–Stokes equations,* J. of Mathematical Fluid Mechanics **2** (2000), 381–399.
[29] G. Prodi, *Un teorema di unicità per el equazioni di Navier–Stokes,* Ann. Mat. Pura Appl. **48** (1959), 173–182.
[30] J. Serrin, *On the interior regularity of weak solutions of the Navier–Stokes equations,* Arch. Rat. Mech. Anal. **9** (1962), 187–195.
[31] H. Sohr, *Zur Regularitätstheorie der Instationären Gleichungen von Navier–Stokes,* Math. Z. **184** (1983), 359–375.
[32] H. Sohr and W. von Wahl, *On the singular set and the uniqueness of weak solutions of the Navier–Stokes equations,* Manuscripta Math. **49** (1984), 27–59.
[33] H. Sohr, W. von Wahl and M. Wiegner, *Zur Asymptotik der Gleichungen von Navier–Stokes,* Nachr. Akad. Wiss. Göttingen **146** (1986), 1–15.
[34] V. A. Solonnikov, *Estimates of solutions of a non–stationary Navier–Stokes system,* Zap. Nauch. Sem. LOMI **38** (1973), 153–231 (in Russian).
[35] Y. Taniuchi, *On generalized energy inequality of the Navier–Stokes equations,* Manuscripta Mathematica **94** (1997), 365–384.
[36] R. Temam, *Navier–Stokes Equations,* North–Holland, Amsterdam–New York–Oxford (1977).
[37] W. von Wahl, *Regularity of weak solutions of the Navier–Stokes equations,* Proc. Symp. Pure Appl. Math. **45** (1986) 497–503.

Jiří Neustupa
Czech Technical University, Faculty of Mechanical Engineering
Department of Technical Mathematics
Karlovo nám. 13
121 35 Praha 2
Czech Republic
E-mail address: `neustupa@marian.fsik.cvut.cz`

Patrick Penel
Université de Toulon et du Var, Mathématique
83957 La Garde, France
E-mail address: `penel@univ-tln.fr`

Appendix

We append references to basic literature on mathematical fluid mechanics where the readers can find explanation of the backgrounds to most of the topics discussed in this book. The list is naturally not exhaustive.

1. V. I. Arnold and B. A. Khesin, *Topological Methods in Hydrodynamics,* Applied Mathematical Sciences 125, Springer–Verlag, New York (1997).
2. S. Chandrasekhar, *Hydrodynamic and Hydromagnetic Stability,* Clarendon Press, Oxford (1961).
3. C. Cercignani, *The Boltzmann Equation and its Applications,* Springer–Verlag, Berlin (1988).
4. P. Constantin and C. Foias, *Navier–Stokes Equations,* Chicago Univ. Press, Chicago (1988).
5. R. Dautray and J. L. Lions, *Analyse Mathématique et Calcul Numérique pour les Sciences et les Techniques,* Masson, Paris (1988).
6. P. G. Drazin and W.H. Reid, *Hydrodynamic Stability,* Cambridge Monographs on Mechanics and Applied Mathematics, Cambridge Univ. Press, Cambridge (1982).
7. M. Feistauer, *Mathematical Methods in Fluid Dynamics,* Pitman Monographs and Surveys in Pure and Applied Mathematics 67, Longman Scientific & Technical, Harlow (1993).
8. U. Frisch, *Turbulence,* Cambridge Univ. Press, Cambridge (1995).
9. G. P. Galdi, *An Introduction to the Mathematical Theory of the Navier–Stokes Equations, Vol. I: Linearized Steady Problems,* Springer Tracts in Natural Philosophy 38, Springer–Verlag, New York (1994).
10. G. P. Galdi, *An Introduction to the Mathematical Theory of the Navier–Stokes Equations, Vol. II: Nonlinear Steady Problems,* Springer Tracts in Natural Philosophy 39, Springer–Verlag, New York (1994).
11. G. P. Galdi, J. G. Heywood and R. Rannacher (editors), *Fundamental Directions in Mathematical Fluid Mechanics,* Advances in Mathematical Fluid Mechanics 1, Birkhauser–Verlag, Basel (2000).
12. V. Girault and P. A. Raviart, *Finite Element Methods for Navier–Stokes Equations,* Series in Computational Mathematics, Springer–Verlag (1986).
13. E. Godlewski and P. A. Raviart, *Hyperbolic Systems of Conservation Laws,* Springer–Verlag, New York (1996).

14. Ch. Hirsch, *Numerical Computation of Internal and External Flows,* Wiley, Chichester (1989).

15. D. Kröner, *Numerical Schemes for Conservation Laws,* Wiley–Teubner, Chichester (1997).

16. O. A. Ladyzhenskaya, *Mathematical Problems in the Dynamics of Viscous Incompressible Fluid,* Gordon and Breach, New York (1963).

17. R. J. LeVeque, *Numerical Methods for Conservation Laws,* Birkhäuser–Verlag, Basel (1990).

18. A. J. Lichtenberg and M. A. Lieberman, *Regular and Chaotic Dynamics,* Applied Mathematical Sciences 88, Springer–Verlag, New York (1992).

19. J. L. Lions, *Quelques Méthodes de Résolution des Problèmes aux Limites Non Linéaires,* Dunod, Paris (1969).

20. P. L. Lions, *Mathematical Topics in Fluid Mechanics I, Incompressible Models,* Oxford Science Publication, Clarendon Press, Oxford (1996).

21. P. L. Lions, *Mathematical Topics in Fluid Mechanics II, Compressible Models,* Oxford Science Publication, Clarendon Press, Oxford (1998).

22. J. Málek, J. Nečas, M. Rokyta and M. Růžička, *Weak and Measure-Valued Solutions to Evolution Partial Differential Equations,* Applied Mathematics and Mathematical Computation 13, Chapman and Hall (1996).

23. J. Málek, J. Nečas and M. Rokyta (editors), *Advanced Topics in Theoretical Fluid Mechanics,* Pitman Research Notes in Mathematics 392, Longman, New York (1998).

24. J. Málek, J. Nečas and M. Rokyta (editors), *Advances in Mathematical Fluid Mechanics,* Springer–Verlag, Berlin–Heidelberg–New York (2000).

25. O. Pironneau, *Méthode des Élémets Finis pour les Fluides,* Collection Recherches en Mathématiques Appliquées 7, Masson, Paris (1988).

26. J. Serrin, *Mathematical Principles of Classical Fluid Mechanics,* in Handbuch der Physik VIII/1, Springer–Verlag, Berlin (1972), 125–262.

27. R. Temam, *Navier-Stokes Equations,* North–Holland, Amsterdam–New York–Oxford (1977).

28. R. Temam, *Infinite Dimensional Dynamical Systems in Mechanics and Physics.* Applied Mathematical Sciences 68, Springer–Verlag, New York (1988).

29. E. F. Toro, *Riemann Solvers and Numerical Methods for Fluid Dynamics,* Springer–Verlag, Berlin (1997).

30. W. von Wahl, *The Equations of Navier-Stokes and Abstract Parabolic Equations,* Aspekt der Mathematik, Vieweg & Sohn, Wiesbaden (1985).

List of Authors

Bardos *Claude* 1
Bause *Markus* 27
Cherfils-Clérouin *Catherine* 47
Feireisl *Eduard* 73
Feistauer *Miloslav* 105
Friedlander *Susan* 143
Fürst *Jiří* 173
Heywood *John* 27, 195
Janda *Michal* 173
Kozel *Karel* 173
Kozono *Hideo* 207
Lafitte *Olivier* 47
Neustupa *Jiří* 237
Novotný *Antonín* 27
Padula *Mariarosaria* 27
Penel *Patrick* 237
Raviart *Pierre-Arnaud* 47
Shnirelman *Alexander* 143

MIX
Papier aus verantwortungsvollen Quellen
Paper from responsible sources
FSC® C105338

FSC
www.fsc.org

If you have any concerns about our products,
you can contact us on
ProductSafety@springernature.com

In case Publisher is established outside the EU,
the EU authorized representative is:
Springer Nature Customer Service Center GmbH
Europaplatz 3, 69115 Heidelberg, Germany

Printed by Libri Plureos GmbH
in Hamburg, Germany